FERMENTATION PROCESSES

EMERGING AND CONVENTIONAL TECHNOLOGIES

FERMENTATION PROCESSES

EMERGING AND CONVENTIONAL TECHNOLOGIES

Oleg Paliy
Katherine Bauer

Kruger Brentt
Publishers

2025

Kruger Brentt Publishers UK. LTD.
Company Number 9728962

Regd. Office: 68 St Margarets Road, Edgware, Middlesex HA8 9UU

ISBN: 978-1-78715-322-6

For information on all our publications visit our website at http://krugerbrentt.com/

PREFACE

"Fermentation Processes: Emerging and Conventional Technologies" delves into the diverse and evolving landscape of fermentation science and technology, offering readers a comprehensive exploration of both traditional and cutting-edge fermentation processes. This book serves as an essential resource for researchers, practitioners, industry professionals, and enthusiasts seeking to understand the principles, applications, and innovations driving advancements in fermentation technology.

Fermentation, the transformative process by which microorganisms convert organic substrates into useful products such as food, beverages, pharmaceuticals, and biofuels, has been practiced by humans for millennia. From ancient techniques like brewing and cheese-making to modern industrial processes such as bioreactor cultivation and enzyme production, fermentation plays a vital role in various sectors of the global economy.

The primary objective of this book is to provide readers with a comprehensive overview of fermentation processes, covering key topics such as microbial physiology, substrate selection, process optimization, downstream processing, and product development. Each chapter offers insights, case studies, and practical guidance to help readers navigate the complexities of fermentation technology and harness its potential for innovation and sustainability.

Contributions from leading experts in the field explore various aspects of fermentation, including microbial diversity, metabolic engineering, bioreactor design, fermentation monitoring and control, and downstream processing techniques. Special emphasis is placed on emerging trends and technologies in fermentation, such as synthetic biology, fermentation of alternative substrates, biorefinery concepts, and bioprocess intensification.

As we confront global challenges such as food security, environmental sustainability, and public health, fermentation technology offers promising solutions for producing sustainable and value-added products while reducing reliance on fossil fuels and chemical processes. By fostering collaboration, knowledge exchange, and innovation in the fermentation community, we can unlock new opportunities for improving human health, protecting the environment, and advancing the bioeconomy.

We extend our sincere gratitude to the contributors whose expertise and dedication have enriched this book. It is our hope that "Fermentation Processes: Emerging and Conventional Technologies" will serve as a valuable resource and catalyst for dialogue, innovation, and collaboration in the dynamic and ever-expanding field of fermentation science and technology.

Oleg Paliy

Katherine Bauer

CONTENTS

01 THE SIGNIFICANCE OF FERMENTATION IN THE PRODUCTION OF PREMIUM SILAGE

1. INTRODUCTION: AN OVERVIEW

Grazing is the predominant and cost-effective method of feeding cattle. However, it is not feasible year-round due to climate constraints that restrict the development of grasses. The presence of pastures in livestock systems is contingent upon the changing seasons, as the various elements that influence plant growth, including as temperature, luminosity, and rainfall, fluctuate throughout seasons. Consequently, there are periods characterized by abundant feed production and periods marked by its scarcity. During the winter season, there is insufficient fodder production to adequately nourish the animals.[1].

The selection of an appropriate forage preservation method to provide a continuous supply of feed primarily relies on the prevailing environmental conditions at harvest. In regions with high temperatures and arid seasons, haymaking is likely the optimal method for preserving forage. This straightforward technique involves drying the freshly cut crop to remove moisture, resulting in a stable and well-preserved substance. Producing high-quality hay in tropical countries with hot and humid climates is challenging due to the combination of high humidity and frequent rainfall during the optimal time of crop maturation, which is when the crop has the highest nutritional value. Ensilage is a significant technique for preserving forage in this situation since it is less reliant on weather conditions compared to haymaking. Furthermore, silage serves as the primary energy source in the overall mixed rations of ruminants in numerous regions across the globe. [2, 3]. The aim of this chapter was to explain the fermentation processes involved in making silage and how they can be manipulated to impact the overall quality of the silage. This includes examining the implications on animal performance and health.

2. THE SIGNIFICANCE OF FERMENTATION IN THE PROCESS OF SILAGE PRODUCTION

According to [2], in short, the silage is made by keeping chopped crop air-tight in a silo, as follows: (1) the crops are harvesting and chopping in a specific length at the better nutritional value and proper moisture content; (2) application of continuously heavy weights to pack at adequate densities; (3) and complete sealing. The preservation of the forage crops as silage depends of anaerobic environment, because it is based on a lactic acid fermentation that decreases the pH and associated with high osmotic pressure that inactivates the microorganisms preserving the nutritive value of the fresh crop (Figure 1). Even the presence of some mycotoxins in the fresh crop may be denatured due to the acid pH of silage.

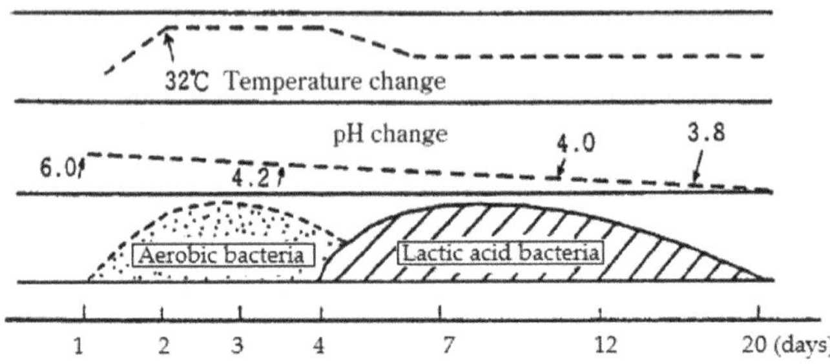

Figure 1. Diagram of silage fermentation process [2].

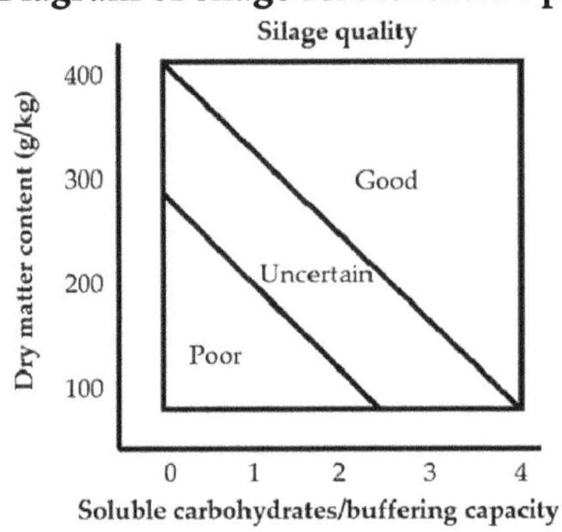

Figure 2. Effects of dry matter content and water-soluble carbohydrates: buffering capacity on silage quality [5].

The main principle of silage is anaerobic environment and fermentation of the water-soluble carbohydrates in the fresh crop by the epiphytic lactic acid bacteria (LAB) and production of lactic acid. However, different fermentation pathways may occur into the silo environment, depending on the availability of substrate, the predominant microbial populations, the dry matter (DM) content, and the buffering capacity of the crop at the ensiling (Figure 2). In addition, the fermentation must be limited to a certain extent, because it alters the chemical composition of the feed. This process may last for days or months, which may result in silage containing high levels of alcohols, butyric acid, ammonia, amines, and acetic acid that represent the major silage losses. Generally, the epiphytic microbial populations found in growing crops include pseudomonas, actinomycetes, listeria, and mainly the LAB that we expect to dominate the fermentation process to produce high-quality silage (Table 1) [4].

High-quality silages are resulted of a fast and efficient fermentation preserving the crop nutrients, which depends if the fresh crop has high nutritional value and good characteristics for the ensiling process, as described before. In addition, the fermentation process cannot improve the crop nutritive value, but in some cases occur an increase on digestibility, always with energy losses. Efficient fermentation ensures a more palatable and digestible feed, which improves the animal performance. As noted above, the most important factors related to the crop characteristics to ensiling are adequate dry matter content, sufficient water-soluble carbohydrates for fermentation, and low buffering capacity.

The dry matter content affects directly the microbial activity, specific density, and effluent losses. Crops with dry matter content below 25% at ensiling show high effluent losses and high activity of undesirable microorganisms such as the genus *Clostridium* [6]. In addition, the LAB are more tolerant to low moisture conditions (low water activity) than other undesirable anaerobic microorganisms. However, dry matter content above 45% difficult the process of forage packing, resulting in high porosity, which may cause losses by the development of aerobic microorganisms [7].

Table 1. Typical microbial populations on crops before ensiling [4].

Group	Population, colony-forming units/g of fresh forage
Total aerobic bacteria	>10,000,000
Lactic acid bacteria	10–1,000,000
Enterobacteria	1000–1,000,000
Yeasts	1000–100,000
Molds	1000–10,000
Clostridia	100–1000
Bacilli	100–1000
Acetic acid bacteria	100–1000
Propionic acid bacteria	10–1000

About the amount of water-soluble carbohydrates, they present a narrow range of optimum values (60–80 g/kg of dry matter), because they are readily available substrates for the LAB and other microorganisms [6]. Furthermore, the excess sugar can stimulate the growth of anaerobic yeasts that are not fully inhibited by the low pH, as occurs in sugarcane silage, which results in high DM losses because the fermentation goes to the ethanol pathway [8].

The silage resistance to the pH lowering is named buffering capacity. This is exerted by compounds present in the crop, as the crude protein, inorganic ions, organic acids, and others. The greater buffering capacity needs more water-soluble carbohydrates content for an effective fermentation by reducing pH and inhibiting undesirable fermentations [9].

The fermentation coefficient (FC) was developed to predict if the crop is suitable to ensiling or not, as follows [10]:

$$FC = DM\ (\%) + 8\ WSC/BC$$

where FC = fermentation coefficient, DM = dry matter content, WSC = water-soluble carbohydrates, and BC = buffer capacity.

The forage crops with FC < 35 can result in undesirable fermentations and high dry matter losses, requiring additive application to control silage fermentation. If the FC ≥ 35, sufficient fermentable substrates are available. However, in high DM, crops are used microbial inoculants to ensure the presence of osmotolerant LAB to dominate the fermentation process.

3. PROCESSES OF SILAGE FERMENTATION

The ensiling process, didactically, is divided into four principal phases [4]:

- Initial aerobic phase since the harvest to the oxygen exhaustion in the silo. This phase is characterized by crop respiration and activity of all obligate and facultative aerobic organisms such as molds, yeasts, and some bacteria until finish up all the oxygen (Figure 1). In addition, the plant enzymes such as proteases and carbohydrases remain active. This phase must be short, because the sugars are converted to CO_2 and water with heat release, representing dry matter losses, increased Maillard products, and drops in the silage quality. This phase is also important because of CO_2, hydrogen peroxide, and other compounds that are produced with antimicrobial effect.

- The main fermentation phase started with a short lag phase followed by rapid growth of facultative and obligate anaerobic microorganisms. The undesirable microorganisms as enterobacteria, clostridia, and yeasts compete with the desirable genera of LAB by the substrates. The main genera of LAB commonly associated with silage are *Lactobacillus, Pediococcus, Leuconostoc, Enterococcus, Lactococcus, Streptococcus*, and *Weissella*. The lactic acid production and the rate of pH decline are responsible for the disappearance of enterobacterial and clostridial secondary fermentations.

Obtaining a well-fermented silage depends on the fresh crop characteristics as the adequate dry matter content (300–500 g/kg of fresh matter), water-soluble carbohydrates (60–120 g/kg of dry matter), and low buffering capacity. In addition to speed of harvesting, length of chop and silage distribution and compaction will be responsible for the successful conservation of feed nutrients.

- Stable phase: In the acidic environment and without oxygen, the activity of microorganisms decreases substantially, and only acid-tolerant enzymes keeps a slow hydrolysis of carbohydrates and protein. The final pH of the ensiled forage depends on the ensiled crop. Theoretically, under ideal conditions, silage can be stored indefinitely if those conditions are maintained, because the losses are minimal. However, in the farm, it is usually stored for a maximum of 1 year or until the next harvest season. In arid and semiarid regions, farmers can store silage for longer periods because the dry period can comprehend two or more years.
- Feed-out phase is very critical, because the undesirable microorganisms consume the compounds that make the silage stable in the silo (lactic acid) in the presence of oxygen and can produce many compounds decreasing the silage quality. Well-fermented silage with high lactic acid content and residual carbohydrates are more susceptible to aerobic deterioration, because they are the main substrates for the yeasts that initiate the deterioration process. The molds, yeasts, and acetic acid bacteria consume the acids, sugars, and protein for growth releasing heat and can cause considerable changes in the chemical composition in addition with rise in pH other microorganisms that were inhibited can proliferate and lead to a massive spoilage. Because this, care must be taken of removing a uniform layer of silage every day to not provide sufficient time for the undesirable microorganism's proliferation. Normally, silage can stay stable when exposed to air for approximately 30–40 h, but it depends of environmental conditions and silage characteristics.

3.1. Substrates

The primary substrates for fermentation are the water-soluble carbohydrates, as well as a variety of amino acids and vitamins found in the crop. Furthermore, following the breakdown of enzymes, plants have the ability to break down starch and hemicelluloses, resulting in an increased supply of hexoses and pentoses for microbial development. Hexose monosaccharides, oligosaccharides, and polysaccharides, such as glucose, fructose, sucrose, and fructans, are the primary types of carbohydrates that may dissolve in water and are easily accessible for fermentation. Another significant carbohydrate is starch, which serves as the primary storage polysaccharide in many crops. However, it is not commonly utilized in fermentation due to its insolubility in water. [9].

3.2. Types of fermentations

In silage fermentation, several pathways occur simultaneously; the fermentation type depends on the environmental conditions, microorganism species, and substrate availability. The LAB show two basic types of hexose fermentation to lactic acid. The most efficient pathway in energy conservation is the obligate homofermentative, which produces almost exclusively lactic acid (>85%). The facultative heterofermentative lactic acid bacteria show besides the homolactic pathway; they present ability to ferment pentoses, because they have both enzymes aldolase and phosphoketolase. The obligate heterofermentative lactic acid bacteria present DM loss from hexose fermentation due the CO_2 production as well as lactic acid, and acetic acid or ethanol [4]. The acetate or ethanol production depends on the fermentation substrate: if the fermentation substrate is a hexose, the end-product is acetic acid, and if it is a pentose, the end-product is ethanol [6]. Although heterolactic pathway causes DM loss, a partial increase in acetic acid concentration improves the aerobic stability of silage, because the acetic acid inhibits the activity of yeasts during the feed-out phase [11]. The end-products of well-fermented silages are presented in Table 2.

Table 2. Amounts of common fermentation end products in various silages [12].

Item, % of dry matter	Silage and dry matter contents				
	Alfafa, 32.5%	Alfafa, 50.0%	Grass, 30.0%	Corn, 37.5%	High moisture corn, 75.0%
pH	4.3–4.5	4.7–5.0	4.3–4.7	3.7–4.2	4.0–4.5
Lactic acid	7.0–8.0	2.0–4.0	6.0–10.0	4.0–7.0	0.5–2.0
Acetic acid	2.0–3.0	0.5–2.0	1.0–3.0	1.0–3.0	<0.5
Propionic acid	<0.5	<0.1	<0.1	<0.1	<0.1
Butyric acid	<0.5	0	<0.5	0	0
Ethanol	0.5–1.0	0.5	0.5–1.0	1.0–3.0	0.2–2.0
Ammonia N[F1]	10.0–15.0	<12	8.0–12.0	5.0–7.0	<10.0

[1]% of total nitrogen.

When the acidification is not fast and adequate and/or with high moisture content, undesirable secondary fermentations can occur by other microorganisms, which are able to compete for nutrients with the LAB. Enterobacterial fermentation pathway is similar to the heterofermentative LAB and ferments glucose to acetic acid, formic acid, and alcohol. In addition, enterobacteria can decarboxylate and deaminate amino acids and reduce NO_3. Other undesirable is the clostridial fermentations, which derive their energy from organic compounds such as carbohydrates and proteins producing butyric acid, acetic acid, propionic acid, ethanol, biogenic amines, and CO_2. Those processes represent major losses that decrease silage quality and increase the production cost because of the low DM recovery. In addition, other smaller fermentations like the *Propionibacterium* can ferment glucose, fructose, glycerol, lactate, lactose, sucrose, xylose, and starch producing propionic acid, acetic acid, $CO_{2,}$ and formic acid or isovaleric acid. The facultative anaerobic yeasts can ferment glucose, maltose, and sucrose with the main products such as ethanol, $CO_{2,}$ and others compounds (alcohols, volatile fatty acids and lactate). The facultative anaerobic bacilli can ferment carbohydrates to organic acids or ethanol, 2,3-butanediol, and glycerol [4].

The secondary fermentations are undesirable because they preserve less energy in its end-products compared to the lactic acid fermentation, which is explained by the production of CO_2. These fermentations can also produce toxic compounds that impair the animal health and performance.

3.3. Efficiency of the fermentation process

The prevalent fermentation pathways in the ensiling process depend on several factors. They are related to the fresh crop and are basically the contents of DM and water-soluble carbohydrates. In addition, there are some characteristics related to the process techniques such as particle size, specific density, and especially the length time until the installation of anaerobic conditions in the silo. According to [13], the homofermentative LAB pathway results in only 0.7% of energy loss and it can be described as follows:

Glucose or fructose + 2 ADP + 2 Pi = 2 lactate + 2 ATP + 2 H_2O.

The heterofermentative LAB pathway from glucose results in 24% of DM loss and 1.7% of energy loss. When they ferment, fructose results in 4.8% of DM loss and 1.0% of energy loss, and it can be described as follows:

Glucose + ADP + Pi = lactate + ethanol + CO_2 + ATP + H_2O, or

Fructose + 2 ADP + 2 Pi = lactate + acetate + 2 mannitol + 2 CO_2 + 2 ATP + H_2O.

In the clostridial fermentations, DM loss is 51.1% and the energy loss is 18.4%, and it can be described as follows:

2 lactate + ADP + Pi = butyrate + 2 CO_2 + 2 H_2 + ATP + H_2O.

In the yeasts' fermentation, the DM loss is 48% and the energy loss is 0.2%, and it can be described as follows:

Glucose + 2 ADP + 2 Pi = 2 ethanol + 2 CO_2 + 2 ATP + 2 H_2O.

4. MANIPULATING SILAGE FERMENTATION

The knowledge about silage fermentation provides technology improvement to produce high-quality silages. In addition, crops that were once considered inappropriate to ensiling, mainly legumes, are routinely ensiled in many farms nowadays. Theoretically, all forage crops can be conserved as silage, if the ensiling techniques such as the finely chopped, well packed in the silo, and complete sealed through of plastic sheet are done carefully to promote adequate anaerobic condition. However, the crop intrinsic characteristics will direct the fermentation pathway and affect the final silage quality.

4.1. Changing the harvest time

Each crop, depending on environment, has the ideal stage of maturity for silage production considering the yield due to the profitability, dry matter, and fermentable sugar contents for bacteria and maximum nutritional value for livestock (Table 3). Practically, all factors involving the fermentation will change with crop maturity stage. In addition, the water-soluble carbohydrates have a diurnal fluctuation cycle, and their concentrations are highest at 18:00 h and lowest at 06:00 h. Generally, advancing crop maturity results in increases in dry matter, carbohydrates, and LAB population as well as total microorganism number. In addition, decreases in buffering capacity and crude protein concentration are observed, and some crops have showed a decrease in digestibility with advancing maturity [9].

Table 3. Harvest and dry matter recommendation for main crops conserved as silage [14].

Crop	Maturity	Dry matter (%)	Cut length (mm)
Corn	Milk line 1/2–2/3 down the kernel	28–37	9.5–12.7
Alfafa	Mid-bud 1/10 bloom, wilt to	30–40	6.4–9.5
Cereal	Milk or soft dough, wilt to	28–37	6.4–9.5
Grasses	When the first stems head out	28–37	6.4–9.5
Clover	1/4–1/2 bloom, wilt to	28–37	6.4–9.5
Sorghum	Grain medium to hard dough	30–35	9.5–12.7

4.2. Wilting

Some crops, like tropical grasses and some legume such as alfalfa and forage soybean (Table 4), have a quite low DM content at the same time when the nutritive value is high. Obtaining a good fermentation and eliminating the effluent losses must increase the dry matter content prior to chopping and ensiling. Generally, those crops need be wilted at harvest with a mower-conditioner to increase DM content and to enhance the lactic fermentation. Mowing and conditioning can increase the leaves losses and affect the microbial populations on the crop. The plant juice released can increase the nutrients losses and bacterial population and cause a shift in the microbial species present [15].

Table 4. Effect of crop vegetative stage and preharvest wilting time on ensiling parameters and *in vitro* rumen degradability of forage soybean silage [16].

	Maturity stage			Wilting	
	R4	R5	R6	20 h	28 h
Crop dry matter, g/kg of fresh matter					
After cutting	244	266	282	–	–
Wilted	449[F2]	471[F2]	529[F2]	438	528[F2]
Ensiled	454[F2]	485[F2]	518[F2]	444	528[F2]
pH	5.19	5.23	5.10	5.11	5.24
Chemical composition, g/kg of dry matter					
Ammonia-N	3.3	2.5	2.4	2.9	2.5
Lactic acid	32.7	29.9	29.3	32.0[F2]	29.3
In vitro rumen degradability					
Fiber	0.319[F2]	0.388[F2]	0.465[F2]	0.399	0.382
Crude protein	0.391[F2]	0.503[F2]	0.548[F2]	0.495[F2]	0.466
[a–c,*] $P < 0.05$.					

The wilting before ensiling is more common in regions with dry weather or with well-defined seasons, because the rainfall during the wilting period may cause significative losses than the ensiling wet crop. During the wilting, the crop remains metabolically active, and the cell respiration and proteolysis cause losses, the most important factor is the time until reaching the desired DM. The fast dehydration decreases plant carbon losses and protein degradation. The respiration loss is unavoidable, and its intensity depends on the oxygen, DM, and water-soluble carbohydrate contents. Depending on environmental conditions, the crop containing high level of crude protein may have high proteolysis during wilting, which decreases the silage quality [15].

4.3. Silage additives

In specific cases, when all ensiling techniques and fermentation process are understood and managed properly, the use of additives is necessary to regulate the fermentation process and to obtain high-quality silages. Silage additives can be used to help fixing some historic problems of the crops (low LAB epiphytic, and low DM and soluble sugars contents), oversized silos, silage storage for prolonged time, or silage moved from silo to another structure [17]. In addition, the additives are used to reduce heating and DM losses improving the silage fermentation quality and profitability. Most commercial additives contain more than one active ingredient in order to enhance efficacy and broad range of applicability [10]. According to [18], the additives, basically, have five functions (Table 5). Once again, it is important to emphasize that the use of additives will never correct or fix failures from poor management of the silage-making process.

Table 5. Silage additives [18].

Functions	Examples
Fermentation stimulators	Homofermentative lactic acid bacteria
	Glucose, sucrose, molasses, cereals, wheat, citrus pulp, and enzymes
Fermentation inhibitors	Formic acid, acetic acid, lactic acid, benzoic acid, acrylic acid, citric acid, and sorbic acid
	Formaldehyde, sodium nitrite, sodium metabisulfite, sodium chloride, antibiotics, and sodium hydroxide
Aerobic deterioration inhibitors	Heterofermentative lactic acid bacteria
	Propionic acid, caproic acid, sorbic acid, and ammonia
Nutrients	Urea, ammonia, biuret, and limestone
Moisture absorbents	Citrus pulp, ground corn, cassava meal, straw, and coffee hulls

4.3.1. Fermentation stimulators

The additives that promote the desirable lactic acid fermentation are called fermentation stimulators, by either providing additional fermentable sugars or increasing the LAB population in the ensiled crop.

Additives containing water-soluble carbohydrates will improve the fermentation in crops containing low sugars such as some legumes and tropical grasses. The use of molasses in the ensiling process was a practice widely used in the past to accelerate and increase the lactic acid fermentation. However, it was recommended to be used in relatively high concentrations (40–50 g/kg of fresh matter) and crops containing low DM content showed increase in effluent losses. Due to the high cost and viscosity, which are difficult to apply the molasses, today it is not too used in the farms. Other products or by-products can also be used for the same purpose, but attention should be paid to the availability and cost [19].

Enzyme additives usually are active enzyme combination (cellulases, hemicellulases, and amylases) used to break down the crop fiber and starch to release water-soluble carbohydrates, which could be fermented by LAB. The best results are improvements in silage fermentation and decreases in fiber content. However, the enzymes require certain conditions for maximum activity such as the pH, temperature, surface area, dry matter content, and crop proteases may inhibit enzyme activity. In addition, their positive effects also depend on the LAB initial population, crop characteristics, and application rate. The most suitable role for enzymes may be in combination with microbial inoculants [17, 19].

Inoculants containing homofermentative LAB are used with the purpose of increasing the initial population of this bacteria ensuring efficient fermentation to produce lactic acid. In addition, the use of homofermentative inoculants may accelerate pH reduction because the lactic acid is a stronger acid (p*K*a 3.86) than acetic acid (p*K*a 4.76) [4]; improving the lactic acid:acetic acid ratio consequently reduces dry matter losses. Homofermentative inoculation would also limit degradation and deamination of crop proteins and reduce ammonia production, which increases silage quality [20]. It was observed by [21], when evaluating the effects of homofermentative inoculants in alfalfa silage; they observed that some of the evaluated inoculants, with faster growing and ability to dominate the epiphytic microflora, decreased the pH since the first day of fermentation (Figure 3).

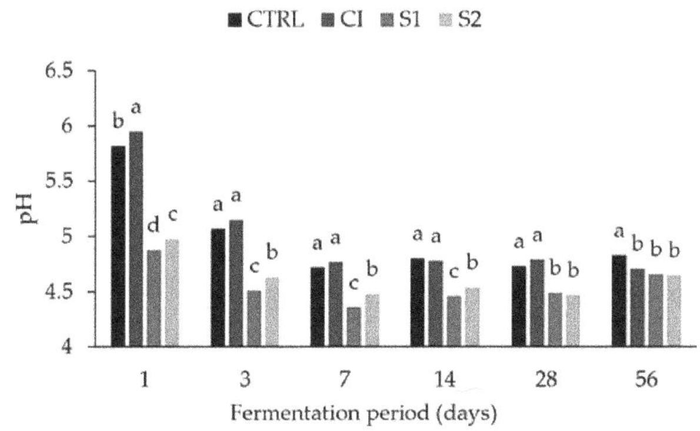

Figure 3. The pH (a), ammonia nitrogen (b) and lactic acid of alfalfa silages as a function of microbial inoculant within each fermentation period. [a–c]**Means followed by different letters in bars are different according to the predicted difference ($P < 0.05$). CTRL = control (without inoculant); CI = commercial inoculant, Sil-All® 4 × 4 W.S. (Alltech, Sao Paulo, Brazil); S1 = Pediococcus acidilactici, Strain 10.6; S2 = P. pentosaceus, Strain 6.16 [21].**

Microbial inoculants include one or more of these bacteria: *Lactobacillus plantarum, L. acidophilus, L. salivarius, Pediococcus acidilactici, P. pentacaceus, Enterococcus faecium*, and *Streptococcus bovis*. Some combinations are used in accordance with the LAB capacity and potential of synergistic actions. For example, the use of *Streptococcus*, which exhibit faster growth and simultaneous drop in pH, combined with *Pediococcus*, which are more tolerant to conditions of temperature, pH, and high dry matter content. However, *Lactobacillus plantarum* is the most common species used [17]. According to [22], the inoculant should be added at a rate that is at least 10% of the epiphytic population to fermentation improvement. For commercial inoculants, recommendation ranges from 1×10^5 to 1×10^6 colony-forming units (cfu)/g of fresh forage.

4.3.2. Fermentation inhibitors

These are all chemical additives that affect the undesirable fermentation and microorganism growth. Based on the same principle of food conservation, several substances are used for this purpose. However, the choice of a suitable additive depends on cost-efficiency and historical occurrence of silage with poor-quality fermentation. Generally, they are used in wet crops with low WSC content and/or high buffer capacity. In addition, in crops containing high WSC, the acid-tolerant yeast can proliferate and decrease the silage quality. Salts of acids have become the most popular fermentation inhibitors, because they are easier and safer to handle [10], and they are effective on controlling yeast growth [23].

4.3.3. Inhibitors of aerobic deterioration

During the feed-out phase, when opening the silo, the presence of oxygen allows the development of molds, yeasts, and aerobic bacteria that consume the silage nutrients. The length of time that silage remains cool and does not spoil after it is exposed to air is called of aerobic stability. There are chemical and biological additives that are used to improve the aerobic stability by inhibit aerobic spoilage, mainly yeasts and acetic acid bacteria, because these microorganisms are responsible to initiate the aerobic deterioration. Generally, the chemical additives are more expensive and difficult to handle than are biological, and successful treatment depends on application rate. However, the variation in the effects when chemical additives are used is lower than the biological additives. Chemical additives with strong antimycotic activity are sorbic and benzoic acid [19, 23]. Besides the use of chemical additives, there is the possibility of using of biological additives based on heterofermentative LAB, such as *Lactobacillus buchneri*, which anaerobically degrade lactic acid to acetic acid and 1,2-propanediol causing a yeast inhibition [10, 23]. Yeast inhibition by organic acids is due to the undissociated form in acid pH. The inhibition effectiveness depends on the dissociation constant (pk) of organic acid; the acids with the highest pk are more effective in inhibiting. The ascending order of pk is formic acid, lactic acid, acetic acid, and propionic acid (3.75, 3.86, 4.76, and 4.87, respectively) [4].

4.3.4. Nutrients

The quality of crop can be improved by supplementation of dietary components that are essential for ruminants through of specific additives at the time of ensiling. In addition, despite of the buffering effect, the urea and ammonia can improve the aerobic stability of silage and increase crude protein content [6]. Grains can be added to increase levels of metabolizable energy in the silage. In other cases, some minerals can be added in order to meet a possible deficiency of the crop to better animal performance [19].

4.3.5. Moisture absorbents

Good results have been obtained in crops with a low DM content (<25%) at the ensiling to prevent excessive effluent losses and clostridial fermentations. Some additives can also improve the nutritive value and final silage quality [6]. Grains can be added to increase moisture absorbent to reduce silage effluent losses [19].

4.4. Using mixed crops

It can be used with several goals always taking advantage of a potential synergistic effect from improvement of soil tillage and fertilization and increased nutritive value, and/or supply the dry matter content and water-soluble carbohydrates to ensure a high-quality silage. Mixing legumes with cereal crops has been to increase grain yields and crude protein of crops while improving soil fertility but can increase the buffering capacity, which can decrease the fermentation efficiency in drops in the pH [24].

5. HOW DOES THE FERMENTATION PROCESS AFFECT SILAGE QUALITY?

All microorganisms present in the silo, crop epiphytic population, and possible contamination primarily consume energy of water-soluble carbohydrates and other compounds for their growth and proliferation. Theoretically, the homolactic fermentation recovers 99% of the energy from glucose. However, in the silage fermentation process, many pathways occur simultaneously with different extensions, beyond the initial cellular respiration and enzymes activity, which are decisive in the final silage quality. Reducing losses by effluent is also important because it contains cellular content with high nutritional value that can contaminate the environment [6, 25]. High-quality silage is the result of adoption of appropriate techniques, starting with soil preparation and fertilization. In addition, the crop must have high DM yield, adequate nutritional value, and good characteristics for fermentation at the ensiling. Actually, even if high-quality crops are harvested efficiently, significant losses in the quality can occur if the ensiling process is inadequate (Table 6).

Table 6. Dry matter losses in silage under good or poor management [26].

Source	Management	
	Good	Poor
Respiration	0–4%	10–15%
Fermentation	4–6%	10–15%
Seepage	0–2%	5–15%
Aerobic storage	5–7%	10–20%
Total	9–17%	20–40%

Forages should be harvested for silage making when they have high nutritional value and the DM content is between 30 and 35%. Therefore, the monitoring of dry matter content at harvest period is essential, because some crops are required to be wilted or ensiled with additives to reach the recommended DM content. The crop must be chopped to about 0.5–1.5 cm length so that the work of packing and taking out is carried out easily. The chopped forage must be well packed in the silo, so less air will be trapped inside the stack, and the peripheral area should have packed more intensely. Filling the silo as quick as possible (within 3 days) limits the forage exposure to air, but each night until it is filled, the stack should be covered. The last step is complete seal with plastic as soon as filling and compaction is completed. In addition, the plastic should be covered, usually with tires or soil to eliminate gases and to prevent damage of the plastic. The packing density at of a good silage should be about 650 kg of fresh silage per cubic meter.

5.1. Chemical composition and nutritive value

Changes are inevitable in chemical composition during the ensiling process; it is due the conversion of soluble carbohydrates into organic acids, as well as degradation of fiber and protein of fresh crop. First, changes in the composition start immediately after cutting, still in an aerobic environment. Early in this phase, enzymes break down fructans, starch, and hemicellulose, releasing simple sugars, and also degrade protein to peptides, amino acids, amides, and ammonia. In addition, during the respiration, soluble carbohydrates are converted to CO_2 and water by releasing heat. If the respiration period is extended, it can increasing losses due the development of molds and yeasts. Also, the heat released by respiration may decrease the digestibility due to the Millard reaction. The heat binds amino acids to the hemicellulose increasing the indigestible fiber and undegradable protein [4].

During LAB fermentation, the soluble carbohydrates are converted to lactic acid, acetic acid, ethanol, $CO_{2,}$ and water, which represents slight losses of DM and energy. However, if there is a clostridial fermentation, which causes major problems in the silage quality, it converts the soluble carbohydrates and amino acids to organic acids, glycine, biogenic amines, ammonia-nitrogen, H_2, and CO_2. The fermentation length is important in the crop preservation. When the fermentation length is extensive, the losses and changes in nutritional value are greater [4].

Another major problem about the silage chemical composition is at the silo opening. With air exposure, the microorganisms, which were inhibited, can proliferate and consume the silage energy. Heating and spoilage during feed-out is one of the greatest contributors to DM losses. In addition, it can produce some substances, like mycotoxins, that may pose risks to animals fed with this silage [26].

5.2. Animal performance

The feed intake is the key constraint limiting performance of ruminant animals fed diets containing forages. Regulation of feed intake in ruminants involves multiple mechanisms and complicated interactions between animal and feed characteristics. Evaluating factors that affect the silage intake of dairy cows, [27] concluded which silage intake can be predicted based on the silage digestibility, total acids, and DM content. Silage intake increased with increasing silage digestibility which was influenced by stage of maturity at harvest. The same authors showed that the total organic acids produced by silage fermentation process depress the silage intake, but it will depend on the proportion of the silage included in the diet. In addition, a positive association between DM content and silage intake, and DM content independently affects the silage fermentation and animal performance.

Feeding spoiled silage can be a big problem, because the deterioration decreases silage digestibility and intake in cattle. In addition, molds in spoiled silage can produce mycotoxins that cause serious health problems in the animals and farmers [26]. Silage additive is one of the ways to try to ensure efficient fermentation and thus obtain high-quality silage. When studies from North America evaluating the effects of silage additives on animal responses were summarized, [28] showed that although not replace good techniques of the ensiling process, the microbial inoculation can improve the silage quality and animal performance. This activity in animal performance is not well understood and might inhibit detrimental microorganisms in both silage and rumen to enhance the animal health and performance [29].

5.3. Animal health

The microorganisms in the microbial epiphytic population are usually nonpathogenic. However, the contamination, especially with the soil, may increase the presence of enterobacteria and spores of clostridium and bacillus in the silage. Therefore, in some cases, the silage can be a contamination source of animal products, such as meat, milk, and cheese, besides affecting the animal health [4]. During the silage fermentation process, a succession of microorganisms and denaturation and production of several compounds occur. However, the main problem is the occurrence of undesirable fermentations, which reduces the nutritive value of silage. Furthermore, the presence of some microorganisms or compounds produced may be a risk to the animal health [26].

Enterobacteria present in the crop may have a small positive effect on the hygienic quality of silage because during the first stage of ensiling, they reduce the nitrate (NO_3^-) to intermediates as nitrite and nitric oxide which inhibit clostridial fermentations. However, enterobacteria are undesirable because they have an endotoxin, which can reduce the silage intake and increase cases of mastitis, besides the less effective fermentation than LAB [4].

The anaerobic environment into the silo is essential for high-quality silage and inhibition of molds that produce mycotoxins. Generally, the mycotoxins in silage are related to molds with high tolerance to CO_2 concentrations. Feeding spoiled silage results in reduced intake, increased abortions, hormonal imbalances, and suppressed immune function. In addition, good ensiling conditions reduce the most of the population of potential pathogens such as *Listeria monocytogenes, Escherichia coli*, and several *Salmonella* species, because they are strongly inhibited by acid pH (<4.5). Actually, the biggest problems are caused by clostridia and bacilli due the ability to form endospore and their presence later in food production systems, requiring special treatment for their elimination [4].

6. FUTURE TRENDS

With the knowledge of the silage process, some techniques are being developed to improve the efficiency of the preservation and production of high-quality silage. The development of monitors for the DM content of the crop at harvest will help farmers to know the crop quality and ensiling characteristics to choice of additives when needed at accurate rates. In addition, the development of specific additives for each culture that are used in the world with ample effect, since the silage fermentation until the animal performance. Furthermore, today many researches are aimed to developing plastic films more resistant and impermeable to oxygen. Through improvements in the plant, breeding is possible to obtain suitable crops for the most different environmental conditions with high quality and productivity, besides the suitable characteristics for silage production and animal performance.

7. CONCISE SUMMARY

Although silage is a widely recognized method, achieving high-quality results is challenging. Commencing with the cultivation of a crop that possesses a significant amount of essential nutrients, which typically comes with a high price tag and demands meticulous attention. Furthermore, the ensiling process requires specialized technology, physical infrastructure such as silos, and plastic sheets for covering. Furthermore, the farmers are unable to bear the potential consequences of losing the entire harvest due to substandard silage. Certain crops, such as corn and sorghum, possess both superior nutritional content and favorable ensiling qualities. However, in order to guarantee the production of high-quality silage, it is often essential to employ procedures such as crop wilting and the application of chemicals, which can increase the overall cost of the operation. The optimal outcome during ensiling is achieved through a rapid and effective fermentation process that results in a decrease in pH. The outcome of LAB fermentation is contingent upon the presence of an anaerobic environment, water activity, and substrate. The homofermentative lactic acid bacteria (LAB) are highly effective in maintaining the crop's properties. Nevertheless, certain heterofermentative LAB are also considered beneficial due to their impact on the aerobic stability of silage.

REFERENCES

1. Doonan, B.M., Kaiser, A.G., Stanley, D.F., Blackwood, I.F., Piltz, J.W., and White, A.K. Silage in the farming system. In: A.G. Kaiser, J.W. Piltz, H.M. Burns, N.W. Griffiths (Eds.), Successful Silage, Chapter 1. New South Wales Dept. of Primary Industry: Orange, 2004, pp. 1–24.
2. Chiba, S., Chiba, H., and Yagi, M. A guide for silage making and utilization in the tropical regions. A publication of the Japanese Livestock Technology Association. Tokyo: Ministry of Agriculture, Forestry and Fisheries, 2005, pp. 29.
3. Adesogan, A.T. Challenges of tropical silage production. In Proc. 15th International Silage Conference. Madison, WI: University of Wisconsin, 2009, pp. 139–154.
4. Pahlow, G., Muck, R.E., Driehuis, F., Oude Elfering, S.J.W.H. and Spoelstra, S.F. Microbiology of ensiling. In: D.R. Buxton, R.E. Muck, J.H. Harrison (Eds.), Silage Science and Technology. 1st ed. Madison, WI: American Society of Agronomy, 2003. pp. 31–94.
5. Woolford, M.K. The silage fermentation. New York, NY: Marcel Dekker. 1984, pp. 23–132.
6. McDonald, P., Henderson, A.R., Heron, S.J.E. Biochemistry of silage. 2nd ed. Marlow: Chalcombe Publication, 1991. 340 p.
7. Pitt, R.E., Muck, R.E., Pickering, N.B. A model of aerobic fungal growth in silage. Grass and Forage Science, v. 46, 1991. pp. 301–312.
8. Kung Jr, L. Aerobic stability of silage. In: Proceedings of the Alfalfa & Forage Symposium and Corn/Cereal Silage Conference, Visalia, California, USA, 2010, pp. 14.
9. Buxton, D.R., and O'Kiely, P. Preharvest plant factors affecting ensiling. In: Silage Science and Technology. D.R. Buxton, R.E. Muck and J.H. Harrison (Eds.), Madison, WI: American Society of Agronomy, 2003. pp. 199–250.
10. Oude Elferink, S.J.W.H., Driehuis, F., Gottschal, J.C. et al. Silage fermentation processes and their manipulation. In: Proceedings of the FAO Eletronic Conference on Tropical Silage, 1999, Rome. Silage making in the tropics with emphasis on smallholders. Rome: FAO, 2000. pp. 17–30.

11. Kleinschmit, D.H. and Kung, L. A meta-analysis of the effects of Lactobacillus buchneri on the fermentation and aerobic stability of corn and grass and small-grain silages. Journal of Dairy Science, v. 89, n. 10, pp. 4005–4013, 2006.

12. Kung Jr, L. a review on silage additives and enzymes. Newark, DE: Department of Animal and Food Sciences University of Delaware, 2010, 19717–1303.

13. Wilkinson, J.M., Bolsen, K.K., Lin, C.J. History of silage. In: D.R. Buston, E.R. Muck, J.H. Herrison (Eds.), Silage Science and Technology. Madison, WI: ASA-CSSA-SSSA, Agronomy, 2003, 42, pp. 1–30.

14. Schroeder, J.W. Silage Fermentation and Preservation. AS1254. Quality Forage. North Dakota State University Fargo. North Dakota: NDSU Extension Service, 2013, 7 p.

15. Muck, R.E.; Moser, L.E.; Pitt, R.E. Postharvest factors affecting ensiling. In: D.R. Buxton, R.E. Muck, J.H. Harrison (Eds.), Silage Science and Technology. Madison, WI: American Society of Agronomy, 2003. pp. 251–304.

16. Spanghero, M., Zanfi, C., Signor, M., Davanzo, D., Volpe, V., Venerus, S. Effects of plant vegetative stage and field drying time on chemical composition and in vitro ruminal degradation of forage soybean silage. Animal Feed Science and Technology. v. 200, pp. 102–106, 2015.

17. Kung Jr, L., Stokes, M.R., Lin, C.J. Silage additives. In: D.R. Buxton, R.E. Muck, J.H. Harrison (Eds.), Silage Science and Technology. 1st ed. Madison, WI: American Society of Agronomy, 2003. pp. 305–360.

18. McDonald, P. The biochemistry of silage. Chichester: John Wiley & Sons, 1981. 218 p.

19. Kaiser, A.G. Silage additives. In: A.G. Kaiser, J.W. Piltz, H.M. Burns, N.W. Griffiths (Eds.), Successful Silage, Chapter 7. New South Wales Dept. of Primary Industry: Orange, 2004, pp. 172–196.

20. Muck, R.E., and Kung, Jr, L. Effect of silage additives on ensiling. Proceedings of the conference on Silage: Field to feedbunk. North American Conference Hershey, PA, 1997. NRAES-99.

21. Silva, V.P., Pereira, O.G., Leandro, E.S., Da Silva, T.C., Ribeiro, K.G., Mantovani, H.C., and Santos, S.A. Effects of lactic acid bacteria with bacteriocinogenic potential on the fermentation profile and chemical composition of alfalfa silage in tropical conditions. Journal Dairy Science. v. 99, n. 3, pp. 1895–1902, 2016.

22. Muck, R.E. Silage microbiology and its control through additives. Revista Brasileira de Zootecnia, Viçosa, v. 39, pp. 183–191, 2010.

23. Da Silva, T.C., Pereira O.G., Kung Jr, L., Silva, L.D., Paula, R.A., Martins, R.M., Silva, V.P., and Ribeiro, K.G. Effect of a chemical additive containing sodium benzoate, potassium sorbate, and sodium nitrite on the microbial populations and aerobic stability of sugar cane. Proceedings of the XVII International Silage Conference, Piracicaba, Brazil, 2015, pp. 148–149.

24. Titterton, M., and Bareeba, F.B. Grass and legume silages in the tropics. In: L. t'Mannetje (Ed.), Silage Making in the Tropics with Particular Emphasis on Smallholders. Proc. FAO Electronic Conference on Tropical Silage. Plant Production and Protection Paper 161, Rome, Italy: Food and Agricultural Organization of the United Nations, 2000.

25. Rotz, C.A. and Muck, R.E. Changes in Forage Quality during Harvest and Storage. In: G.C. Fahey, Jr. et al (Eds.), Forage Quality, Evaluation, and Utilization, Madison, WI: American Society of Agronomy, 1994. pp. 828–868.

26. Adesogan, A.T. Improving forage preservation with additives. Covington, KY: Annual Meeting of the American Forage and Grassland Council, 2013.

27. Huhtanen, P., Rinne, M., and Nousiainen, J. Evaluation of the factors affecting silage intake of dairy cows: a revision of the relative silage dry-matter intake index. Animal, v. 1, pp. 758–770, 2007.

28. Kung Jr., L. and Muck, R.E. Animal Response to Silage Additives. Proceedings from the Silage: Field to Feedbunk North American Conference, Hershey, 1997, NRAES-99, pp. 200–210.

29. Gollop, N., Zakin, V. and Weinberg, Z.G. Antibacterial activity of lactic acid bacteria included in inoculants for silage and in silages treated with these inoculants. Journal of Applied Microbiology, v. 98, pp. 662–666, 2005.

02 THE RELATIONSHIP BETWEEN FERMENTATION AND REDOX POTENTIAL

1. INTRODUCTION: AN OVERVIEW

The fermentation sector has a lengthy historical background, as our human ancestors would occasionally create wine, yogurt, and pickled food. The majority of these fermentation products are associated with the glycolysis and TCA cycle pathways, which necessitate microaerobic or anaerobic conditions in order to prevent the desired products from being oxidized by oxygen. Accurately managing microaerobic or anaerobic conditions is difficult when utilizing a conventional dissolved oxygen electrode due to the probe's limited detection threshold. Hence, the assessment of redox potential, also known as oxidoreduction potential (ORP), is regarded as an optimal alternative method due to its prompt reaction time and exceptional sensitivity to oxidation reactions. In addition, redox potential is also associated with the metabolic network, which includes genes, proteins, and metabolites. Given that cells require a balanced intracellular redox potential, controlling either the intracellular or intercellular redox potential could be useful strategies for redirecting metabolic flux towards certain products. This concept has been utilized to create a diverse array of fermented items. This chapter provides an overview of the fundamental concept of redox potential and its impact on genes, proteins, and metabolites within cells. In addition, this article demonstrates how metabolic modification and process engineering can be used to manage redox potential in different metabolite fermentations, with a specific focus on ethanol generation.

2. BASIC THEORY OF REDOX POTENTIAL

The oxidation-reduction potential (ORP), also known as redox potential, is the measure of a molecule's inclination to gain electrons. The electron transfer process contains two components called a redox pair, in which the oxidizing component

(Ox) attracts electrons and eventually transforms into the reducing component (Red). This relationship is illustrated below:

$$Ox + ne^{-} = Red\ 1 \quad (1)$$

Electrons are exchanged during a redox reaction, in which a pair of oxidation reaction and reduction reaction must be involved. As an illustration, when oxidizing iodide by ferric iron to form iodine, the iodine ion loses two electrons to from iodine (known as oxidation), concurrently ferric ion receives the same amount electrons to form ferrous ion (known as reduction). As a result, a complete redox reaction is established.

Oxidation: $2I^{-} = I_2 + 2e^{-}$

Reduction: $2Fe^{3}+ + 2e^{-} = 2Fe^{2}+$

Redox reaction: $2Fe^{3}+ + 2I^{-} = 2Fe^{2}+ + I_2$

In an aqueous system, the redox potential is related to the capacity of releasing or accepting electrons from all redox reactions. Similar to pH where it indicates the availability of hydrogen ions, the overall redox potential portrays a relative state of gaining or losing electrons. However, the net changes of redox potential are caused by all oxidizing and reducing agents in the aqueous system, not just alkalis and acids that determine pH values.

In 1889, Walter Hermann Nernst (1864–1941; Nobel Prize: 1920) developed an equation to interpret the theory of galvanic cells by taking the changes of Gibbs free energy (ΔG) and the mass ratio into account. The Gibbs free energy is a thermodynamic potential, a reduction of G is a necessary condition for the spontaneity of processes at constant pressure and temperature. The chemical reaction can occur only if the ΔG is negative.

$$E_h = E^0 + \frac{RT}{nF} In \frac{[Ox]}{[Red]} 2 \quad (2)$$

$$\Delta G = -nF\Delta E_h 3 \quad (3)$$

E^0 is the standard redox potential of a system obtained at standard state. Every chemical pair has its own intrinsic redox potential. The greater affinity for electrons, the higher standard redox potential could be. Generally, NAD+/NADH, NADP+/NADPH, GSSG/2GSH, ubiquinone (ox/red), and oxygen/water are some of the most common chemical pairs in cells, whose E^0 were -320, -315, -240, +100, and +820 mV, respectively.

R is the universal gas constant; T is the absolute temperature; F, Faraday constant (96,485 C/mol), is the number of coulombs per mole of electrons, and n is the number of transferred electrons. The equation implies the concentration of species and temperature plays the key roles for redox potential change.

For instance, the reaction of NADH oxidized by oxygen is the final step of electron transport chain during aerobic respiration in mitochondrion. Usually, the reaction involves two redox pairs, just like oxygen/water (+820 mV) and NAD+/NADH (320 mV), thus $\Delta E_h = 820$ mV - (-320 mV) = 1140 mV, $\Delta G = -125$ kJ/mol, which indicates that this process occurs spontaneously due to the negative value of ΔG.

While the Nernst equation is commonly applied in biological systems due to its relevance to electron transfer chains, it is important to note that the redox potential measured by a platinum electrode is not a thermodynamically derived number. It quantifies the redox status in a liquid system by measuring electrical potentials. A live biosystem revolves around cellular development and metabolism, yet it is an open system that does not always reach a state of intracellular equilibrium. However, the importance of redox potential in the operation of biological systems was anticipated about a century ago by two esteemed British scientists at the University of Cambridge [1]. Subsequently, numerous scientists have effectively investigated different connections between the redox potential outside of cells, as measured by an electrode, and the internal biological characteristics.

3. EXTRACELLULAR REDOX POTENTIAL

The extracellular redox potential differs from the intracellular redox state due to the separation of cytomembranes and the maintenance of cellular redox homeostasis. Environmental influences have a crucial role in indirectly altering the cellular redox potential. According to the Nernst Equation, the redox potential is defined solely by the ratio of the oxidative state to the reductive state at a constant temperature, which remains unchanged in most biological processes. Figure 1 depicts three overarching methods for regulating the extracellular redox potential in biological devices.

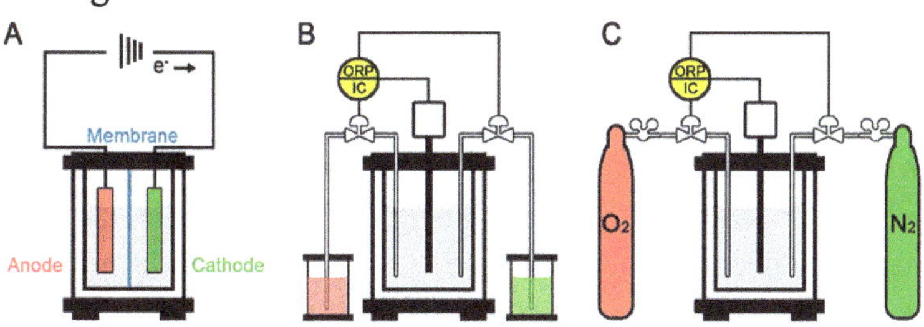

Figure 1. Approaches to control extracellular redox potential. (A) energy input, (B) redox reagents, and (C) gas sparging.

3.1. Regulate the redox potential of the extracellular environment by the input of energy

Bioelectrical reactors (BERs), equipped with anodic and cathodic electrodes, were developed to regulate extracellular redox state in the medium through an external power source. It was used to replace chemical electron donor and acceptor in biosystem. BERs control redox potential at a certain level as easy as tuning a radio. It has been applied to microorganism cultivation and metabolites production [2]. Nevertheless, BERs have been implemented in a laboratory setting or for the production of high-value products in order to compensate for its complicated equipment requirement and extra electrical energy consumption.

3.2. Regulate the redox potential outside of cells using redox reagents

Numerous chemicals with higher or lower standard redox potential than common metabolic components are supplemented into fermentation broth in order to alter environmental redox potential. Some commonly used reductants and oxidants to control extracellular redox potential include $FeCl_3$, Na_2S, potassium ferricyanide, dithiothreitol, cysteine, methyl viologen, neutral red, H_2O_2, and even directly NADH and NAD+ as additives. Unlike BERs requiring the design of a specific reactor, supplementing redox reagents can be employed in any type of bioreactor. However, the disadvantages are obvious: (a) extra chemicals added in media potentially interfere with intended bioprocessing and (b) some chemicals are too costly for industrial fermentation.

Those problems could be solved using substrates with different reducing degree. Girbal and Soucaille [3] used mixed substrates (glucose, glycerol, and pyruvate) to interfere with the intracellular NADH/NAD+ ratio in *Clostridium acetobutylicum*. Snoep et al. [4] chose some energy source substrates, such as mannitol, glucose, and pyruvate, to govern cellular redox potential in *Enterococcus faecalis*.

3.3. Regulate the redox potential outside the cell by introducing gas by sparging

Oxygen and nitrogen are commonly used in aerobic and anaerobic fermentation, respectively. Thus, sparging pure or mixed gases into fermentation broth is one of the desired approaches to avoid unwanted reactions caused by redox salts. Generally speaking, oxygen elevates redox potential and hydrogen depresses it, whereas nitrogen and helium as inert gases remove dissolved oxygen or hydrogen from the medium. Furthermore, by adjusting the ratio of mixed gases, a different redox potential level can be maintained. Carbon monoxide and SO_2 were also utilized to reduce the redox potential sometimes [5]. However, aerating a fermenter during fermentation is considered cost-effective only when air is used. As a mix of nitrogen, hydrogen and helium were applied to regulate redox potential in the above settings, these methods become too luxurious for industrial applications.

3.4. Extracellular redox potential and dissolved oxygen levels

Controlling the level of dissolved oxygen in a fermenter is essential for microorganisms to propagate under optimum physiological condition, not only because oxygen is involved in maintaining cell membrane integrity and function by synthesizing unsaturated fatty acid and sterol, but also for keeping metabolic flux channeling toward the production of desired products.A number of bioreactions toward the syntheses of intended metabolites requires maintaining dissolved oxygen at a proper level. For most microaerobic and anaerobic fermentations, conventional oxygen probe has trouble in distinguishing trace level dissolved oxygen from background noise, and its response time is not sufficient for the purpose of regulating dissolved oxygen level. Even for aerobic fermentation, redox potential still offers much more details about gaseous conditions than that collected from dissolved oxygen measurement [6]. The standard redox potential for the O_2/H_2O pair has the highest value among typical metabolites related to microbial metabolism during fermentation. If electrons were transferred to acceptors, oxygen must be the preferable choice even though its concentration is lower than other metabolites. Therefore, redox potential is much more sensitive in monitoring the presence of a trace amount of dissolved oxygen under microaerobic and anaerobic conditions.

4. INTRACELLULAR REDOX POTENTIAL

Currently, advanced technologies, such as a nanosensor that can embed into individual cells, have been developed to measure intracellular redox potential directly for in-depth understanding on intracellular redox balance and its impact on cell physiology and metabolism. However, the indirect approaches, such as the measurement of NAD(P)H pools, NAD(P)+/NAD(P)H, GSH/GSSG, and the total oxidization power, are still commonly adopted to monitor the distribution of intracellular redox potential.

4.1. Universal redox pairs in a cell

A conjugate pair that constitutes a complete redox reaction is the fundamental of metabolic network in a cell. Many metabolic functions are realized through keeping intracellular redox balance with the main redox pairs, such as glutathione (GSH)/ glutathione disulfide (GSSG), thioredoxin (TrxSS/Trx$(SH)_2$), nicotinamide adenine dinucleotide (NAD), and nicotinamide adenine dinucleotide phosphatase (NADP). These redox systems, such as NADP+/NADPH, GSSG/2GSH, and TrxSS/Trx$(SH)_2$, are not isolated systems. Both the Trx and GSH systems use NADPH as a source of reducing equivalents; thus, they are thermodynamically connected to each other. The role of NAD(P)+/NAD(P)H in redox reaction is illustrated in Figure 2.

Both glutathione (GSH) and thioredoxin are important reducing agents in all organisms, involved in cell oxidative stress response where they play an antioxidant role. Glutathione is a tripeptide (glutamine, cysteine, and glycine) that prevents damage to cellular components caused by reactive oxygen species such as free radicals and peroxides, lipid peroxides, and heavy metals. Thioredoxin is another class of small redox proteins with thiol system in the cell, which appears in many crucial biological processes, including redox signaling.

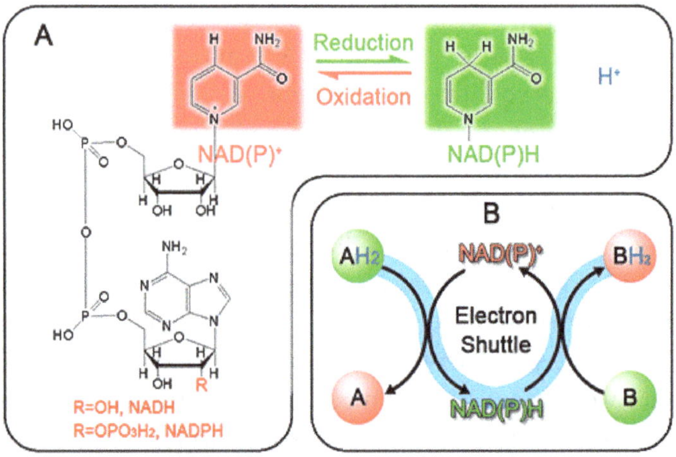

Figure 2. The structure (A) and function (B) of NAD(P)H.

The coenzymes are essential electron carriers in cellular redox reactions with the oxidized form NAD(P)+ and the reduced form NAD(P)H. The reduction reaction requires an input of energy and the oxidation reaction is exergonic. During carbohydrate metabolism, NADH plays as a notable reducing substance in catabolism, whereas NADPH, the other reducing component connected to anabolism, favors formation of amino acids, fatty acids, and nucleic acids. There are 129 enzymes that need NAD+ as cofactor in order to serve 931 redox reaction and 108 enzymes that require the involvement of NADP+ as cofactor in order to catalyze 1099 redox reaction (KEGG, 2016-3).

4.2. Oxidation-reduction action occurring across the membrane

Cytosol is isolated from the extracellular environment by a selectively permeable cytomembrane, which not only prevents the main redox pair escaping from the plasma freely but also conditionally allows the external redox chemicals to enter into the cytoplasm. As shown in Figure 3, chemicals with different reduction degrees, such as dithiothreitol (DDT), diamine, hydrogen peroxide, and oxygen, can unrestrictedly cross the membrane bilayer, causing the changes to the intracellular redox potential. However, most of these chemicals are prohibited to across the membrane. In another scenario, membrane proteins, such as oxidoreductase, involved in electron transport will respond and change the extracellular redox potential. For example, ferric reductase assists ferrous iron transport across the cell membrane [7]. Hydrogenase facilitates electron flow through the membrane with the conversion of NADH and NAD+ [8]. A low redox potential level results in the changes of thiol and disulfide balance on membrane proteins, making the membrane more permeable to protons [9]. A thiol-rich membrane protein transduces external GSH reducing power across the erythrocyte membrane, which can be explained as a thiol/disulfide exchange mechanism [10].

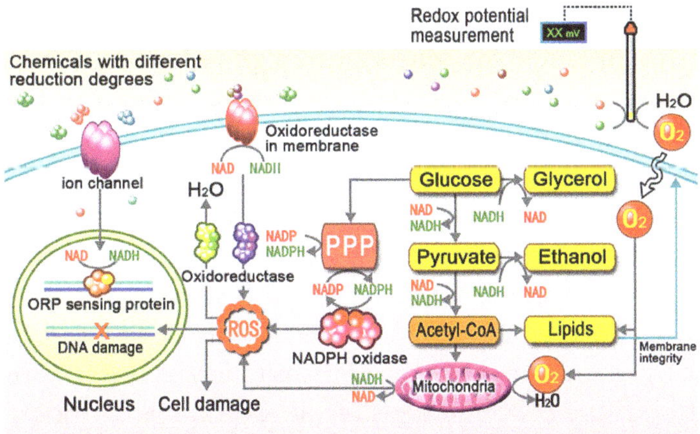

Figure 3. Intracellular redox response to extracellular redox potential and effects of redox potential on cellular metabolism and stress response.

4.3. The impact of redox potential on a cell

The influences of redox potential on enzymes activity have also been reported. Almost all enzymes related to oxidation–reduction reaction are redox potential sensitive, such as alcohol dehydrogenase, D-glyceraldehyde-3-phosphate dehydrogenase, quinone reductase (involved in quinone detoxification), NADH diphosphatase (involved in peroxisomal function), ubiquinone oxidoreductase (catalyzing the oxidation of NADH in the respiratory chain or in cytoplasm), mitochondrial NADH kinase (response to oxidative stress), and so on. The above-mentioned proteins have been investigated in *Saccharomyces cerevisiae* in the past decades. Numerous proteins contain sulfhydryl groups (PSH) due to their cysteine content. In fact, the concentration of PSH groups in cells and tissues is much greater than that of GSH. These groups can be present as thiols (-SH), disulfides (PS-SP), or mixed disulfides; Hsp33 as a possible chaperone and cysteine protease in heat shock protein families is regulated by redox potential, whose conformation changes from reduced state to oxidized state with the exposure of hydrophobic surface [11]. Being a key regulator of glutathione and, in turn, of redox potential, the identification of GSTp as, a JNK regulator, provides an important link between cellular redox potential and the regulation of stress kinase activities [12].

Gene expression is controlled by redox states as well. It has been reported that overexpressing genes related to redox process in *Escherichia coli* resulted in the decrease of NADH/NAD+ ratio, which improve the cell growth profiles, because sufficient NAD+ is required to oxidize carbohydrate substrate during cell growth [13]. *GPD2* encodes NAD-dependent glycerol 3-phosphate dehydrogenase, the key enzyme of glycerol synthesis, and is essential for cell survival under osmotic and low redox potential conditions. Unlike its homologous gene *GPD1* controlled by high osmolality glycerol response pathway, *GPD2* is regulated under anoxic conditions or, more accurately, oxygen-independent reducing environment [14]. *YAP1*, a transcription factor for sensing the high redox state (e.g. H_2O_2), usually exists in the cytoplasm but is transferred into nucleus to activate the transcription of antioxidant genes *SOD1*, *TWF*, *TRX2*, *GLR1*, and *GSH1*, when Yap1p C-terminal region with three conserved cysteine residues is oxidized in response to oxidative stress [15]. A redox sensing protein (RSP) binds transcriptional regulation regions located upstream from *adh*A, *adh*B, and *adh*E as a transcriptional repressor. The structure of RSP was changed from α-helix to β-sheet rich conformation when redox potential declined by adding NADH. Meanwhile, the repression of an alcohol dehydrogenase transcription caused by RSP was reversed [16]. Thioredoxin reduces cysteine moieties in the DNA-binding sites of several transcription factors and is therefore important in gene expression [17].

External redox potential correlates the net balance of intracellular reducing equivalents and the changes in the cellular redox environment can alter signal transduction, DNA and RNA synthesis, protein synthesis, enzyme activation, and even regulation of the cell cycle. Thus, monitoring and controlling environmental redox potential helps to elucidate cellular physiology and intracellular metabolic interaction.

5. REDOX POTENTIAL AND METABOLIC FLUX ARE TWO INTERCONNECTED CONCEPTS IN BIOCHEMISTRY.

Strategies to control intracellular redox potential can be developed by altering intracellular redox potential pools, consequently resulting in redistribution of metabolic profiles. However, cells have a series of built-in mechanisms to adjust their own intercellular redox balance by cofactor regeneration through the oxidoreductase-harboring genes, including mitochondrial alternative oxidase (AOX), formate dehydrogenase (FDH), cytoplasmic H_2O-forming NADH oxidase (NOX), and mitochondrial NADH kinase (POS5). Therefore, modification of these genes is a promising strategy to "design" a robust strain subjected to redox regulation through extracellular manipulation, although such an alternation may result in unexpected outcomes.

5.1. Alternative oxidase

The alternative oxidase (AOX, EC: 1.10.3.11), also named ubiquinol oxidase, forms a part of the electron transport chain in mitochondria. The function of this oxidase is believed to dissipate excess reducing power. The reaction catalyzed by AOX oxidase (ubiquinol oxidase) is shown in Reaction (4).

$$\text{ubiquinone} + H^+ + \text{NADH} = \text{ubiquinol} + \text{NAD}^+ \quad (4)$$

When a cell subjected to increasing glycolytic fluxes under aerobic conditions, a decrease in respiratory capacity is caused by the presence of excess glucose that repressed respiratory pathways. Introducing a heterologous alternative oxidase into *S. cerevisiae*, increased metabolic flux toward respiration and reduced aerobic ethanol formation [18]. In other investigation, the introduction of AOX pathway improved reactive oxygen species and pyruvate levels simultaneously under stressful conditions, such as suboptimal temperature and hyperosmotic pressure [19].

5.2. Formate dehydrogenase

Formate dehydrogenases (FDH, EC: 1.2.1.2) are a set of enzymes that catalyze the oxidation of formate to carbon dioxide (see Reaction 6), donating electrons to a second substrate, such as NAD+ or cytochrome. NAD+-dependent formate dehydrogenases are important in methylotrophic yeast and bacteria and are vital in the catabolism of C1 compounds, such as methanol.

$$\text{formate} + NAD^+ = CO_2 + NADH + H^+ \qquad (5)$$

As the FDH gene from *Candida boidinii* was introduced into *Paenibacillus polymyxa*, highly expressed exogenous FDH increased NADH/NAD+ and the titers of NADH-dependent products such as lactic acid and ethanol, while resulting in significantly decreased acetoin and formic acid [20]. In addition, the increased capacity of a FDH gene in *Bacillus subtilis* efficiently enhanced the production of 2,3-butanediol and decreased the formation of acetoin through increasing the availability of NADH [21]. In another case, an engineered strain for the conversion of D-fructose to allitol was developed by constructing a multienzyme coupling pathway and cofactor recycling system in *E. coli*. FDH gene was used to support the cofactor recycling system for the availability of NADH [22].

5.3. NADH oxidase

NADH oxidase (NOX, EC: 1.6.3.4) is a membrane-associated enzyme that catalyzes the production of superoxide, a reactive free radical, by transferring one electron from NADH to oxygen as the electron acceptor (see Reaction 7). It is considered one of the major sources of producing superoxide anions in humans as well as bacteria, subsequently used in oxygen-dependent killing mechanisms for invading pathogens.

$$2H^+ + 2NADH + O_2 = 2H_2O + 2NAD^+ \qquad (6)$$

Glycerol is a main by-product in the 2,3-butanediol metabolic pathways. To minimize glycerol accumulation by an engineered *S. cerevisiae*, the *Lactococcus lactis* NOX gene was inserted and expressed, resulting in substantial decreases in intracellular NADH/NAD+ ratio. As a result, the carbon flux was redistributed from glycerol to 2,3-butanediol [23]. NADH oxidase was also expressed with l-arabinitol dehydrogenase in *E. coli* to efficiently produce l-xylulose. Thus, the efficiency above 96% for the conversion of l-arabinitol into l-xylulose was achieved under optimized conditions [24].

5.4. NADH kinase

NADH kinase (like POS5, EC: 2.7.1.86) catalyzes the replacement reaction with two substrates ATP and NADH and two products ADP and NADPH (see Reaction 8). It provides a key source of the important cellular antioxidant NADPH.

$$ATP + NADH = ADP + NADPH \quad (7)$$

NADPH is a key cofactor for carotenoid biosynthesis. *Corynebacterium glutamicum* was always used for the production of amino acids, such as L-isoleucine. By implementing NADPH-supplying strategies based on NAD kinase (PpnK), NADH kinase, glucose-6-phosphate dehydrogenase (Zwf), and PpnK coupling with Zwf, the expression of all genes increased both the intracellular NADPH concentration and the L-isoleucine production [25]. Researchers constructed the NADPH regenerators of heterologous NADH kinase to increase the availability of NADPH and resulted in a superior S-adenosylmethionine production in *E. coli* without requiring L-methionine addition [26]. When a *S. cerevisiae* strain-producing carotenoid was constructed by overexpressing glucose-6-phosphate dehydrogenase and NADH kinase individually, the final product β-carotene yield increased by 18.8% and 65.6%, respectively. Thus, NADPH supply improved by overexpression of NADH kinase is more important than glucose-6-phosphate dehydrogenase [27].

6. UTILIZATION OF REDOX POTENTIAL IN FERMENTATION PROCESSES

Controlling redox potential at a desired level alters the intracellular metabolic flow in order to favor the formation of desired product(s). Many researches have been conducted in this regard with a large number of examples for enhanced production of metabolites under redox potential–controlled conditions. Most studied metabolites using redox potential–controlled approaches are hydrogen, pyruvate, 1,3-propanediol, butanol, and 2,3-butanediol, and the following metabolites are reviewed but provided with references: acetoin [28], succinic acid [29], xylitol [30], and so on.

6.1. Hydrogen

Hydrogen, as a clean and high-combustion energy in widespread areas, can be generated by fermentative anaerobes. Hydrogen production from anaerobic fermentation by bacteria demands reducing level because the standard redox potential of $H_2/H+$ is low. Zhang et al. [8] showed that the addition of NAD+ during hydrogen fermentation by *Enterobacter aerogenes* resulted in the increase of overall hydrogen. Nakashimada et al. [31] investigated *E. aerogenes* for its hydrogen production under different intracellular redox state through the utilization of different substrates bearing various reduction degrees. Low redox potential accelerated the NAD(P)H-dependent hydrogenase activity in membrane and favors high H_2 evolution capability. Ren et al. [32] assessed H_2 production during butyric acid fermentation, propionic acid fermentation, and ethanol fermentation by controlling redox potential and pH simultaneously. Besides, the NAD+ synthetase encoded by nadE gene was homologously overexpressed in *E. aerogenes* to decrease the NADH/NAD+ ratio and thus enhanced hydrogen yield [33].

6.2. Pyruvate

Pyruvate, which is produced during the process of glycolysis, is a valuable precursor for the production of several pharmaceuticals and agrochemicals. Additionally, it is currently utilized in the food sector. Through the integration of adaptive evolution and cofactor engineering techniques, a collection of genetically modified yeasts capable of synthesizing pyruvate exclusively from glucose as the primary carbon source was successfully generated. As a result, the engineered strains were capable of producing 75.1 g/L of pyruvate, which represents a 21% increase compared to the natural strain. The strain achieved a production yield of 0.63 g pyruvate per gram of glucose. [34].

6.3. Propanediol

1,3-propanediol, made from glycerol under anaerobic condition, is a monomer for producing various industrial polymers. Du et al. [35] demonstrated that controlling redox potential at -190 mV was preferable for *Klebsiella pneumoniae* to ferment glycerol into 1,3-propanediol. They further developed a redox potential-based strategy for screening high productivity strain using the correlation between redox potential level and growth rate [36]. Zheng et al. [37] regulated redox potential under low levels (-200 and -400 mV) during 1,3-propanediol fermentation in order to avoid the accumulation of by-product. Wu et al. [38] engineered the pathways of 2,3-butanediol and formic acid in a recombinant *K. pneumonia* to improve 1,3-propanediol production. The intracellular metabolic flux was redistributed pronouncedly by shrinking all nonvolatile by-products and supplying the availability of NADH. Jain et al. [39] established novel metabolic pathways for 1,2-propanediol in *E. coli* by disrupting the major competing pathways for acetate production as well as the ubiquinone biosynthesis pathway that conserved more NADH.

6.4. Butanol

Butanol attracts public attentions due to its favorable physicochemical properties for blending with or for directly substituting for gasoline. Fermentation of butanol by *C. acetobutylicum* is generally a biphasic process consisting of acidogenesis and solventogenesis. It has been reported that an earlier initiation of solvent genesis under redox potential control at -290 mV could increase solvent production by 35% [40]. Li et al. [41] supplemented nicotinic acid, the precursor of NADH and NADPH, into the growth medium, and led to a significant increase of NADH and NADPH levels for a wild-type *Clostridium sporogenes* strain. As a result, the metabolic pattern was shifted toward the production of more reduced metabolites, in which butanol production was then enhanced. Bui et al. [42] constructed the recombinant *K. pneumoniae* by overexpressing the genes *kivD*, *leuABCD*, and *adhE1*, with several NADH regeneration strategies to overcome redox imbalance, including the introduction of NAD+-dependent enzymes or elimination of the NADH competition pathway (1,3-propanediol synthesis). The NADH/NAD+ ratio was increased resulting in butanol titer increase [42].

6.5. Butanediol

2,3-butanediol (2,3-BD) is a promising bulk chemical with extensive industry applications. In order to enhance the production of 2,3-BD, various strategies for increasing the NADH availability were developed through regulation of low dissolved oxygen, supplement of reducing substrates and gene modification. An *udhA* encoding transhydrogenase was introduced and more NADH from NADPH was provided to allow the enhancement of production [43]. For the same reason, two NADH regeneration enzymes, glucose dehydrogenase and formate dehydrogenase, were introduced into *E. coli* with 2,3-butanediol dehydrogenase, respectively [44]. In other case, an engineered *S. cerevisiae* harboring NADH oxidase gene (noxE) from *L. lactis* minimized glycerol accumulation, because intracellular NADH/NAD+ ratio was decreased substantially and carbon flux was redirected to 2,3-BD from glycerol [23].

7. DESIGNING THE PROCESS OF REDOX POTENTIAL: A CASE STUDY ON THE FERMENTATION OF ETHANOL

Fuel ethanol, the most successful renewable energy so far, is produced worldwide and applied in transportation as alternative to fossil fuel. However, the high cost associated with bioethanol production urges researchers to innovate new fermentation technologies like redox potential–controlled ethanol fermentation. In this section, the role of redox potential in *S. cerevisiae* pathways, the correlation between yeast growth and redox potential, and the application of redox potential to very high gravity fermentation will be reviewed.

7.1. The role of redox potential in yeast pathway

S. cerevisiae has been considered as a model microorganism, whose genome, proteome, and relevant pathway information are almost unveiled. As illustrated in Figure 4, glucose is converted into small molecules through the coupling of redox reactions, in which NADH plays an essential role in key metabolites production such as ethanol, glycerol, and lactate. In this process, glucose is oxidized by NAD+ to make pyruvate and NADH. The surplus of reducing power is then balanced by the formation of glycerol and ethanol, where NAD+ is restored. When the growth environment favors the production of acetic acid, the implementation of redox potential control can alter the trend, leading to a more reduced state toward ethanol production.

Compared with other control parameters, such as temperature, pH, and the ingredients of medium, redox potential has less influence on improving fermentation results. Hence, the implementation of redox potential control in ethanol fermentation was not popular until the new concept of "very high gravity (VHG)" was proposed. VHG is generally regarded as the final ethanol concentration is greater than 15% (v/v) or initial glucose concentration is greater than 250 g/L. VHG is a promising technology to reduce energy consumption and labor cost, as well as elevate the efficiency of the fermenter. However, high sugar concentration depresses cell growth and bioconversion. Redox potential control helps cells survive from osmotic pressure and ethanol toxicity by constructing healthier membranes or other potential mechanisms. Yeast grown under VHG condition without redox potential control requires much longer fermentation times in order to completely utilize substrate [45]; therefore, the improvement of ethanol production by redox potential control would be expected.

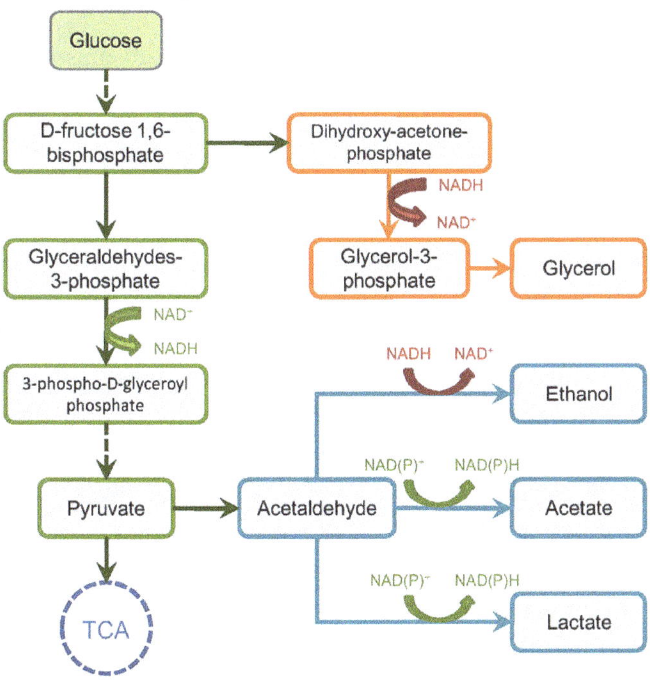

Figure 4. Metabolic pathway of glucose degradation in Saccharomyces cerevisiae.

Lin et al. [45] controlled redox potential under -150 mV, -100 mV, and no control conditions and demonstrated that VHG ethanol fermentation under -150 mV resulted in the highest final ethanol concentration and the highest ethanol-to-glucose yield. Compared with the case of 200g glucose/L, the effect of redox potential control becomes significant under VHG conditions [45]. Jeon and Park [46] cultivated *Zymomonas mobilis* and *S. cerevisiae* to produce ethanol in two separate compartments of an electrochemical bioreactor. The results showed that *Z. mobilis* favors the reducing environment, but *S. cerevisiae* produced more ethanol under higher redox potential conditions [46]. Na et al. [47] observed that ethanol production was enhanced in the anode compartment than in the cathode one, although the reduced environment would be better for fermentation process.

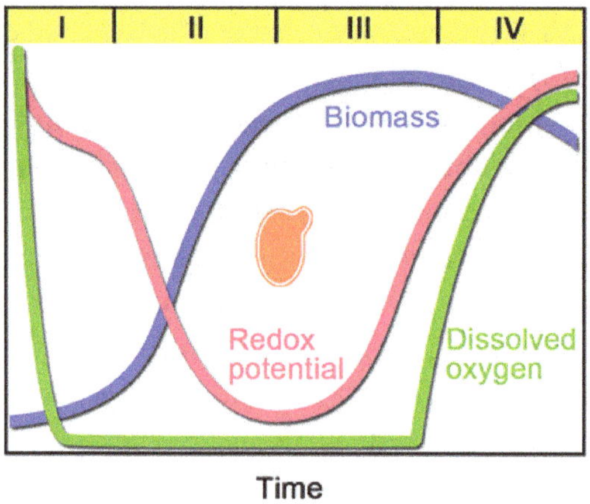

Figure 5. Profiles of redox potential, biomass, and dissolved oxygen.

The correlation between cell development and redox potential: During ethanol fermentation, fluctuations in redox potential occur due to two primary substances: the electron donor NAD(P)H, which is produced during dissimilatory activities such as glycolysis, and assimilatory processes such as biomass synthesis; and the electron acceptor oxygen, which is dissolved by sparging and/or agitation. The redox potential profiles are therefore associated with cellular activity and oxygen levels. A typical redox potential profile exhibits a shape similar to that of a bathtub curve. Initially, yeast was introduced into the sterilized medium with a redox potential equivalent to the typical oxygen levels. Yeast utilizes oxygen as the ultimate electron acceptor during the process of respiration to rapidly reproduce, resulting in a significant decrease in redox potential (Stage I, Figure 5). Yeast undergoes a shift in its respiratory needs from aerobic to anaerobic phases when there is a significant decrease in dissolved oxygen. This transition is accompanied by changes in gene expression and pathways, as seen in Figure 5 between Stage I and II. Following adjustment, yeast cells see an increase in their growth rate during the exponential phase, accompanied by fast glucose consumption. While ethanol generation is theoretically a redox neutral process, the inclusion of reducing substrates such as sugar tends to decrease the redox potential of fermentation. The redox potential continues to decrease as fermentation progresses and may reach a minimum of -300 mV if there are no additional substances that can oxidize present in the fermentation broth (Stage II-III, Figure 5). The lowest dip in redox potential level (Stage III, Figure 5) is observed as a result of substrate depletion and the reduction in cell viability caused by ethanol toxicity. During the last stages of fermentation, a sudden rise in redox potential is caused by continuous aeration or thorough agitation. From a technical standpoint, the emergence of an upward slope indicates that the fermentation process is nearing completion. (Stage IV, Figure 5).

7.3. Process design using redox potential

The performance of VHG ethanol fermentation can be further improved by (1) searching for the optimal redox potential setting and (2) extending redox potential control period to prolong the exponential growth phase. Three redox potential control schemes are collected [48]. The simple aeration-controlled scheme (ACS) has a short redox potential–controlled period. For glucose-controlled feeding scheme (GCFS), glucose was supplemented along with dissolved oxygen presented in the feed stream. For combined chemostat and aeration-controlled scheme (CCACS), a constant glucose was fed along with air supply determined by redox potential–controlled device. The GCFS extends the redox potential–controlled period by offering enough glucose for yeast propagation and maintaining the low residual glucose. As a result, the ethanol yield is increased noticeably. The operation of GCFS as a fed batch, as such the buildup of ethanol causes yeast cessation, resulting in incomplete fermentation. The CCACS is a set of continuous equipment that feeds the fresh medium into a fermenter and discharge spent broth into aging vessels at a constant dilution rate. Sterilized air was used to adjust the fermentation redox potential at a predetermined level. In the chemostat fermenter, both intracellular and extracellular factors should reach their respective steady states. Thus, constant growth rate and yeast viability are sustained under a preset redox potential level, which is helpful to prolong the redox potential–controlled duration and to maximize the benefits from redox potential control. The CCACS achieved the longest controlled period and the highest ethanol yield among all three schemes. However, a chemostat device alone could not result in zero glucose discharge. The incorporation of aging vessel design into fermentation operation thus was developed [49].

8. PROSPECTIVE RESEARCH ON REDOX POTENTIAL AND FERMENTATION

While numerous fermentation processes have been successfully established for long-term operation, achieving cost savings is an ongoing endeavor, especially when it comes to producing biofuels and bio-based compounds in large quantities.

Each cent saved is guaranteed to yield substantial economic benefits. Given that redox reactions and homeostasis serve as the foundation for intracellular metabolism, the ability to monitor and regulate the redox potential status within a cell has the potential to redirect the flow of metabolic substances and energy. Multiple studies have demonstrated and verified that precise regulation of redox potential can modify cellular metabolism, hence improving the conversion of specific compounds.

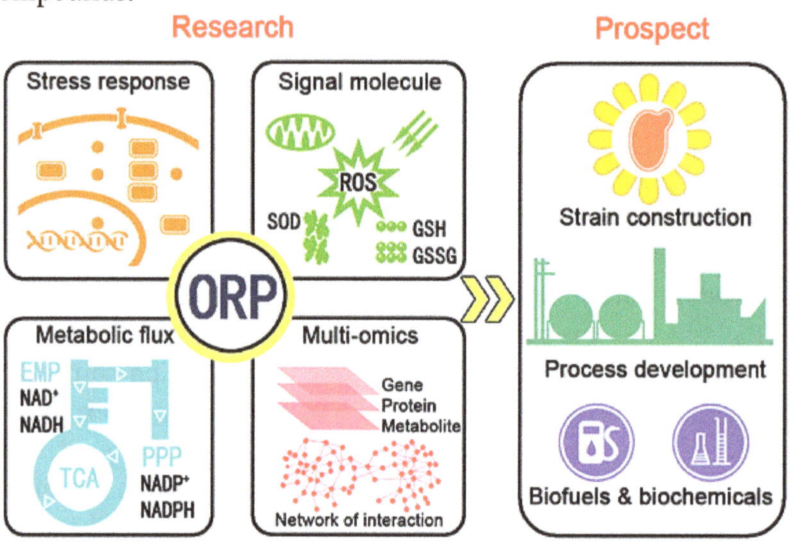

Figure 6. Research and prospect in redox potential-controlled fermentation.

To identify the effects of redox potential control on various levels, such as gene expression,proteinbiosynthesis,andbiomolecularinteractingnetwork,itisnecessary to use an integrated approach that includes technologies capable of detecting intracellular redox potential levels (Figure 6). The fundamental mechanism of this phenomena can be further clarified by studying it at the molecular and bioprocess engineering levels. Greater acquisition of specific information leads to enhanced utilization of redox potential control in various applications. As a result, strong strains and efficient procedures can be designed to achieve high levels of production. The future outlook for controlling redox potential is really appealing. The fermentation process will be conducted using genetically modified strains that have been engineered to have a specific redox potential balance. The strain will undergo stringent management by maintaining a precise redox potential level. The metabolic flow profiles will be evaluated at various redox potential levels in order to optimize the generation of desired metabolites and identify potential bottlenecks for strain enhancement. Anticipated benefits are expected to arise from the development of new fermentation technology that is controlled by redox potential.

REFERENCES

1. Needham J G. Curtis gates lloyd. Science. 1926;64:569-570. DOI: 10.1126/science.64.1667.569-a
2. Thrash J C, Coates J D. Review: Direct and indirect electrical stimulation of microbial metabolism. Environmental Science Technology. 2008;42:3921-3931. DOI: 10.1021/es702668w
3. Girbal L, Soucaille P. Regulation of Clostridium acetobutylicum metabolism as revealed by mixed-substrate steady-state continuous cultures - role of NADH/NAD ratio and ATP pool. Journal of Bacteriology. 1994;176:6433-6438.
4. Snoep J L, Joost M, Demattos T. Effect of the energy-source on the NADH/NAD ratio and on pyruvate catabolism in anaerobic chemostat cultures of Enterococcus faecalis nctc-775. FEMS Microbiology Letters. 1991;81:63-66.
5. Kukec A, Berovic M, Celan S. The role of on-line redox potential measurement in Sauvignon blanc fermentation. Food Technology and Biotechnology. 2002;40:49-55.
6. Kwong S C W, Rao G. Effect of reducing agents in an aerobic amino-acid fermentation. Biotechnology and Bioengineering. 1992;40:851-857. DOI: 10.1002/bit.260400713
7. Dancis A, Klausner R D, Hinnebusch A G. Genetic-evidence that ferric reductase is required for iron uptake in Saccharomyces cerevisiae. Molecular and Cellular Biology. 1990;10:2294-2301. DOI: 10.1128/MCB.10.5.2294
8. Zhang C, Ma K, Xing X H. Regulation of hydrogen production by Enterobacter aerogenes by external NADH and NAD(+). International Journal of Hydrogen Energy. 2009;34:1226-1232. DOI: 10.1016/j.ijhydene.2008.11.070
9. Riondet C, Cachon R, Wache Y. Changes in the proton-motive force in Escherichia coli in response to external oxidoreduction potential. European Journal of Biochemistry. 1999;262:595-599. DOI: 10.1046/j.1432-1327.1999.00429.x
10. Ciriolo M R, Paci M, Sette M. Transduction of reducing power across the plasma membrane by reduced glutathione. European Journal of Biochemistry. 1993;215:711-718. DOI: 10.1111/j.1432-1033.1993.tb18083.x
11. Raman B, Kumar L V S, Ramakrishna T. Redox-regulated chaperone function and conformational changes of Escherichia coli Hsp33. FEBS Letters. 2001;489:19-24. DOI: 10.1016/s0014-5793(01)02074-9
12. Adler V, Yin Z, Tew K D. Role of redox potential and reactive oxygen species in stress signaling. Oncogene. 1999;18:6104-6111. DOI: 10.1038/sj.onc.1203128

13. Singh A, Lynch M D, Gill R T. Genes restoring redox balance in fermentation-deficient E. coli NZN111. Metabolic Engineering. 2009;11:347-354. DOI: 10.1016/j.ymben.2009.07.002

14. Ansell R, Granath K, Hohmann S. The two isoenzymes for yeast NAD(+)-dependent glycerol 3-phosphate dehydrogenase encoded by GPD1 and GPD2 have distinct roles in osmoadaptation and redox regulation. The EMBO Journal. 1997;16:2179-2187. DOI: 10.1093/emboj/16.9.2179

15. Estruch F. Stress-controlled transcription factors, stress-induced genes and stress tolerance in budding yeast. FMES Microbiology Reviews. 2000;24:469-486. DOI: 10.1111/j.1574-6976.2000.tb00551.x

16. Pei J, Zhou Q, Jing Q. The mechanism for regulating ethanol fermentation by redox levels in Thermoanaerobacter ethanolicus. Metabolic Engineering. 2011;13:186-193. DOI: 10.1016/j.ymben.2010.12.006

17. Schafer F Q, Buettner G R. Redox environment of the cell as viewed through the redox state of the glutathione disulfide/glutathione couple. Free Radical Biology and Medicine. 2001;30:1191-1212. DOI: 10.1016/S0891-5849(01)00480-4

18. Vemuri G N, Eiteman M A, McEwen J E. Increasing NADH oxidation reduces overflow metabolism in Saccharomyces cerevisiae. Proceedings of the National Academy of Sciences of the United States of America. 2007;104:2402-2407. DOI: 10.1073/pnas.0607469104

19. Dinakar C, Vishwakarma A, Raghayendra A S. Alternative oxidase pathway optimizes photosynthesis during osmotic and temperature stress by regulating cellular ROS, malate valve and antioxidative systems. Frontiers in Plant Science. 2016;7:68. DOI: 10.3389/fpls.2016.00068

20. Zhang L, Xu Y Y, Gao J. Introduction of the exogenous NADH coenzyme regeneration system and its influence on intracellular metabolic flux of Paenibacillus polymyxa. Bioresource Technology. 2016;201:319-328. DOI: 10.1016/j.biortech.2015.11.067

21. Yang T W, Rao Z M, Hu G Y. Metabolic engineering of Bacillus subtilis for redistributing the carbon flux to 2,3-butanediol by manipulating NADH levels. Biotechnology for Biofuels. 2015;8:129. DOI: 10.1186/s13068-015-0320-1

22. Zhu Y M, Li H Y, Liu P P. Construction of allitol synthesis pathway by multi-enzyme coexpression in Escherichia coli and its application in allitol production. Journal of Industrial Microbiology & Biotechnology. 2015;42:661-669. DOI: 10.1007/s10295-014-1578-1

23. Kim J W, Seo S O, Zhang G C. Expression of Lactococcus lactis NADH oxidase increases 2,3-butanediol production in Pdc-deficient Saccharomyces

cerevisiae. Bioresource Technology. 2015;191:512-519. DOI: 10.1016/j.biortech.2015.02.077

24. Gao H, Kim I W, Choi J H. Repeated production of L-xylulose by an immobilized whole-cell biocatalyst harboring L-arabinitol dehydrogenase coupled with an NAD(+) regeneration system. Biochemical Engineering Journal. 2015;96:23-28. DOI: 10.1016/j.bej.2014.12.017

25. Shi F, Li K, Huan X J. Expression of NAD(H) kinase and glucose-6-phosphate dehydrogenase improve NADPH supply and l-isoleucine biosynthesis in Corynebacterium glutamicum ssp lactofermentum. Applied Biochemistry and Biotechnology. 2013;171:504-521. DOI: 10.1007/s12010-013-0389-6

26. Chen Y W, Xu D B, Fan L H. Manipulating multi-system of NADPH regulation in Escherichia coli for enhanced S-adenosylmethionine production. RSC Advances. 2015;5:41103-41111. DOI: 10.1039/c5ra02937f

27. Zhao X, Shi F, Zhan W. Overexpression of ZWF1 and POS5 improves carotenoid biosynthesis in recombinant Saccharomyces cerevisiae. Letters in Applied Microbiology. 2015;61:354-360. DOI: 10.1111/lam.12463

28. Zhang X, Zhang R, Bao T. The rebalanced pathway significantly enhances acetoin production by disruption of acetoin reductase gene and moderate-expression of a new water-forming NADH oxidase in Bacillus subtilis. Metabolic Engineering. 2014;23:34-41. DOI: 10.1016/j.ymben.2014.02.002

29. Balzer G J, Thakker C, Bennett G N. Metabolic engineering of Escherichia coli to minimize byproduct formate and improving succinate productivity through increasing NADH availability by heterologous expression of NAD(+)-dependent formate dehydrogenase. Metabolic Engineering. 2013;20:1-8. DOI: 10.1016/j.ymben.2013.07.005

30. Kastner J R, Eiteman M A, Lee S A. Effect of redox potential on stationary-phase xylitol fermentations using Candida tropicalis. Applied Microbiology and Biotechnology. 2003;63:96-100. DOI: 10.1007/s00253-003-1320-9

31. Nakashimada Y, Rachman M A, Kakizono T. Hydrogen production of Enterobacter aerogenes altered by extracellular and intracellular redox states. International Journal of Hydrogen Energy. 2002;27:1399 - 1405. DOI: 10.1016/S0360-3199(02)00128-3

32. Ren N Q, Chua H, Chan S Y. Assessing optimal fermentation type for bio-hydrogen production in continuous-flow acidogenic reactors. Bioresource Technology. 2007;98:1774-1780. DOI: 10.1016/j.biortech.2006.07.026

33. Wang J, Yu W, Xu L. Effects of increasing the NAD(H) pool on hydrogen production and metabolic flux distribution in Enterobacter aerogenes

mutants. International Journal of Hydrogen Energy. 2013;38:13204-13215. DOI: 10.1016/j.ijhydene.2013.07.121

34. Wang Z, Gao C, Wang Q. Production of pyruvate in Saccharomyces cerevisiae through adaptive evolution and rational cofactor metabolic engineering. Biochemical Engineering Journal. 2012;67:126-131. DOI: 10.1016/j.bej.2012.06.006

35. Du C Y, Yan H, Zhang Y P. Use of oxidoreduction potential as an indicator to regulate 1,3-propanediol fermentation by Klebsiella pneumoniae. Applied Microbiology and Biotechnology. 2006;69:554-563. DOI: 10.1007/s00253-005-0001-2

36. Du C, Zhang Y, Li Y. Novel redox potential-based screening strategy for rapid isolation of Klebsiella pneumoniae mutants with enhanced 1,3-propanediol-producing capability. Applied and Environmental Microbiology. 2007;73:4515-4521. DOI: 10.1128/aem.02857-06

37. Zheng Z M, Cheng K K, Hu Q L. Effect of culture conditions on 3-hydroxy-propionaldehyde detoxification in 1,3-propanediol fermentation by Klebsiella pneumoniae. Biochemical Engineering Journal. 2008;39:305-310. DOI: 10.1016/j.bej.2007.10.001

38. Wu Z, Wang Z, Wang G. Improved 1,3-propanediol production by engineering the 2,3-butanediol and formic acid pathways in integrative recombinant Klebsiella pneumoniae. Journal of Biotechnology. 2013;168:194-200. DOI: 10.1016/j.jbiotec.2013.04.022

39. Jain R, Huang J, Yuan Q. Engineering microaerobic metabolism of E. coli for 1,2-propanediol production. Journal of Industrial Microbiology & Biotechnology. 2015;42:1049-1055. DOI: 10.1007/s10295-015-1622-9

40. Wang S, Zhu Y, Zhang Y. Controlling the oxidoreduction potential of the culture of Clostridium acetobutylicum leads to an earlier initiation of solventogenesis, thus increasing solvent productivity. Applied Microbiology and Biotechnology. 2012;93:1021-1030. DOI: 10.1007/s00253-011-3570-2

41. Li T, Yan Y, He J. Reducing cofactors contribute to the increase of butanol production by a wild-type Clostridium sp. strain BOH3. Bioresource Technology. 2014;155:220-228. DOI: 10.1016/j.biortech.2013.12.089

42. Bui L M, Lee J Y, Geraldi A. Improved n-butanol tolerance in Escherichia coli by controlling membrane related functions. Journal of Biotechnology. 2015;204:33-44. DOI: 10.1016/j.jbiotec.2015.03.025

43. Fu J, Wang Z W, Chen T. NADH plays the vital role for chiral pure D-(-)-2,3-Butanediol production in Bacillus subtilis under limited oxygen conditions.

Biotechnology and Bioengineering. 2014;111:2126-2131. DOI: 10.1002/bit.25265

44. Wang Y, Li L, Ma C. Engineering of cofactor regeneration enhances (2S,3S)-2,3-butanediol production from diacetyl. Scientific Reports. 2013;3:2643. DOI: 10.1038/srep02643

45. Lin Y H, Chien W S, Duan K J. Correlations between reduction-oxidation potential profiles and growth patterns of Saccharomyces cerevisiae during very-high-gravity fermentation. Process Biochemistry. 2010;45:765-770. DOI: 10.1016/j.procbio.2010.01.018

46. Jeon B Y, Park D H. Improvement of ethanol production by electrochemical redox combination of Zymomonas mobilis and Saccharomyces cerevisiae. Journal of Microbiology and Biotechnology. 2010;20:94-100. DOI: 10.4014/jmb.0904.04029

47. Na B K, Hwang T S, Lee S H. Effect of electrochemical redox reaction on growth and metabolism of Saccharomyces cerevisiae as an environmental factor. Journal of Microbiology and Biotechnology. 2007;17:445-453.

48. Liu C G, Lin Y H, Bai F W. Development of redox potential-controlled schemes for very-high-gravity ethanol fermentation. Journal of Biotechnology. 2011;153:42-47. DOI: 10.1016/j.jbiotec.2011.03.007

49. Liu C G, Lin Y H, Bai F W. Ageing vessel configuration for continuous redox potential-controlled very-high-gravity fermentation. Journal of Bioscience and Bioengineering. 2011;111:61-66. DOI: 10.1016/j.jbiosc.2010.09.003

03

SOLID-STATE FERMENTATION IN A BAG BIOREACTOR: THE IMPACT OF COMBINING CORN COB WITH PHYTOPATHOGEN BIOMASS ON THE PRODUCTION OF SPORES AND CELLULASE BY TRICHODERMA ASPERELLUM

1. INTRODUCTION: AN OVERVIEW

Phytopathogens are responsible for causing diseases that result in significant losses in crops globally. It is widely recognized that the management of these diseases using chemical pesticides is ineffective, since it leads to the development of resistant strains of plant pathogens. Additionally, these pesticides produce toxic residues and have carcinogenic consequences [1, 2]. In recent years, there has been a growing interest in using antagonist microorganisms to control phytopathogens. This approach offers a highly effective and environmentally benign method of biological management [2]. In order to achieve this, it is required to produce high concentrations of spores from biological control agents (BCA). Currently, there are multiple production procedures available for various bacteria. The solid-state fermentation (SSF) is the most effective method for this purpose, as it results in spores with a longer shelf life compared to those obtained in liquid cultures [3, 4]. Alternatively, spore generation in solid-state fermentation (SSF) is quite simple, making it accessible to those without prior knowledge. This facilitates the dissemination of this technology to farmers [3]. The synthesis of spores under solid-state fermentation (SSF) conditions utilizing polyethylene bioreactors, specifically bag-type bioreactors, is a recent area of research. There is limited information available on the design of bioreactors and the optimal conditions for spore generation [1, 5, 6]. SSF typically investigates the potential of utilizing waste from other production processes, making it a popular practice in poor nations [1]. Various agricultural byproducts such as rice wastes, maize meal, corn cob, rice husk, banana husk, wheat bran, and tea leaves have been utilized as substrates for spore generation through solid-state fermentation (SSF) [3, 5, 6]. Certain substances can be introduced in small quantities to the substrate of solid-state fermentation (SSF) to stimulate the production of specific metabolites. For instance, there have been reports of incorporating casein and gluten to enhance the production of proteases, as well as utilizing waste shrimp silage to increase chitinase production, among other examples [7, 8]. The primary objective of this study was to utilize maize cobs as a substrate in solid-state fermentation (SSF) to generate spores of the beneficial fungus Trichoderma asperellum, which acts as a biological control agent (BCA), in a polyethylene bioreactor. The biomass of the phytopathogenic fungus Colletotrichum gloeosporioides and Phytophthora capsici was included in the procedure to stimulate the production of hydrolase enzymes, specifically endoglucanases, exoglucanases, and chitinases.

2. MATERIALS AND METHODS

2.1. Microorganism and culture conditions

The *T. asperellum* (T2-10) and *P. capsici* were kindly proportioned by the Agricultural Parasitology Department of the UAAAN (Universidad Autónoma Agraria Antonio Narro, Saltillo, México). *C. gloeosporioides* was proportioned by the Food Research Department of the UAdeC (Universidad Autónoma de Coahuila). The fungi were cultivated and conserved in a milk-glycerol 8.5% solution. Potato dextrose agar (PDA) was used to reactivate the fungi. In HACH® tubes, 5 mL of PDA was taken, then they were closed and sterilized at 121°C for 15 min. The tubes in slant were inoculated with the fungal strains and incubated at 30°C for 5 days. The conservation was at ± 4°C.

2.2. Phytopathogen biomass production

A cornmeal medium (17 g/L) was used to produce phytopathogen biomass. This medium was maintained under shaking for 1 h at 58°C. Then, it was filtrated and sterilized (15 at 115°C). The inoculation of phytopathogens was as follows: *C. gloeosporioides* (1×10^6 spores/mL) and *P. capsici* (10 PDA plugs from a culture of 7days old), and incubated at 28°C for 7 days under shaking (200 rpm).

2.3. Substrates

In this work, we evaluate as substrate corn cob (CC) proportioned by the Mexican Institute of Maize, UAAAN Coahuila, México. The material was dried, ground, fractioned (300–1680 μm) and stored under low moisture conditions for further evaluation. This material was used as a substrate on SSF without pretreatment.

2.4. Solid-state fermentation

Polyethylene bags were used as bioreactor in all experiments. Sporulation and cellulase production were evaluated. Plackett-Burman design (PBD) was used in this experiment to determine the most influential factors on spore and enzyme production by *T. asperellum* under SSF conditions on a bag bioreactor. The factors such as temperature (°C), pH, substrate (g), inoculum (spores/g), moisture (%), phytopathogen biomass (%) and incubation time (days) were evaluated, one maximum (+1) and one minimum (-1) (Table 1). Spore counting was done at the end of SSF process using a hemocytometer. The fermented material was placed in a Falcon® tube with 10 mL of distilled water. The enzymatic extract was homogenized in a vortex (1 min) for further determination of enzyme activity.

Table 1. PBD matrix used to determine the influence of different variables (A, B, C, D, E, F and G) on spore and enzyme activity in SSF by *T. asperellum*.

Run	A	B	C	D	E	F	G
1	-1	-1	-1	1	1	1	-1
2	1	-1	-1	-1	-1	1	1
3	-1	1	-1	-1	1	-1	1
4	1	1	-1	1	-1	-1	-1
5	-1	-1	1	1	-1	-1	1
6	1	-1	1	-1	1	-1	-1
7	-1	1	1	-1	-1	1	-1
8	1	1	1	1	1	1	1
Code	Factors			High value	Low value		
A	Substrate (g)			30		15	
B	pH			8.0		6	
C	Inoculum (spores/g)			1×10^7		1×10^5	
D	Temperature (°C)			30		24	
E	Moisture (%)			66		50	
F	Inducer (%)			3		1	
G	Time (days)			7		5	

2.5. Enzyme activity determination

After SSF each sample was analyzed to determine cellulase activity [9], chitinase activity [10] and reducing sugars [11]. The carboxymethylcellulose activity (CMCA) was carried out at 50°C for 30 min. Sample (1 mL) and substrate (1 mL of carboxymethylcellulose 1%) were the mix reaction. Citrate buffer (1 mL at 50 mM, pH 4.8) and substrate (1 mL) were the substrate control. The enzyme control was the mix of sample (1 mL) and citrate buffer (1 mL).

The filter paper activity (FPA) was carried out at 50°C for 1 h. Sample (1 mL) and substrate (filter paper Whatman No.1 (1 cm×5 cm) in 1 mL of citrate buffer at 50 mM, pH 4.8) were the reaction mix. The control substrate was the mix of citrate buffer (2 mL) and filter paper. Sample (1 mL) and citrate buffer (1 mL) were the enzyme control.

Chitinase activity was carried out at 37°C for 1 h. The reaction mix, enzyme and substrate control were done similar to carboxymethylcellulose activity. In this case, substrate (chitin oligosaccharides) and buffer solution (acetate 50 mM, pH 4.0) were replaced.

Sugar concentration was determined after each enzyme reaction. An enzyme activity (U) was defined as the amount of enzyme that catalyze the release of 1 μmol of glucose per minute.

2.6. Design and statistical analysis

A PBD was used to SSF. Spore and enzyme production were the response variables. Data were analyzed by ANOVA using STATISTICA 7.0 software; when needed mean treatments were compared using Tukey's multiple range procedure. A *p*-value of less than 0.05 was regarded as significantly different.

3. RESULTS

3.1. Screening of significant factors using Plackett-Burman design

Studies were performed in eight runs each one to identify the combination of factors which allow us to obtain a significant level of spore, cellulase and chitinase production by *T. asperellum* on corn cob under SSF conditions using phytopathogen biomass (*C. gloeosporioides* and *P. capsici*).

3.2. Solid-state fermentation with C. gloeosporioides biomass

Table 2. Enzyme production and sporulation index of *T. asperellum* on a mixture of corn cob and *C. gloeosporioides* biomass under SSF conditions.

Treatment	Sporulation index (Spores/g)	Enzyme activities		
		CMCA	FPA	CA
A	4.2E+08 ± 6.25E+07^{cd}	1.847 ± 0.02^{b}	1.183 ± 0.17^{b}	3.671 ± 0.46^{b}
B	4.0E+08 ± 4.30E+07^{d}	0.132 ± 0.02^{d}	0.165 ± 0.06^{cd}	1.795 ± 0.05^{f}
C	4.8E+08 ± 5.10E+07^{cd}	2.582 ± 0.08^{a}	1.549 ± 0.05^{a}	5.118 ± 0.28^{a}
D	2.3E+08 ± 6.65E+07^{e}	0.204 ± 0.10^{d}	0.006 ± 0.00^{d}	2.149 ± 0.06^{ef}
E	2.3E+08 ± 7.35E+07^{e}	0.529 ± 0.10^{d}	0.443 ± 0.05^{c}	2.996 ± 0.08^{cd}
F	7.3E+08 ± 9.55E+07^{a}	1.958 ± 0.17^{b}	1.381 ± 0.23^{ab}	3.692 ± 0.13^{b}
G	6.2E+08 ± 8.15E+07^{ab}	1.243 ± 0.02^{c}	0.390 ± 0.01^{c}	2.538 ± 0.01^{de}
H	5.2E+08 ± 3.50E+06^{bc}	2.474 ± 0.47^{a}	1.193 ± 0.25^{b}	3.418 ± 0.29^{bc}

Table 2 summarizes the results obtained in SSF using the biomass of *C. gloeosporioides* blended with corn cob. The sporulation index is favored by the treatments F and G (7.3×10^8 and 6.2×10^8 Spores/g CS, respectively), with no significant difference among the values. The conditions of treatment C allow the best production to CMCA, FPA and CA (2.582, 1.549 y 5.118 U/g), respectively).

Numbers in each column followed by a common letter are not significantly different ($P<0.05$).

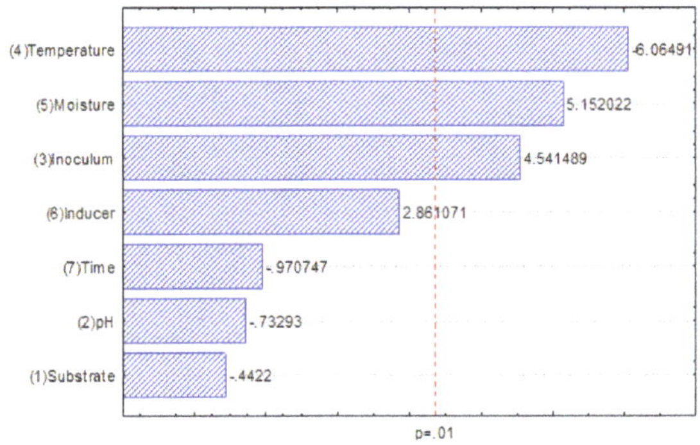

Figure 1. Pareto plot of the standardized effects on the spore production of T. asperellum using corn cob in SSF with C. gloeosporioides as inducer.

In the first case, the spore production was influenced by the temperature in a negative way. Between the range of the values (24 and 30°C), the study shows that 24°C is the best to produce a better sporulation index and possibly if we reduce the value, the sporulation can be major. The moisture, inoculum and inducer are the other factors that also have influence on spore production, just in a positive way. It means, it is necessary to increase the value of each factor (Figure 1). The moisture, pH and inoculum were the factors more determining endoglucanase production (CMCA). These factors had positive values, which mean that high values allow high enzyme activity. A significant effect was observed with the substrate concentration, but this effect was negative, so low amount of substrate is needed to obtain high enzyme yields (Figure 2).

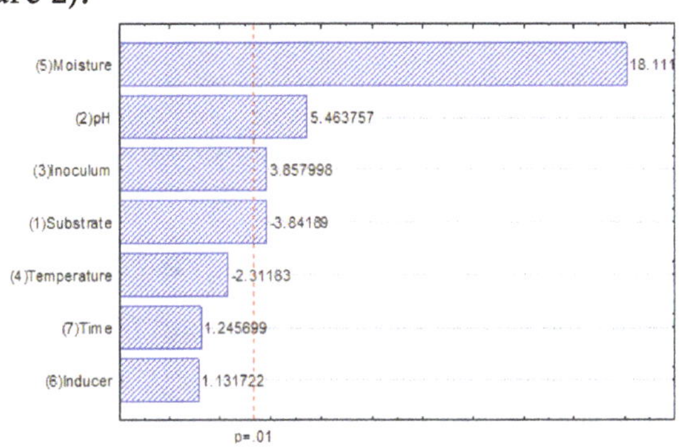

Figure 2. Pareto plot of the standardized effects on CMCA from an extract of T. asperellum using corn cob in SSF with C. gloeosporioides as inducer.

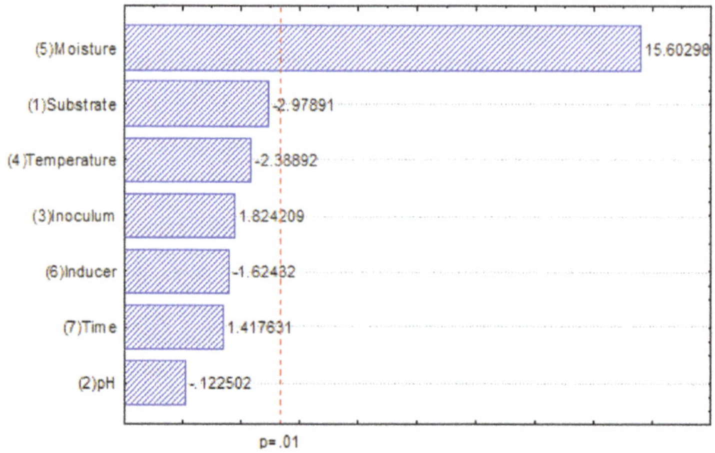

Figure 3. Pareto plot of the standardized effects on FPA from an extract of T. asperellum using corn cob in SSF with C. gloeosporioides as inducer.

The exoglucanase (FPS) in the same way to CMCA was influenced by the moisture (Positive). Low levels in the substrate and temperature show the best enzymatic yields (Figure 3). The moisture was the factor with major influence on the chitinase production. The substrate and inducer were also significant, but in negative way, it is necessary to use low values to increase the yield. The time and pH were important, so these factors must be in high levels (Figure 4).

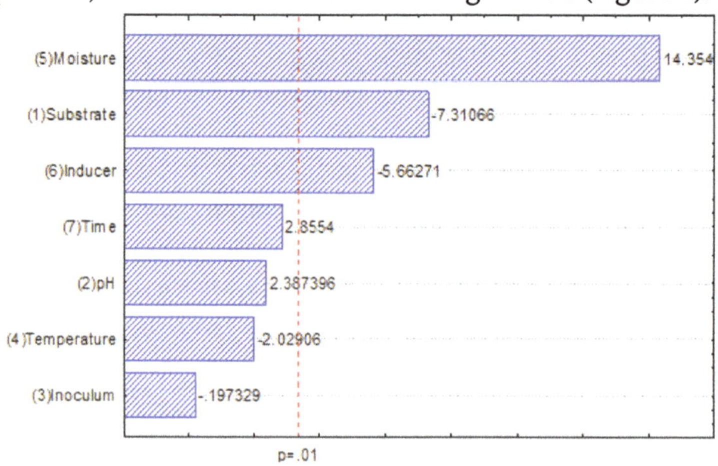

Figure 4. Pareto plot of the standardized effects on chitinase from an extract of T. asperellum using corn cob in SSF with C. gloeosporioides as inducer.

3.3. Solid-state fermentation with P. capsici as inducer

Now, Table 3 shows the results obtained in SSF using the biomass of *P. capsici* as inducer. The conditions of treatment G allow the best production to all dependent variables evaluated. The values obtained were sporulation index (1.2×10^9 Spores/g CS), CMCA (7.825 U/g), FPA (2.764 U/g) and CA (3.609 U/g).

Table 3. Enzyme production and sporulation index of *T. asperellum* grown on a mixture of corn cob with *P. capsici biomass under SSF conditions*.

Treatment	Sporulation index (Spores/g)	Enzyme activities		
		CMCA	FPA	CA
A	3.1E+08 ± 2.0E+06[d]	4.238 ± 0.32[c]	1.660 ± 0.02[d]	2.848 ± 0.22[b]
B	1.4E+09 ± 1.5E+08[a]	4.861 ± 0.13[bc]	2.466 ± 0.14[b]	3.216 ± 0.15[ab]
C	2.6E+07 ± 1.0E+06[f]	0.062 ± 0.02[e]	0.121 ± 0.00[g]	1.527 ± 0.25[d]
D	8.1E+08 ± 9.0E+06[b]	5.012 ± 0.40[b]	2.272 ± 0.01[c]	2.978 ± 0.10[b]
E	4.4E+08 ± 6.0E+07[c]	4.814 ± 0.29[bc]	2.200 ± 0.02[c]	3.082 ± 0.04[b]
F	3.7E+08 ± 3.0E+07[d]	1.538 ± 0.56[d]	0.696 ± 0.01[e]	2.144 ± 0.24[c]
G	1.2E+09 ± 1.1E+08[a]	7.825 ± 0.21[a]	2.764 ± 0.01[a]	3.609 ± 0.18[a]
H	1.1E+08 ± 3.0E+06[e]	1.168 ± 0.21[d]	0.310 ± 0.13[f]	1.956 ± 0.22[cd]

Numbers in each column followed by a common letter are not significantly different ($P<0.05$).

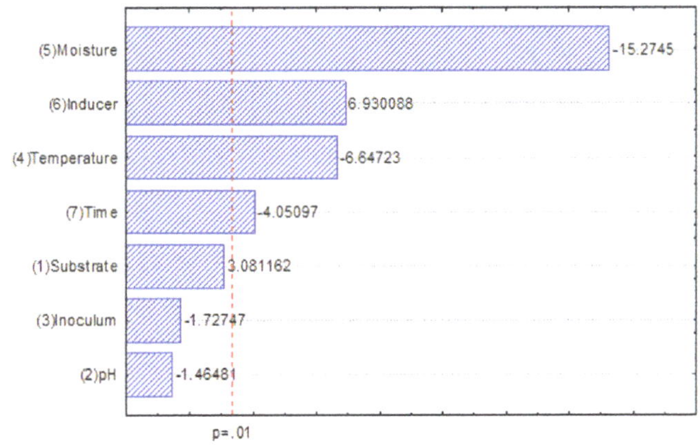

Figure 5. Pareto plot of the standardized effects on the spore production of T. asperellum using corn cob in SSF with P. capsici as inducer.

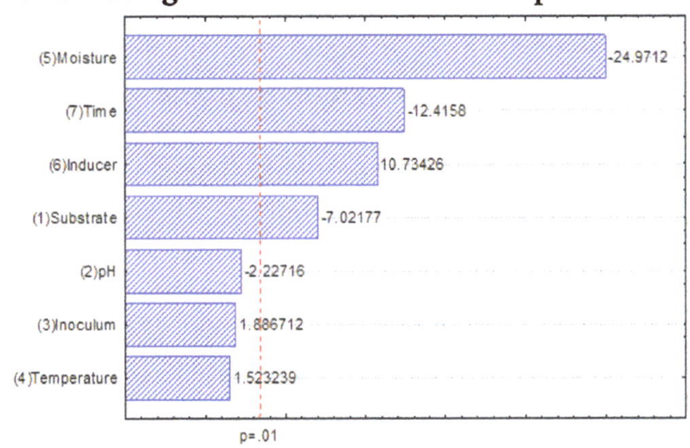

Figure 6. Pareto plot of the standardized effects on CMCA from an extract of T. asperellum using corn cob in SSF with P. capsici as inducer.

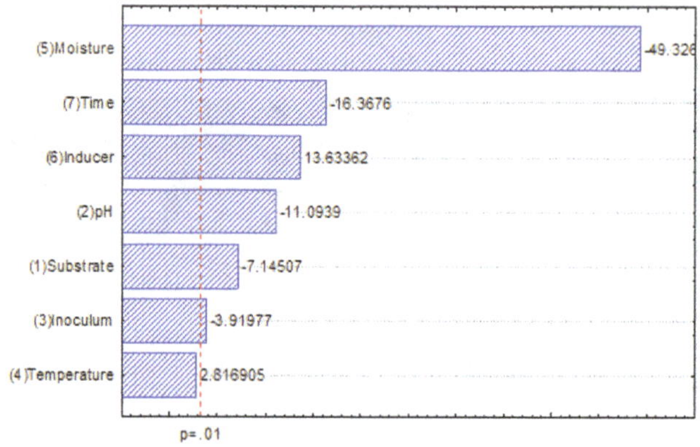

Figure 7. Pareto plot of the standardized effects on FPA from an extract of T. asperellum using corn cob in SSF with P. capsici as inducer.

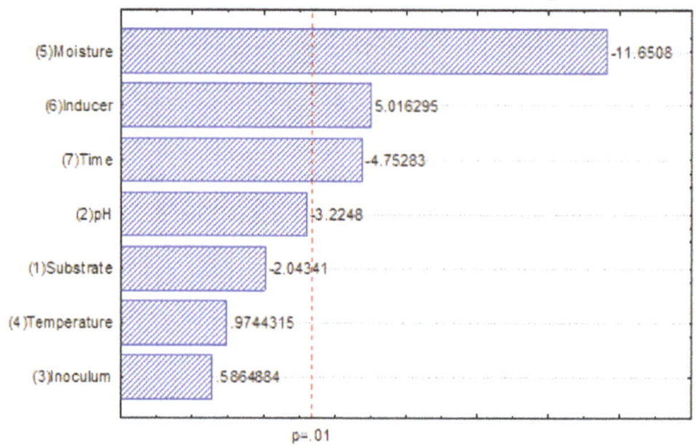

Figure 8. Pareto plot of the standardized effects on chitinase from an extract of T. asperellum using corn cob in SSF with P. capsici as inducer.

The sporulation of *T. asperellum* was not influenced by the pH and inoculum. But low levels (moisture, temperature and time) and high levels (inducer and substrate) show high spore production (Figure 5). In endoglucanase production, four factors are important under the SSF conditions evaluated. Low levels of moisture, time and substrate and high levels of the inducer show the major levels of enzymatic activity (Figure 6). All factors evaluated were significant to the exoglucanase production. Low levels of moisture, time, pH, substrate ad inoculum and high levels of inducer and temperature show the best enzyme yields (Figure 7). Finally, low levels of moisture, time and pH and high levels of inducer shows the major chitinolytic activity (Figure 8).

4. DISCUSSION

Multiple investigations have documented the generation of various enzymes using SSF [12, 13]. At present, the SSF is widely utilized due to the cost-effectiveness of raw materials such as sugarcane bagasse, wheat bran, and others [14]. Factors such as temperature, pH, moisture, culture purity, and process time pose challenges in maintaining strict control over the fermentation process [8]. Occasionally, it can be challenging to identify a single set of SSF conditions that will result in a high yield for all examined response variables, including sporulation, cellulases, and chitinases. The treatment F yielded the most favorable results for spore generation in the instance of SSF with biomass of C. gloeosporioides, while the treatment C showed the highest enzyme activity. Nevertheless, treatment F also exhibits significant enzyme yields. The treatment F enables the acquisition of elevated levels of spores, cellulases, and chitinases. The treatment F settings consisted of a substrate weighing 30 g, a pH level of 6, an inoculum containing 1×107 spores per gram, a temperature of 24°C, a moisture level of 66%, an inducer concentration of 1%, and a duration of 5 days. The treatment G yielded the most favorable outcomes in the solid-state fermentation (SSF) with P. capsici biomass. These results indicate that the production of spores, cellulase, and chitinase was significantly elevated under certain conditions: substrate quantity of 15 g, pH level of 8, inoculum concentration of 1×107 spores per gram, temperature of 24°C, moisture level of 50%, inducer concentration of 3%, and an incubation period of 5 days. Initially, we hypothesize that introducing a specific concentration of phytopathogen biomass could stimulate the production of certain hydrolase enzymes. Chitinases were produced when C. gloeosporioides was utilized, while cellulases were produced when P. capsici was used. The composition of the phytopathogen, specifically chitin and cellulose, has a direct influence on this action.

In the SSF with *C. gloeosporioides*, the induction of enzymes did not happen maybe because the chitinase is a constitutive enzyme result of the natural metabolism of the microorganism. Previously, the same effect in the chitinase production with *Meyerozyma caribbica* in liquid culture using *C. gloeosporioides* as inducer [15] was observed. A similar study was also reported evaluating the β-*N*-acetylhexosaminidase (a chitinase) by *Verticillium lecanii* using shrimp waste silage as inducer and sugarcane bagasse as support [8].

In the case of SSF with *P. capsici*, the inductor was effective and shows an important effect in all enzyme activities evaluated. We did not find reports of the use of biomass to induce some types of cellulase.

In this study, the moisture and temperature are the two important factors. Among the values evaluated, a level of 66% of moisture and 24°C of temperature shows the best yields in spore and enzymes production on SSF with *C. gloeosporioides* as inducer. In the SSF with *P. capsici* as inducer, the level of moisture was 50%. The two values of moisture used in this study are low which is reported in the literature [8]. They mentioned that the inducer is very important, but also the moisture because they observed that above 75% it can affect the porosity, oxygen diffusion and favor the bacterial contamination. In other hand, low moisture percentage reduces the microbial growth.

This study demonstrated that the biomass addition of any one phytopathogen shows an increment in the spore production by *T. asperellum*. The fungi sporulation starts when the environmental and nutritional conditions become hard to life support. The chemical composition of the inducer possibly causes some stress on *T. asperellum* which accelerate the sporulation process. The experimental stage suggests that high levels of biomass of the inducer increase the sporulation.

Currently, there researches are aimed at the high biomass production of the biological control agents using several systems to produce it. Kancelista *et al.* [16] reported the use of corn cob under SSF by *T. asperellum* obtaining a yield of 3.13×10^9 spores/g. Motta and Santana [17] who working the SSF with empty fruit bunch and a *Trichoderma* spp. The sporulation index was 4.4×10^9 spores/g in a Raimbault columns.

There are few works that report the use of polyethylene bioreactors to produce spores in SSF using some biological control. The use of this kind of bioreactor needs to utilize special plastic bags which allow the gas exchange and microorganism respiration [18]. In some cases, we can use a cotton tap on the bag to allow gas exchange. In this study, the maximal spore production obtained was 7.3×10^8 and 1.4×10^9 spores/g CS to the SSF with *C. gloeosporioides* and *P. capsici*, respectively. Singh *et al.* [6] used a *Trichoderma harzianum* strain and a similar bioreactor, obtaining a production of 8×10^8 y de 4.4×10^6 spores/g CS using tea leaves and sawdust, respectively. Viccini *et al.* [3] did a study of the spore production of *Clonostachys rosea* under SSF conditions using a polyethylene bioreactor and rice grains as substrate. The yield obtained was 1.8×10^8 spores/g CS.

5. CONCISE SUMMARY

It is possible to obtain high levels of spores, cellulases and chitinases using a polyethylene bioreactor under SSF conditions by *T. asperellum* and corn cob as substrate. Under the SSF conditions evaluated, the biomass of *C. gloeosporioides* has an inducer effect just on the spore production. However, *P. capsici* have effect on all response variables evaluated. The spore production was twice when used *P. capsici* as inducer. The most influential factor under SSF was the moisture. Levels of 66 and 50% of this factor increase the yield in all response variables evaluated (sporulation, cellulases and chitinases), *C. gloeosporioides* and *P. capsici*, respectively. When the biomass of *C. gloeosporioides* was used as a inducer, the best SSF conditions with corn cob and *T. asperellum* are as follows: substrate (30 g), pH (6), inoculum (1×10^7 Spores/g), temperature (24°C), moisture (66%), inducer (1%) and time (5 days). In the case of *P. capsici*, the conditions are: substrate (15 g), pH (8), inoculum

(1×10^7 Spores/g), temperature (24°C), moisture (50%), inducer (3%) and time (5 days). Further research on SSF with agroindustrial wastes using polyethylene bioreactors, mainly to the reduction of cost in the process, is necessary. Also, it must be make more analyses to determine the optimal production conditions, as well as, the use of inducers.

REFERENCES

1. Millner RJ. 2000. Current status of Metarhizium as mycoinsecticide in Australia. BiocontrolNewsandInformation.21(2):47-50.doi:10.1080/03235400601160065.
2. Abo-Elyousr KAM, Abdel-Hafez SII, Abdel-Rahim IR. 2014. Isolation of Trichoderma and evaluation of their antagonistic potential against Alternaria porri. Journal of Phytopathology. 162: 567–574. doi:10.1111/jph.12228.
3. Viccini G, Mannich M, Fontana-Capalbo DM, Valdebenito-Sanhueza R, Mitchell DA. 2007. Spore production in solid-state fermentation of rice by Clonostachys rosea, a biopesticide for gray mold of strawberries. Process Biochemistry. 42:275-278. doi:10.1016/j.procbio.2006.07.006.
4. Hanada RE, Pomella AWV, Soberanis W, Loguercio LL, Pereira JO. 2009. Biocontrol potential of Trichoderma martiale against the black-pod disease (Phytophthora palmivora) of cacao. Biological Control. 50:143-149. doi:10.1016/j.biocontrol.2009.04.005.
5. Singh HB, Singh BN, Singh SP, Nautiyal CS. 2010. Solid-state cultivation of Trichoderma harzianum NBRI-1055 for modulating natural antioxidants in soybean seed matrix. Bioresource Technology. 101:6444-6453. doi:10.1016/j.biortech.2010.03.057
6. Singh A, Srivastava S, Singh HB. 2007. Effect of substrates on growth and shelf life of Trichoderma harzianum and its use in biocontrol of diseases. Bioresource Technology. 98:470-473. doi:10.1016/j.biortech.2006.01.002.
7. Battaglino RA, Huergo M, Pilosof AMR, Bartholomai GB. 1991. Culture requirements for the production of protease by Aspergillus oryzae in solid state fermentation. Applied Microbiology and Biotechnology. 35:292-296. doi:10.1007/BF00172714.
8. Matsumoto Y, Saucedo-Castañeda G, Revah S, Shirai K. 2004. Production of β-N-acetylhexosaminidase of Verticillium lecanii by solid state and submerged fermentations utilizing shrimp waste silage as substrate and inducer. Process Biochemistry. 39:665-671. doi:10.1016/S0032-9592(03)00140-7
9. Galindo, A. I. G., Medina-Morales, M. A., Romanini, D., Pico, G., & Aguilar, C. N. (2016). Cellulase production by solid state fermentation and its separation with polyacrylate. New Biotechnology, 33(3):429.

10. Saito, A., Fujii, T., Yoneyama, T., Redenbach, M., Ohno, T., Watanabe, T., & Miyashita, K. (1999). High-multiplicity of chitinase genes in Streptomyces coelicolor A3 (2). Bioscience, Biotechnology, and Biochemistry. 63(4):710-718. doi:10.1271/bbb.63.710
11. Miller, G. L. (1959). Use of dinitrosalicylic acid reagent for determination of reducing sugar. Analytical Chemistry. 31(3):426-428. doi:10.1021/ac60147a030
12. Buenrostro-Figueroa, J., Ascacio-Valdés, A., Sepúlveda, L., De la Cruz, R., Prado-Barragán, A., Aguilar-González, M. A., Rodriguez, R., Aguilar, C. N. 2014. Potential use of different agroindustrial by-products as supports for fungal ellagitannase production under solid-state fermentation. Food and Bioproducts Processing. 92(4):376-382. doi:10.1016/j.fbp.2013.08.010
13. Pirota RDPB, Delabona PS, Farinas CS. 2014. Simplification of the biomass to ethanol conversion process by using the whole medium of filamentous fungi cultivated under solid-state fermentation. BioEnergy Research. 7:744–752. doi:10.1007/s12155-013-9406-4.
14. Shahzadi T, Ikram N, Rashid U, Afroz A, But HI, Anwar Z, Irshad M, Mahmood S. 2013. Optimization of physical and nutritional factors for induced production of cellulase by co-culture solid-state bio-processing of corn stover. WSEAS Transactions on Environment and Development. 4(9):263-267. Available at: http://www.wseas.org/multimedia/journals/environme...
15. Bautista-Rosales PU, Calderon-Santoyo M, Servín-Villegas R, Ochoa-Álvarez NA, Ragazzo-Sánchez JA. Action mechanisms of the yeast Meyerozyma caribbica for the control of the phytopathogen Colletotrichum gloeosporioides in mangoes. Biological Control. 65:293-301. doi:10.1016/j.biocontrol.2013.03.010
16. Kancelista A, Tril U, Stempniewickz R, Piegza M, Szczech M, Witowska D. 2013. Application of lignocellulosic waste materials for the production and stabilization of Trichoderma biomass. Polish Journal of Environmental Studies. 22(4):1083-1090. doi:bwmeta1.element.agro-6ccb71f0-b739-49a9-a9ea-2b515c9f6d5a.
17. Motta FL, Santana MHA. 2014. Solid-state fermentation for humic acid production by Trichoderma reesei strain using an oil palm empty fruit bunch as the substrate. Applied Biochemistry and Biotechnology. 172:2205-2217. doi:10.1007/s12010-013-0668-2.
18. Krishna C. 2005. Solid-state fermentation an overview. Critical Reviews in Biotechnology. 25(1): 1-30. doi:10.1080/07388550590925383.

04 INVESTIGATION OF SOLID-STATE AND LIQUID FERMENTATION FOR LACCASE PRODUCTION BY PLEUROTUS OSTREATUS

1. INTRODUCTION: AN OVERVIEW

Laccases are oxygen oxidoreductase enzymes that are synthesized by plants, insects, bacteria, and fungus. The laccases that have been extensively researched are derived from fungus, particularly those belonging to the white rot fungi group. These laccases have been studied utilizing various culture systems, with a focus on solid-state fermentation (SSF) and liquid fermentation (SmF). Overall, solid-state fermentation has been proposed as a more effective method for producing metabolites and enzymes compared to submerged fermentation (SmF) [1]. However, recent studies have shown that the basidiomycete Pleurotus ostreatus, when grown in SmF, actually exhibited higher laccase values compared to when the fungus was grown using solid-state fermentation. The P. ostreatus strain ATCC 32783 has been examined for its ability to produce intracellular laccases in both peripheral and central vegetative mycelium [2]. The solid-state fermentation (SSF) and submerged fermentation (SmF) systems have also been assessed for laccase production, revealing varying levels of activity and number of isoenzymes depending on the culture system [3, 4]. When comparing certain strains of P. ostreatus with ATCC 32783, it was found that the latter exhibited superior qualities compared to the others [5]. In a separate investigation, the impact of the pH of the culture medium on the activity of laccases in P. ostreatus cultivated in SmF was assessed. The activity was found to be seven times greater when the pH of the medium remained constant [6]. A recent research describes the identification of a new gene, lacP83, in the fungus P. ostreatus ATCC 32783. This gene encodes a laccase enzyme that is mostly produced under conditions of liquid culture. The study examined the impact of different initial pH levels (ranging from 3.5 to 8.5) on laccase activity. It analyzed the production pattern and expression profile of five laccase genes in P. ostreatus ATCC 32783 grown in SmF. The results indicated that a pH of 8.5 promoted biomass production but not enzymatic activity, and it also led to repression of gene expression. On the other hand, a pH of 4.5 resulted in higher laccase enzyme activity, with levels reaching up to 78,500 U/L [8].

2. *PLEUROTUS OSTREATUS*

2.1. *Pleurotus ostreatus*: phases of growth and composition

The genus *Pleurotus* (Jacq.: Fr.) Kummer (Pleurotaceae, higher Basidiomycetes) comprises a group of edible ligninolytic mushrooms, which have two phases of growth: one is called vegetative or mycelial and is seen as strands of hyphae, which colonize the substrate and the other, the reproductive or fruit body, is represented by the mushroom itself that in basidiomycetes, is called the basidiome. *P. ostreatus* is characterized by a white spore print, with an eccentric stipe and a fan or oyster-shared pileus or cap (5–25 cm). The Latin *Pleurotus* means "beside the ear" and *ostreatus* means "oyster shaped." This fungus is commonly named "oyster mushroom" for the resemblance of its fruiting body a white shell [9]. In this case, the spores are located in a special structure called basidium. In the *P. ostreatus* growth, after spore germination (or inoculation of *in vitro*-grown mycelia), the substrate is invaded by microscopic filaments called hyphae. Hyphae continually grow and branch to form a network of hyphae. Mycelial growth is generally coupled with increased enzyme production and respiration. Hyphae absorb digestive products, penetrating the substrate until its complete hyphal invasion. The vegetative growth is in direct contact with the support (substrate), providing the nutritive materials required for mushroom growth (Figure 1)[10].

The growth of all parts of a fungus occurs at the hyphal tips. The mycelial structure grows by synthesizing new wall at the hyphal apex, as they increase in length, additional sites for wall synthesis are formed in the subapical region, originating lateral branches that synthetize wall again confined to the hyphal tip [11]. The formation of a new branch requires the production of a new apex from the existing mature hyphal cell. It has been reported that some enzymes such as proteinases could create weakened zones in the cell wall, which could be pushed out by cytoplasmic flow to initiate branching. So in the mycelial growth, in addition to activation of the cell wall synthesis, enzymes such as chitin synthetase and proteinases, which might weaken the lateral cell walls, are also important. During growth of vegetative hyphae, a large amount of nutrient absorbed from the substrate is stored. Polysaccharide may be stored in the form of glucans in secondary wall layers and/or as glycogen granules in the cytoplasm in the cell. The cells in a hypha are separated by a cross-wall called septum. Septa placement has been reported to depend on the position of nuclear division. The septum is formed by chitin deposition on a preformed ring of actin microfilament. Dolipore septa divide hyphae into compartments or cells where movement of cytoplasmic material between them is carefully regulated. It gives rigidity to the hyphae and it can help avoid further injury if damage occurs at the hyphal tip [12]. In a developed colony, the hyphae tip growth or peripheral growth zone forms continuity between hyphae because of the presence of dolipore septa. The growing tips have a constant forward advance, so the mycelium that is left behind seems no longer involved in the growth. At the beginning of the fructification, the characteristic invasive growth of the vegetative mycelium in the substrate is modified. The fruit body initial is formed by increase in mycelial mass, with the formation of additional hyphal branching between the hyphae. The fruit body formation begins with the aggregation of hyphae to form a "knot" that will develop into a primordium and then a mature fruit body with differentiated stem and cap. It has been reported that the most important quantitative change in the cell wall during fruiting is the almost total loss of water-soluble glucan, instead, chitin has been reported important in fruit body development, which is essential for elongation of the stem hyphal walls. The precursor of chitin is N-acetylglucosamine and it is incorporated in the elongation of the hyphae of fruit bodies during expansion. Glycogen is accumulated in the base of fruit bodies at very earlier stages of growth and then disappears from the base as it is accumulated in the cap. In fruit body development, carbohydrates from the culture medium are temporarily store in R-glucan (alkali insoluble glucan) in the wall of mycelia and fruit body primordium hyphae, which is utilized for pileus development in the growing fruit bodies. *Pleurotus* species are cosmopolitan reported mainly as subtropical mushrooms. The optimal temperatures for growth of the mycelium are around 25–28°C and the range of pH is about 6.0–7.0. For fruit body formation, optimal temperature, relative humidity, CO_2, and light are 10-21°C, 85–90%, <1000 ppm, and 1000–1500 lx, respectively [13].

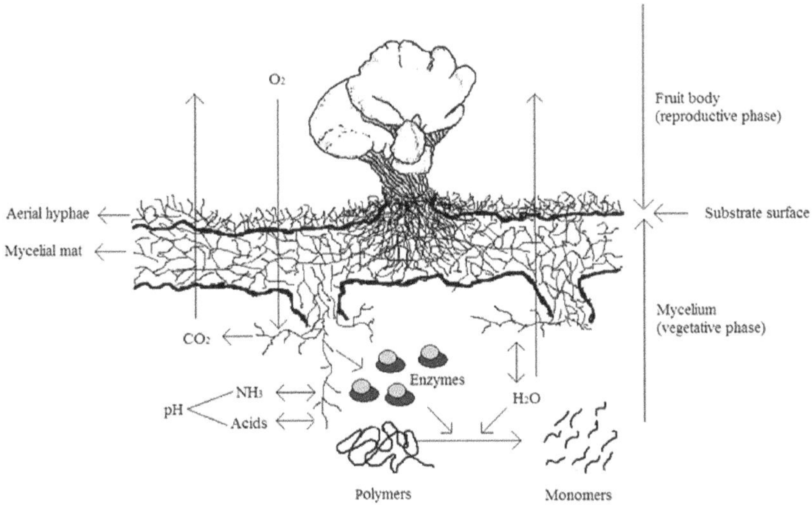

Figure 1.Schematic representation of the growth of Pleurotus ostreatus [10].

P. ostreatus can be considered as functional food with nutritional and health benefits in addition to nutritional value [14]. This mushroom contains vitamins as well as an abundance of essential amino acids. It also has proteins, lipids, ash, glycosides, tocopherols, phenolic compounds, flavonoids, carotenoids, folates, organic acids, etc. [15, 16]. In general, mushrooms contain 90% water and 10% dry matter, and their nutritional value can be compared to those of eggs, milk, and meat [17]. The total energetic value of cultivated species of *P. ostreatus* is 151 J in 100 g of fresh mushrooms [18].

Pleurotus ostreatus is the second most widely farmed edible fungus globally, following Agaricus bisporus [13]. The advancement of technology has enabled the global cultivation of mushrooms. Due to its capacity to secrete a diverse array of hydrolyzing and oxidizing enzymes [19], it is capable of breaking down various lignocellulosic substrates. Additionally, it may be cultivated on natural materials derived from agriculture, woodland, animal husbandry, and manufacturing industries [13].

2.2. Ligninolytic enzymes of *Pleurotus* spp.

Fromanecophysiologicalpointofview,white-rotbasidiomycetesaremicroorganisms able to degrade lignin efficiently. However, the degree of lignin degradation with respect to other wood components depends on the environmental conditions as well as the fungal species involved. *Pleurotus* species cause white rot of wood and other lignocellulosic materials, due to their oxidative and extracellular ligninolytic system. The fungal degradation occurs exocellularly, either in association with the outer cell envelope layer or extracellularly, because of the insolubility of lignin, cellulose, and hemicellulose. Three ligninolytic enzyme families have been reported as the enzymatic complex from *Pleurotus* species; manganese peroxidase (EC 1.11.1.13), versatile peroxidase (EC 1.11.1.16) and laccase (EC 1.10.3.2) but lack lignin peroxidase. Recently, was reported that in *Pleurotus ostreatus*, the role that generally played the lignin peroxidase, has been assumed by versatile peroxidase. [20]. Studies on the enzymes secreted by *P. ostreatus* have shown that the concerted action of laccase and aryl-alcohol oxidase produce significant reduction in the molecular mass of soluble lignosulphonates [21].

Additional peroxidases, such as dye decolorizing peroxidases have also been detected in *P. ostreatus* [22, 23]. Lignin biodegradation is an oxidative process, as a consequence *Pleurotus* enzymes can be involved in such processes. The manganese peroxidase gene family (mnps) of *P. ostreatus* is composed of five Mn^2+-dependent peroxidases (mnp3, 6, 7, 8, and 9) and four versatile peroxidases (mnp1, 2, 4, and 5), all having related gene and protein structure [24]. Mn^2+-dependent peroxidases (MnP) catalyze the H_2O_2-dependent oxidation of lignin and its derivatives [25]. Mn is an obligatory cosubstrate for these enzymes, as it is required to complete the catalytic cycle. In fact, the oxidation of lignin and other phenols by MnP is dependent on free Mn^{2+} ions. This peroxidase does not oxidize nonphenolic lignin structures. It lacks sufficient oxidative potential to cleave the major nonphenolic units of lignin. MnP contains Mn^{2+}-binding catalytic site that is formed by three acidic residues (two Glu and one Asp) and generates Mn^{3+}, which acts as a diffusible oxidizer on phenolic or nonphenolic lignin units through lipid peroxidation reactions [26, 27]. In many fungi, MnP thought to play a crucial role in the primary attack of lignin because it generates a diffusible and strong oxidant (Mn^{3+}). Organic acids such as oxalate and malonate are secreted by white-rot fungi, stimulating the MnP reaction throughout the stabilizing of Mn^{3+} [28, 29]. Versatile peroxidases feature Mn-binding residues as well as conserved Trp involved in the electron transfer that enables oxidation of nonphenolic compounds. Versatile peroxidases possess two catalytic sites, one for the direct oxidation of low- and high-redox potential compounds, and the other for oxidation of Mn in a preferred manner [23, 30–33]. This dual activity mode of action enables versatile peroxidases to modify a wide range of substrates. It has been suggested a role for versatile peroxidases of *P. ostreatus* in the transformation of azo dyes [23, 30, 34] and carbamazepine [35].

Laccases are enzymes that belong to the class of blue copper oxidases. They facilitate the process of one-electron oxidation of ortho- and para-diphenols, as well as aromatic amines. This oxidation involves the removal of an electron and a proton from a hydroxyl group, resulting in the formation of a free radical. Phenolic units in lignin undergo oxidation, resulting in the production of phenoxy radicals. Laccases also facilitate the demethoxylation of certain lignin model compounds [36–38]. This oxidation activity is accompanied by the reduction of molecular oxygen to water. Within laccase, the binding of phenolic substances is facilitated by histidine and aspartic residues, while the binding of nonphenolic substrates is mostly mediated by histidine residue [39]. The addition of various aromatic compounds to the fungal cultures can enhance the laccase activity, resulting in the production of diverse types of laccase. This is attributed to the supplementation of aromatic compounds [40, 41]. The ligninolytic system of P. ostreatus renders this organism valuable in various practical applications involving the utilization of cell-free or purified forms of peroxidases for the purposes of bioremediation and biotransformation of persistent organic pollutants.

2.3. Laccases

Laccases, also known as benzenediol: oxygen oxidoreductases (EC 1.10.3.2), are a type of enzyme that falls under the category of multicopper oxidases. These glycoproteins possess the capacity to use copper ions to facilitate the oxidation of various aromatic substrates, resulting in the production of water as a by-product through the reduction of molecular oxygen [42, 43]. Laccases were first time reported in the Japanese lacquer tree (*Rhus vernicifera*) [44]. Laccases has been observed in plants, insects, bacteria, but the most studied are from the fungi classified as of rot-white, where are considered as ligninolytic enzymes because lignin sources are the best substrate for the growth of these fungi. These enzymes occur mainly in basidiomycetes, deuteromycetes, and ascomycetes, but their production in lower fungi has never been observed [45]. There exists a wide diversity of laccases including isoenzymes produced by fungi that have very different physicochemical properties. Numbers of isoenzymes depend on the fungal species [2, 7, 46–48]. In general, laccases show molecular weight between 40 and 100 kDa with 10–50% of their total weight of glycosylation and with isoelectric point (p*I*) around pH 4.0. It has been reported that the glycosylation in fungal laccase plays a role in secretion, copper retention, susceptibility to proteolytic degradation, and thermal stability [49, 50]. The growth conditions of fungi and their physiological states are responsible for the expression of different laccase isoenzymes, which are coded by gene families and differentially regulated [4, 7, 41, 51, 52]. In *P. ostreatus*, 12 possible genes encoding laccases have been reported and only described and characterized 7 isoenzymes laccase: lacc2 [47], lacc4 [53], lacc6 [54], lacc9 [55], lacc10 [56, 57], lacc12 [58] and lacP83 [7]. The characteristics of some purified enzymes from *P. ostreatus* using 2,6-dimethoxyphenol (DMP), 2,2'-azino-bis(3-ethylbenzothiazoline-6-sulphonic acid) (ABTS), syringaldazine (SYR) and guaiacol (GUA) as substrate are reported in Table 1[59–65].

Table 1. Characteristics of laccases from *Pleurotus* spp.

Fungal species	Optimum pH of activity using different substrates				pI	MW (kDa)	Optimum temperature (°C)	Reference
	DMP	ABTS	SYR	GUA				
P. ostreatus			5.8		3.6	67	50	[59]
POXA1b P. ostreatus	4.5	3.0	6.0		6.9	62		[54]
POXA1w P. ostreatus	3.0–5.0	3.0	6.0		6.7	61	45–65	[60]
POXA2 P. ostreatus	6.5	3.0	6.0	6.0	4.0	67	25–35	[61]
POXA3a P. ostreatus	5.5	3.6		6.2	4.1	83–85	35	[47]
POXA3b P. ostreatus	5.5	3.6		6.2	4.3	83–85	35	[47]
POXC P. ostreatus	3.0–5.0	3.0	6.0	6.0	2.9	59	50–60	[56–61]

Fungal species	Optimum pH of activity using different substrates				pI	MW (kDa)	Optimum temperature (°C)	Reference
	DMP	ABTS	SYR	GUA				
Lcc2 *P. pulmonarius*		4.0–5.5	6.2–6.5	6.0–8.0		46	50	[62]
P. florida					4.1	77		[63]
P. sajor-caju IV		2.1			3.6	55		[64]
P. eryngii I		4.5			4.1	65	55	[65]
P. eryngii II		4.5			4.2	61	55	[65]

Laccases have a high capacity and nonspecific oxidation which allow their use in many biotechnology applications, such as detoxification of wastewater produced in pulp bleaching process [66] and from industrial plants [67], treatment of elimination of phenolic compounds in beer and processed fruit juices [68], in effluent discoloration and modification of textile fibers [69], as biosensors [70], as drug testing (to distinguish morphine from codeine) [71]. Another important application is in environmental remediation; laccases have shown ability of degrading hazardous compounds that have carcinogenic and/or mutagenic effects, including polycyclic aromatic hydrocarbons (PAH), polychlorinated biphenyls (PCB), 1,1,1-trichloro-2,2-bis (4-chlorophenyl) ethane (DDT), pentachlorophenols (PCP), toluene, benzene, xylene (BTEX), ethylbenzene, and trinitrotoluene (TNT) [72].

2.4. Manufacturing laccases from Pleurotus using solid-state and submerged fermentation methods

SSF has been defined as the bioprocess carried out in the absence or near absence of free water by the use a solid matrix with high water adsorption; the solid matrix could be biodegradable or inert, but in both cases must possess enough moisture to support growth and metabolism of the microorganism. Biodegradable solid matrix acts as support and source of nutrients and in the inert matrix, it is only the support and the culture liquid medium must be added [73, 74]. SSF has been a very efficient process for the production of enzymes by filamentous fungi [75, 76], possibly, they reproduce the natural living conditions [77].

SSF is the best culture system to study the morphological and metabolic differences between aerial hyphae and those that penetrate in the solid matrix [78]. It has been reported that SSF is better system than SmF for production of fungal enzymes, because it provides higher volumetric productivities, is less sensible to catabolite repression and yields enzymes with a higher stability at temperature and/or pH [1]. The fermentation could be carried out in less time and the productions of undesirable proteases that degrade enzymes of interest are minimized [78, 79]. Several studies in this field have determined the physiological differences during the growth of microbial cells in the two types of processes. The use of an adequate support for performing SSF is essential, since the success of the fermentation depends on it [80]. Castanera et al. [81] reported that laccase gene transcription is upregulated in an induced SmF but downregulated in the SSF, when were determined the laccases expression profiles in different fungal strains under SmF and SSF using wheat straw extract as inducer.

In a study was found that *P. ostreatus* grown in SSF on wheat bran and vinasse produced twice laccase activity (20 U/ml and three isoforms) than those reported in SmF (10 U/ml and two isoforms) [82]. It has been reported the intracellular activity and isoenzymes number of laccase of 10 strains of *Pleurotus* spp. grown on agar without addition of inducers. Differences in the *in vitro* activities using different substrates (2,6-dimethoxyphenol, *p*-anisidine, or *o*-tolidine) were observed between all the strains; zymogram patterns were similar for strains within same species, independently of any of the three substrate used [2]. Similar results were found in the extracellular extracts obtained of the same strains grown as above was mentioned [7]. In other study was used the same composition of the culture medium to grow *P. ostreatus* (ATCC 32783) in both SmF and SSF using polyurethane foam (PUF) as inert support. Atypical behavior was observed, since in SmF the fungus reported a laccase activity of 13,000 U/L with a biomass production of 5.6 g/L and four laccase isoforms, while SSF had a much lower laccase activity (2430 U/L), with biomass production of 4.5 g/L and three laccase isoforms. These results show that *P. ostreatus* performs much better in SmF than in SSF [4].

P. ostreatus (ATCC 32783) was grown at different initial pH of the culture medium in SSF using PUF. In general, the fungus showed high values of specific growth rate at all pH tested, the higher were at pH 3.5 and 8.5 (0.078 and 0.082 h^{-1}, respectively), whereas at pHs of 4.5, 6.5, and 7.5 were 0.047, 0.034, and 0.047 h^{-1}. Furthermore, the maximum biomass values were low, about 3.7 g/L in all cases. The maximum values of laccase activity were approximately 40,000 U/L observed in fermentation development at pH 4.5 and 6.5. The largest number of isoenzymes was observed in fermentations carried out at pH 7.5 and 8.5 [83].

Pleurotus pulmonarius (Fr) Quélet was cultivated on SSF using corn cob as substrate to produce laccase. The addition of 25 mM $CuSO_4$ increased from 270 to 1420 U/L the laccase production [84]. In other study, *Pleurotus pulmonarius* was grown on SSF using several natural supports, obtaining high laccase activities in wheat bran (2860 ± 250 U/L), pineapple peel (2450 ± 230 U/L), and orange bagasse (2100 ± 270 U/L) [85]. Recently, the growth of *Pleurotus eryngii* on SSF using different agricultural wastes was reported and its laccase activity was evaluated in mycelium, primordium, and fruiting body. Laccase activities were comparably low in mycelial and primordium. The highest laccase activity was obtained in fruiting body developed on both wheat straw and cotton stalk. The laccase activities of 125.65 and 205.83 U/L of fruiting body were observed on wheat straw +5% of rice bran and on cotton stalk +5 % of rice bran, respectively [86]. Maximum activity of laccase during vegetative phase of growth of *P. ostreatus* can be directly correlated with degradation of lignin in this stage [87].

Different strains of *Lentinula edodes* and *Pleurotus* species were compared for the first time for their ability to produce lignocellulolytic enzymes in SmF and SSF using various plant raw materials. Two strains of *Lentinula edodes* (IBB 123 and IBB 363) appeared to be better producers of laccase than oyster mushrooms. In SSF, *Lentinula edodes* IBB 123 reached laccase activity of 57 U/flask on day 7 of fermentation. *Pleurotus tuber-regium* IBB 624 showed 20 U/flask of laccase activity after 10 days of fermentation, other fungi of this genus produced only 7–16 U/flask of laccase during 7 or 10 days of SSF [88].

P. ostreatus and *Pleurotus sajor-caju* were grown in SSF and their ability to produce laccase and carboxymethylcellulase (CMCase) on different agricultural wastes was studied. *Pleurotus* was inoculated on viticulture wastes, wheat straw, paddy straw, sesame straw, sawdust, and the mixtures of these wastes with wheat bran. Different mycelial growth times were related with different patterns of enzyme activities. During the incubation period, *P. ostreatus* showed its highest values of laccase activity at 10th day and decreased gradually until the first harvest. The highest laccase activity was observed on mixture of wheat straw:bran (2:1) (5.48 U/mg), followed by on paddy straw:bran (2:1) (4.36 U/mg) and on viticulture wastes:bran (2:1) (3.51 U/mg) at 10th day of mycelial growth. The lowest laccase activity was obtained on viticulture wastes (0.30 U/mg) [89]. It has been reported that laccase activity could be regulated, increasing the activity in morphogenesis during the mycelial growth and then the enzyme level decreases rapidly [90]. The laccase production of an indigenous strain of *P. ostreatus* (HP-1) was studied on SSF. Culture parameters, including type and concentration of substrate, moisture content, inoculum size, temperature, pH, surfactant presence, and nitrogen source, were optimized by conventional one factor at a time methodology. Maximum laccase activity of 3952 U/g of dry substrate was obtained with wheat straw as substrate, incubation temperature 28°C, five agar plugs as inoculum, pH 5.0, 60% moisture content, surfactant concentration 0.015 g/L, and combination of L-asparagine and NH_4NO_3 at 10 mM concentration each as nitrogen source. Laccase activity was increased with the use of various aromatic inducers and $CuSO_4$. Highest laccase activity of 14189 U/g of dry substrate was obtained using 0.28 mM $CuSO_4$ under optimized conditions [91]. *P. ostreatus* was grown in SSF conditions for production of laccase, manganese peroxidase, and lignin peroxidase. Highest enzymes levels (laccase 455.11, manganese peroxidase 210.77, and lignin peroxidase 54.50 U/ml) were observed at 7 days in a medium containing 5 g wheat straw (66% w/w moisture), 4 ml inoculum at pH 4.5 and 30°C, using 1% (v/v) glycerol as carbon source, 0.2% (w/w) urea as nitrogen source, 1% (w/v) 2,2-azinobis 3-ethylbenzthiazoline 6 sulphonate as laccase inducer and 1% (w/v) $MnSO_4$ for manganese peroxidase, 1% (w/v) $CuSO_4$ as metal ion for laccase, and Mn+ for manganese peroxidase [92].

Different concentrations of apple pomace were evaluated on laccase production by *P. ostreatus*. During the first four days of fermentation, there was not laccase production. The maximum laccase activity (114.64 U/ml), was observed at 9 days of fermentation in the medium with 2.5% (w/v) apple pomace. This activity was approximately 2.8 times (30.24 U/ml) and 0.9 times (60.49 U/ml) higher than that of *P. ostreatus* with 5% (w/v) and without apple pomace, respectively. These results suggest that *P. ostreatus* might use the nutrient content of apple pomace (rich in carbohydrates, dietary fiber, and minerals) without laccase activity in the initial stages of cultivation (approximately 4 days) [93].

Recently, optimization of the laccases production of *P. ostreatus* grown on sugarcane bagasse in SSF was worked. Water activity, pH, temperature, and concentrations of $CuSO_4$, $(NH_4)_2SO_4$, KH_2PO_4, asparagine, and yeast extract were variables used in the optimization. The concentrations of $CuSO_4$ and $(NH_4)_2SO_4$ had a significant influence on the production of laccase, but the use of yeast extract and the addition of ferulic acid as inducer provided increases of laccase activity of 5.7 times and 2.0 times, respectively. The highest laccase activity of 151.6 U/g was produced at the 5th day of SSF [94].

Cocultivation of *P. ostreatus* MTCC 1804 and mutant *Penicillium oxalicum* SAUE-3.510 was studied for the production of xylanase-laccase mixture under SSF condition. Growth compatibility between both fungi was analyzed by growing them on potato dextrose agar plate, obtaining 58% and 33% higher levels of xylanase and laccase production, respectively. A mixture of sugarcane bagasse and black gram husk (3:1) was the best solid substrate and support for fungal colonization and enzyme production during co cultivation. Maximum activity values of xylanase (8205.31 IU/g) and laccase (375.53 IU/g) during SSF were observed using 4 g of solid support with 80% of moisture content. The coculture system was efficient in the production of xylanases and laccases that may be employed in agroindustrial waste degradation [95].

P. ostreatus (ATCC 32783) was grew under SmF conditions and found the production profile of laccases as well as the isoenzymes patterns through zymograms. Four laccase isoenzymes were produced, one throughout the fermentation time and three only during growth stationary phase. The maximum laccase activity (12200 U/L) was observed in the growth stationary phase. Kinetic parameters of a purified isoenzyme (enzyme produced throughout the fermentation), such as the apparent molecular weight of 43.7 kDa, K_m 90 μM, V_{max} 1.18 ΔAbs/min, and p*I* of 2.3, were obtained [3].

On the other hand, the growth and activity of laccases from of five different strains of *P. ostreatus* developed under SmF conditions with and without copper added to the culture medium was studied. It was observed that the concentration of $CuSO_4 \cdot 5H_2O$ (0.25 g/L) did not affect the growth of the strain ATCC 32783, however, other strains showed lower growth rates and less biomass, the ATCC 201216 strain almost was inhibited. ATCC 32783 strain showed the highest values of laccase activity in the presence of copper reaching up to 37490 U/L, whereas in the culture without copper was obtained 1086 U/L; ATCC 201216 strain in the presence and absence of copper produced 1400 and 1000 U/L, respectively. These results suggest that not all strains have the same answer to the presence of Cu in the culture medium, and the sensitivity to Cu be could use to select strains with high laccase production for commercial exploitation [5].

The activity and isoenzymes number of laccase from *P. ostreatus* ATCC 32783 grown in SmF conditions using a buffered and nonbuffered media were studied. For both culture media, the initial pH was 3.5. Laccase activity was around 100–500 U/L during the 100–400 h (approximately) of fermentation in both media. Buffered culture medium showed minimal pH changes, while the pH in nonbuffered medium changed drastically, reached a value of 6.5 after 240 h of fermentation. The highest laccase activity (3200 U/L) at 500 h of fermentation was obtained in the buffered medium and in nonbuffered culture medium was only of 450 U/L. One laccase isoenzyme was observed during the entire fermentation process in both media, but in the nonbuffered medium, an additional isoenzyme was produced when the pH reached a value of 6.5. These results suggest that some laccase isoenzymes are regulated by pH signals and also observed that the fungus produces metabolites to regulate the pH of the medium [6].

A gene called lacP83 that encode a laccase isoenzyme of *P. ostreatus* ATCC 32783 grown in SmF was described. Using the PCR inverse strategy, a 2887 bp sequence was obtained from a genomic library of *P. ostreatus*. The coding sequence was of 1527 bp long with 17 exons and the protein encoded had 509 amino acids, shows a putative signal peptide and conserved Cu binding domains. In the promoter region (466 bp upstream of ATG), putative binding transcription factors such as metal response element, xenobiotic response element, a stress response element, and a defense response element were found. The gene and protein sequences of lacP83 had 85–94% and 90–96%, respectively, of similarity with laccases of *Pleurotus* previously reported. This laccase showed differences in its promoter sequence and apparent molecular weight [7].

A recent study investigated the impact of pH on the expression of five specific genes (lacc1, lacc4, lacc6, lacc9, and lacc10) and the resulting profiles of isoenzymes generated by P. ostreatus ATCC 32783 under SmF conditions. The pH of the culture media was initially altered to 3.5, 4.5, 6.5, and 8.5. This research demonstrated that pH plays a crucial role in the growth, development, and generation of enzymes and metabolites in this fungus. The specific growth rate exhibited a positive correlation with the beginning pH of the culture medium. The highest biomass values were observed at pH levels of 6.5 and 8.5. Similarly, the highest laccase activity was observed at initial pH levels of 4.5 and 6.5, reaching a maximum of 77,500 U/L. The isoenzyme patterns varied based on the initial pH of the culture medium. At acidic pH, three isozymes (29, 47, and 65 kDa) were observed. At pH close to neutral, four isoenzymes (29, 38, 47, and 65 kDa) were observed. At alkaline pH, three isozymes (29, 47, and 65 kDa) were observed. Based on the observation of gene expression, it is believed that lacc6, lacc10, lacc4, and lacc1 correspond to isoenzymes with molecular weights of 65, 47, 38, and 29 kDa, respectively. According to the scientists, the pH plays a crucial function as a transcriptional factor in determining the expression profile and production pattern of laccase enzymes under SmF circumstances [8].

REFERENCES

1. Viniegra-González G, Favela-Torres E, Aguilar NC, Romero-Gómez SJ, Díaz-Godínez G, Augur C. Advantages of fungal enzyme production in solid state over liquid fermentation systems. Biochem Eng J. 2003;13:157–167. DOI: 10.1016/S1369-703X(02)00128-6.

2. Téllez-Téllez M, Sánchez C, Montiel-González AM, Díaz-Godínez G. Laccase activities of the peripheral and central zones of the vegetative mycelium of colonies of Pleurotus species. Agro Food Ind Hi-Tech. 2005;16:38–40.

3. Tlecuitl-Beristain S, Sánchez C, Loera O, Robson GD, Díaz-Godínez G. Laccases of Pleurotus ostreatus observed at different phases of its growth in submerged fermentation: production of a novel laccase isoforms. Mycol Res. 2008;112:1080–1084. DOI:10.1016/j.mycres.2008.03.001.

4. Téllez-Téllez M, Fernández JF, Montiel-González AM, Sánchez C, Díaz-Godínez G. Growth and laccase production by Pleurotus ostreatus in submerged and solid-state fermentation. Appl Microbiol Biotechnol. 2008;81:675–679. DOI: 10.1007/s00253-008-1628-6.

5. Díaz R, Alonso S, Sánchez C, Tomasini A, Bibbins-Martínez MD, Díaz-Godínez G. Characterization of the growth and laccase activity of strains of Pleurotus ostreatus in submerged fermentation. Bioresources 2011;6(1):282–290. DOI: 10.15376/biores.6.1.282-290.

6. Díaz R, Sánchez C, Bibbins-Martínez M, Díaz-Godínez G. Effect of medium pH on laccase zymogram patterns produced by Pleurotus ostreatus in submerged fermentation. Afr J Microbiol Res. 2011;5:2720–2723. DOI: 10.5897/AJMR11.526.

7. Téllez-Téllez M, Díaz-Godínez G, Aguilar MB, Sánchez C, Fernández F. Description of a laccase gene from Pleurotus ostreatus expressed under submerged fermentation conditions. Bioresources 2012;7(2):2038–2050. DOI: 10.15376/biores.7.2.2038-2050.

8. Díaz R, Téllez-Téllez M, Sánchez C, Bibbins-Martínez MD, Díaz-Godínez G, Soriano-Santos J. Influence of initial pH of the growing medium on the activity, production and genes expression profiles of laccase of Pleurotus ostreatus in submerged fermentations. Electron J Biotechnol. 2013;16(4):fulltext 6. DOI: 10.2225/vol16-issue4-fulltext-6.

9. Chang ST. Global impact of edible and medicinal mushroom on human welfare in the 21st century: non-green revolution. Int J Med Mushroom. 1999;1:1–7. DOI: 10.1615/IntJMedMushrooms.v1.i1.10.

10. Sánchez C. Morphogenesis of mushroom fungi: ultrastructural, physiological and histological study of Pleurotus ostreatus [thesis]. Manchester: University of Manchester; 1998.

11. Deshpande MV. Proteinases in fungal morphogenesis. World J Microbiol Biotechnol. 1992;8:242–250. DOI: 10.1007/BF01201871.

12. Moore D. Fungal morphogenesis. New York: Cambridge University Press; 1998.

13. Sánchez C. Cultivation of Pleurotus ostreatus and other edible mushrooms. Appl Microbiol Biotechnol. 2010;85:1321–1337. DOI: 10.1007/s00253-009-2343-7.

14. Deepalakshmi K, Mirunalini S. Pleurotus ostreatus: an oyster mushroom with nutritional and medicinal properties. J Biochem Tech. 2014;5(2):718–726.

15. Sánchez C. Modern aspects of mushrooms culture technology. Appl Microbiol Biotechnol. 2004;64:756–762. DOI: 10.1007/s00253-004-1569-7.

16. Patel S, Goyal A. Recent developments in mushrooms as anticancer therapeutics: a review. Biotechnology. 2012;2:1–15. DOI: 10.1007/s13205-011-0036-2.

17. Oei P. Manual on mushroom cultivation: techniques species and opportunities for commercial application in developing countries. Amsterdam: TOOL Publications; 2003.

18. Manzi P, Aguzzi A, Pizzoferrato L. Nutritional value of mushrooms widely consumed in Italy. Food Chem. 2001;73:321–325. DOI: 10.1016/S0308-8146(00)00304-6.

19. Sánchez C. Lignocellulosic residues: biodegradation and bioconversion by fungi. Biotechnol Adv. 2009;27:85–194. DOI: 10.1016/j.biotechadv.2008.11.001.

20. Fernández-Fueyo E, Ruiz-Dueñas FJ, Martínez MJ, Romero A, Hammel KE, Medrano FJ, Martínez AT. Ligninolytic peroxidase genes in the oyster mushroom genome: heterologous expression, molecular structure, catalytic and stability properties, and lignin-degrading ability. Biotechnol Biofuel. 2014;7:1–23. DOI: 10.1186/1754-6834-7-2.

21. Marzullo L, Cannio R, Giardina P, Santini MT, Sannia G. Veratryl alcohol oxidase from Pleurotus ostreatus participates in lignin biodegradation and prevents polymerization of laccase-oxidized substrates. J Biol Chem. 1995;270:3823–3827. DOI: 10.1074/jbc.270.8.3823.

22. Faraco V, Piscitelli A, Sannia G, Giardina P. Identification of a new member of the dye-decolorizing peroxidase family from Pleurotus ostreatus. World J Microbiol Biotechnol. 2007;23:889–893. DOI: 10.1007/s11274-006-9303-5.

23. Hofrichter M, Ullrich R, Pecyna MJ, Liers C, Lundell T. New and classic families of secreted fungal heme peroxidases. Appl Microbiol Biotechnol. 2010;87:871–897. DOI: 10.1007/s00253-010-2633-0.

24. Salame TM, Knop D, Levinson D, Yarden O, Hadar Y. Redundancy among manganese peroxidases in Pleurotus ostreatus. Appl Environ Microbiol. 2013;79(7):2405–2415. DOI: 10.1128/AEM.03849-12.

25. Pérez J, Jeffries TW. Roles of manganese and organic acid chelators in regulating lignin degradation and biosynthesis of peroxidases by Phanerochaete chrysosporium. Appl Environ Microbiol. 1992;58:2402–2409.

26. Jensen KA, Bao W, Kawai S, Srebotnic E, Hammel KE. Manganese-dependent cleavage of nonphenolic lignin structures by Ceriporiopsis subvermspora in absence of lignin peroxidase. Appl Environ Microbiol. 1996;62:3679–3686.

27. Cullen D, Kersten PJ. Enzymology and molecular biology of lignin degradation. In: Brambl R, Marzluf GA, editors. The mycota III. Biochemistry and molecular biology. Berlin-Heidelberg: Springer-Verlag; 2004. DOI: 10.1007/978-3-662-06064-3_13.

28. Wariishi H, Valli K, Gold MH. Manganese(II) oxidation by manganese peroxidase from the basidiomycete Phanerochaete chrysosporium: kinetic mechanism and role of chelators. J Biol Chem. 1992;267:23688–23695.

29. Gutierrez A, Caramelo L, Prieto A, Martinez MJ, Martinez AT. Anisaldehyde production and aryl-alcohol oxidase and dehydrogenase activities in ligninolytic fungi of the genus Pleurotus. Appl Environ Microbiol. 1994;60:1783–1788.

30. Kamitsuji H, Honda Y, Watanabe T, Kuwahara M. Production and induction of manganese peroxidase isozymes in a white-rot fungus Pleurotus ostreatus. Appl Microbiol Biotechnol. 2004;65:287–294. DOI: 10.1007/s00253-003-1543-9.

31. Salame TM, Yarden O, Hadar Y. Pleurotus ostreatus manganese-dependent peroxidase silencing impairs decolourization of Orange II. Microb Biotechnol. 2010;3:93–106. DOI: 10.1111/j.1751-7915.2009.00154.x

32. Ruiz-Dueñas FJ, Fernández E, Martínez MJ, Martínez AT. Pleurotus ostreatus heme peroxidases: an in-silico analysis from the genome sequence to the enzyme molecular structure. Crit Rev Biol. 2011;334:795–805. DOI: 10.1016/j.crvi.2011.06.004.

33. Garcia-Ruiz E, González-Pérez D, Ruiz-Dueñas FJ, Martinez AT, Alcalde M. Directed evolution of a temperature-, peroxide- and alkaline pH-tolerant versatile peroxidase. Biochem J. 2012;441:487–498. DOI: 10.1042/BJ20111199.

34. Salame TM, Knop D, Tal D, Levinson D, Yarden O. Predominance of a versatile-peroxidase-encoding gene, mnp4, as demonstrated by gene replacement via a gene targeting system for Pleurotus ostreatus. Appl Environ Microbiol. 2012;78:5341–5352. DOI: 10.1128/AEM.01234-12.

35. Golan-Rozen N, Chefetz B, Ben-Ari J, Geva J, Hadar Y. Transformation of the recalcitrant pharmaceutical compound carbamazepine by Pleurotus ostreatus: role of cytochrome P450 monooxygenase and manganese peroxidase. Environ Sci Technol. 2011;45:6800–6805. DOI: 10.1021/es200298t.

36. Kirk TK, Farrell RL. Enzymatic "combustion": the microbial degradation of lignin. Annu Rev Microbiol. 1987;41:465–505. DOI: 10.1146/annurev.mi.41.100187.002341.

37. Eriksson KEL, Blanchette RA, Ander P. Microbial and enzymatic degradation of wood and wood components. Berlin, Heidelberg: Springer; 1990.

38. Hammel KE. Organopollutant degradation by ligninolytic fungi. In: Young LY, Cerniglia CE, editors. Microbial transformation and degradation of toxic organic chemicals.New York: Wiley-Liss; 1995. p. 331–346.

39. Garzillo AM, Colao MC, Buonocore V, Oliva R, Falcigno L, Saviano M, Santoro AM, Zappala R, Bonomo RP, Bianco C, Giardina P, Palmieri G, Sannia G. Structural and kinetic characterization of native laccases from Pleurotus ostreatus, Rigidoporus lignosus, and Trametes trogii. J Prot Chem. 2001;20:191–201. DOI: 10.1023/A:1010954812955.

40. Platt MW, Hadar Y, Chet I. Fungal activities involved in lignocellulose degradation by Pleurotus. Appl Microbiol Biotechnol. 1984;20:150–154. DOI: 10.1007/BF00252594.

41. Bollag JM, Leonowicz A. Comparative studies of extracellular fungal laccase. Appl Environ Microbiol. 1984;48:849–854. DOI: 10.1128/AEM.71.4.1775.

42. Thurston CF. The structure and function of fungal laccases. Microbiology. 1994;140: 19–26.

43. Solomon EI, Sundaram UM, Machonkin TE. Multicopper oxidases and oxygenases. Chem Rev. 1996;96:2563–2605. DOI: 10.1021/cr950046o.

44. Yoshida H. LXIII.—Chemistry of lacquer (Urushi). Part I. Communication from the Chemical Society of Tokio. J. Chem. Soc., Trans. 1883;472. DOI: 10.1039/CT8834300472.

45. Messerschmidt A, Huber R. The blue oxidases, ascorbate oxidase, laccase and ceruloplasmin- Modelling and structural relationships. Eur J Biochem. 1990;187:341. DOI: 10.1111/j.1432-1033.1990.tb15311.x.

46. Yaver DS, Xu F, Golightly EJ, Brown KM, Brown SH, Rey MW, Schneider P, Halkier T, Mondorf K, Dalboge H. Purification, characterization, molecular cloning, and expression of two laccase genes from the white rot basidiomycete Trametes villosa. Appl Environ Microbiol. 1996;62: 834–841.

47. Palmieri G, Cennamo G, Faraco V, Amoresano A, Sannia G, Giardina P. Atypical laccase isoenzymes from copper supplemented Pleurotus ostreatus cultures. Enzyme Microb Technol. 2003;33: 220–230. DOI: 10.1016/S0141-0229(03)00117-0.

48. Rodríguez E, Ruiz-Dueñas FJ, Kooistra R, Ram A, Martínez AT, Martínez MJ. Isolation of two laccase genes from the white-rot fungus Pleurotus eryngii and heterologous expression of the pel3 encoded protein. J Biotechnol. 2008;134:9–19. DOI: 10.1016/j.jbiotec.2007.12.008.

49. Li K, Xu F, Eriksson KEL. Comparison of fungal laccases and redox mediators in oxidation of a nonphenolic lignin model compound. Appl Environ Microbiol. 1999;65:2654–2660.

50. Duran N, Rosa MA, D'Annibale A, Gianfreda L. Applications of laccases and tyrosinases (phenoloxidases) immobilized on different supports: a review. Enzyme Microb Technol. 2002;31:907–931. DOI: 10.1016/S0141-0229(02)00214-4.

51. Rogalski J, Hatakka A, Longa B, Wojtas-Wasilewska M. Hemicellulolytic enzymes of the ligninolytic white-rot fungus Phlebia radiata: influence of phenolic compounds on the synthesis of hemicellulolytic enzymes. Acta Biotechnol. 1993;13: 53–57. DOI: 10.1002/abio.370130110.

52. Mansur M, Suárez T, González AE. Differential gene expression in the laccase gene family from basidiomycete I-62 (CECT 20197). Appl Environ Microbiol. 1998;64:771–774.

53. Pezzella C, Autore F, Giardina P, Piscitelli A, Sannia G, Faraco V. The Pleurotus ostreatus laccase multi-gene family: isolation and heterologous expression of new family members. Curr Genomics. 2009;55:45–57. DOI: 10.1007/s00294-008-0221-y.

54. Giardina P, Palmieri G, Scaloni A, Fontanella B, Faraco V, Cennamo G, Sannia G. Protein and gene structure of a blue laccase from Pleurotus ostreatus. Biochem J. 1999;341:655–663. DOI: 10.1042/bj3410655.

55. Giardina P, Cannio R, Martirani L, Marzullo L, Palmieri G, Sannia G. Cloning and sequencing of a laccase gene from the lignin-degrading basidiomycete Pleurotus ostreatus. Appl Environ Microbiol. 1995;61(6):2408–2413.

56. Palmieri G, Giardina P, Marzullo L, Desiderio B, Nittii G, Cannio R, Sannia G. Stability and activity of a phenol oxidase from the ligninolytic fungus Pleurotus ostreatus. Appl Microbiol Biotechnol. 1993;39:632–636. DOI: 10.1007/BF00205066.

57. Giardina P, Aurilia V, Cannio R, Marzullo L, Amoresano A, Siciliano R, Pucci P, Sannia G. The gene, protein and glycan structures of laccase from Pleurotus ostreatus. Appl Environ Microbiol. 1996;61:2408–2413. DOI: 10.1111/j.1432-1033.1996.00508.x.

58. Lettera V, Piscitelli A, Leo G, Birolo L, Pezzella C, Sannia G. Identification of a new member of Pleurotus ostreatus laccase family from mature fruiting body. Fungal Biol. 2010;114:724–730. DOI: 10.1016/j.funbio.2010.06.004.

59. Hublik G, Schinner F. Characterization and immobilization of the laccase from Pleurotus ostreatus and its use for the continuous elimination of phenolic pollutants. Enzyme Microb Technol. 2000;27:330–336. DOI: 10.1016/S0141-0229(00)00220-9.

60. Palmieri G, Giardina P, Bianco C, Scaloni A, Capasso A, Sannia G. A novel white laccase from Pleurotus ostreatus. J Biol Chem. 1997;272:31301–31307. DOI: 10.1074/jbc.272.50.31301.

61. Sannia G, Giardina P, Luna M, Rossi M, Buonocore V. Laccase from Pleurotus ostreatus. Biotechnol Lett. 1986;8:797–800. DOI: 10.1007/BF01020827.

62. De Souza CGM, Peralta RM. Purification and characterization of the main laccase produced by the white-rot fungus Pleurotus pulmonarius on wheat bran solid state medium. J Basic Microbiol. 2003;43:278–286. DOI: 10.1002/jobm.200390031.

63. Das N, Chakraborty TK, Mukherjee M. Purification and characterization of laccase-1 from Pleurotus florida. Folia Microbiol. 2000;45:447–451. DOI: 10.1007/BF02817619.

64. Lo SC, Ho YS, Buswell JA. Effect of phenolic monomers on the production of laccases by the edible mushroom Pleurotus sajor-caju, and partial characterization of a major laccase component. Mycologia. 2001;93:413–421. DOI: 10.2307/3761726.

65. Muñoz C, Guillen F, Martinez AT, Martinez MJ. Laccase isoenzymes of Pleurotus eryngii: characterization, catalytic properties, and participation in activation of molecular oxygen and Mn2+ oxidation. Appl Environ Microbiol. 1997;63:2166–2174.

66. Bajpai P. Application of enzymes in the pulp and paper industry. Biotechnol Prog. 1999;15:147–157. DOI: 10.1021/bp990013k.

67. Durán N, Esposito E. Potential applications of oxidative enzymes and phenoloxidase-like compounds in wastewater and soil treatment: a review. Appl Catal B: Environ. 2000;28:83–99. DOI: 10.1016/S0926-3373(00)00168-5.

68. Minussi R, Pastore GM, Duran N. Potential applications of laccase in the food industry. Trends Food Sci Technol. 2002;13:205–216. DOI: 10.1016/S0924-2244(02)00155-3.

69. Abadulla E, Tzanov T, Costa S, Robra KH, Covaco-Paulo A, Gübitz A. Decolorization and detoxification of textile dyes with a laccase from Trametes hirsute. Appl Environ Microbiol. 2000;66:3357–3362. DOI: 10.1128/AEM.66.8.3357-3362.2000.

70. Amir L, Tam TK, Pita M, Meijler MM, Alfonta L. Katz E. Biofuel cell controlled by enzyme logic systems. J Am Chem Soc. 2009;131:826–832. DOI: 10.1021/ja8076704.

71. Mayer AM, Staples RC. Laccase: new functions for an old enzyme. Phytochemistry. 2002;60:551–565. DOI: 10.1016/S0031-9422(02)00171-1.

72. Desai SS, Nityanand C. Microbial laccases and their applications: a review. Asian J Biotechnol. 2011;3:98–124. DOI: 10.3923/ajbkr.2011.98.124.

73. Pandey A, Soccol CR, Mitchell D. New developments in solid state fermentation. I. Bioprocesses and products. Proc Biochem. 2000;35:1153–1169. DOI: 10.1016/S0032-9592(00)00152-7.

74. Singhania RR, Patel AK, Soccol CR, Pandey A. Recent advances in solid-state fermentation. Biochem Eng J. 2009;44(1):13–18. DOI: 10.1016/j.bej.2008.10.019.

75. Moo-Young M, Moreira AR, Tengerdy RP. Principles of solid state fermentation, in:Smith JE, Barry DR, Kristiansen B, editors. The filamentous fungi. London: Edward Arnold Publishers; 1983.p. 117–144.

76. Pandey A, Azmi W, Singh J, Banerjee UC. Types of fermentation and factors affecting it, in: Joshi VK, Pandey A, editors. Biotechnology: food fermentation. New Delhi: Educational Publishers; 1999. p. 383–426.

77. Pandey A, Selvakumar P, Soccol CR, Nigam P. Solid state fermentation for the production of industrial enzymes. Curr Sci. 1999;77:149–162.

78. Hölker U, Lenz J. Solid-state fermentation are there any biotechnological advantages?. Curr Opin Biotech. 2005;8(3):301–306. DOI: 10.1016/j.mib.2005.04.006.

79. Díaz-Godínez G, Soriano J, Augur C, Viniegra-González G. Exopectinases produced by Aspergillus niger in solid-state and submerged fermentation: a comparative study. J Ind Microbiol Biotechnol. 2001;26:271–275. DOI: 10.1038/sj.jim.7000113.

80. Couto SR, Sanromán MA. Application of solid-state fermentation to ligninolytic enzyme production. Biochem Eng J. 2005;22(3):211–219. DOI: 10.1016/j.bej.2004.09.013.

81. Castanera R, Pérez G, Omarini A, Alfaro M, Pisabarro AG, Faraco V, Amore A, Ramírez L. Transcriptional and enzymatic profiling of Pleurotus ostreatus laccase genes in submerged and solid-state fermentation cultures. Appl Environ Microbiol. 2012;78(11):4037–4045. DOI: 10.1128/AEM.07880-11.

82. Ramírez NE, Vargas MC, Ariza JC, Martínez C. Characterization of laccase obtained by two production methods with Pleurotus ostreatus. Rev Colombiana Biotecnol. 2003;2:64–72.

83. Velázquez L, Téllez-Téllez M, Díaz R, Bibbins-Martínez MD, Loera O, Sánchez C, Tlecuitl-Beristain S, Díaz-Godínez G. Laccase isoenzymes of Pleurotus ostreatus grown at different pH in solid-state fermentation using polyurethane foam as support. Annu Res Rev Biol. 2014;4(16):2566–2578. DOI: 10.9734/ARRB/2014/10016.

84. Tychanowicz GK, Souza DFD, Souza CG, Kadowaki MK, Peralta RM. Copper improves the production of laccase by the white-rot fungus Pleurotus pulmonarius in solid state fermentation. Braz Arch Biol Technol. 2006;49(5):699–704. DOI: 10.1590/S1516-89132006000600002.

85. dos Santos Bazanella GC, de Souza DF, Castoldi R, Oliveira RF, Bracht A, Peralta RM. Production of laccase and manganese peroxidase by Pleurotus pulmonarius in solid-state cultures and application in dye decolorization. Folia Microbiol. 2013;58(6):641–647. DOI: 10.1007/s12223-013-0253-7.

86. Yildirim N, Yildirim NC, Yildiz A. Laccase enzyme activity during growth and fruiting of Pleurotus eryngii under solid state fermentation medium containing agricultural wastes. IJPAS. 2015; 1:64–71.

87. Pandey VK, Singh MP. Biodegradation of wheat straw by Pleurotus ostreatus. Cell Mol Biol. 2014;60(5):29–34. DOI: 10.14715/cmb/2014.60.5.6.

88. Elisashvili V, Penninckx M, Kachlishvili E, Tsiklauri N, Metreveli E, Kharziani T, Kvesitadze G. Lentinus edodes and Pleurotus species lignocellulolytic enzymes activity in submerged and solid-state fermentation of lignocellulosic wastes of different composition. Bioresour Technol. 2008;99(3):457–462. DOI: 10.1016/j.biortech.2007.01.011.

89. Kurt S, Buyukalaca S. Yield performances and changes in enzyme activities of Pleurotus spp. (P. ostreatus and P. sajor-caju) cultivated on different agricultural wastes. Bioresour Technol. 2010;101(9):3164–3169. DOI: 10.1016/j.biortech.2009.12.011.

90. Tan YH, Wahab MN. Extracellular enzyme production during anamorphic growth in the edible mushroom, Pleurotus sajor-caju. World J Microb Biot. 1997;13(6):613–617.

91. Patel H, Gupte A, Gupte S. Effect of different culture conditions and inducers on production of laccase by a basidiomycete fungal isolate Pleurotus ostreatus HP-1 under solid state fermentation. BioResources 2009;4(1):268–284. DOI: 10.15376/biores.4.1.268-284.

92. Aslam S, Asgher M. Partial purification and characterization of ligninolytic enzymes produced by Pleurotus ostreatus during solid state fermentation. Afr J Biotechnol. 2011;10(77):17875–17883. DOI: 10.5897/AJB11.2233.

93. Park YJ, Yoon DE, Kim HI, Kwon O, Yoo YB, Kong WS, Lee CS. Overproduction of laccase by the white-rot fungus Pleurotus ostreatus using apple pomace as inducer. Mycobiology. 2014;42(2):193–197. DOI: 10.5941/MYCO.2014.42.2.193.

94. Karp SG, Faraco V, Amore A, Letti LAJ, Soccol VT, Soccol CR. Statistical optimization of laccase production and delignification of sugarcane bagasse by Pleurotus ostreatus in solid-state fermentation. Biomed Res Int. 2015;12(16):19. DOI: 10.1155/2015/181204.

95. Dwivedi P, Vivekanand V, Pareek N, Sharma A, Singh RP. Co-cultivation of mutant Penicillium oxalicum SAU E-3.510 and Pleurotus ostreatus for simultaneous biosynthesis of xylanase and laccase under solid-state fermentation. N Biotechnol. 2011;28(6):616–626. DOI: 10.1016/j.nbt.2011.05.006.

05 VARIABLES INFLUENCING RUMEN FERMENTATION UTILIZING BATCH CULTURE TECHNIQUE

1. INTRODUCTION: AN OVERVIEW

Rumen fermentation significantly influences feed digestion and microbial generation in ruminant animals. The feed value, nutrient provision, and animal productivity are ultimately determined by the rate and extent of feed digestion in the rumen, rumen fermentation pattern, and amount of microbial protein production. Hence, it is imperative to assess the digestibility of feed in the rumen in order to forecast animal productivity and formulate the most effective ration. Furthermore, significant feed additives are currently utilized to enhance or alter rumen fermentation, and it is necessary to ascertain their effects. Utilizing animals to assess feed digestibility or the effectiveness of feed additives is a dependable method. However, there are some drawbacks associated with this strategy, including its time-consuming nature, high cost, the need for substantial amounts of feed (or feed additives), and its unsuitability for evaluating feed on a wide scale. Consequently, numerous biological techniques that mimic the rumen fermentation process have been devised. The batch culture method has been extensively utilized for the purpose of screening and comparing different diets and feed additives, such as feed enzymes. The benefits of utilizing in vitro techniques, in contrast to in vivo procedures involving animals, are cost-effectiveness, simplicity, minimal feed requirements, and notably, the capacity to analyze a large number of samples under consistent experimental circumstances [1]. Nevertheless, several variables employed in the batch culture technique, such as the source of the inoculum, the recording system for gas generation, the method of substrate dispersal in the bottle, the sample size, and the method of substrate preparation, have the potential to impact fermentation outcomes [2–4]. One notable concern is the selection of venting methods for gas measurement. Gas accumulates in a closed system, and the increase in pressure in the headspace might impact the speed at which substrate fermentation occurs [5]. Various venting devices for gas pressure relief have been examined, however, the impact on feed digestion remains uncertain [2, 6]. Tagliapietra et al. [3] found that precise management of manual pressure measurements, headspace volume, venting frequency, and fermentable substrate quantity is crucial to prevent excessive headspace pressures that may disrupt fermentation kinetics. Previous studies have documented the practice of inserting substrates in porous bags within incubation vials [1, 7] or directly adding them to the inoculum [8, 9]. Higher concentrations of methane were detected in samples that were immediately dispersed in vials, as opposed to those that were enclosed in bags [4]. The bags may establish a microenvironment that is different from the free inoculum and this difference may shift depending on the pore size of the bags [10]. Prior to incubation, the substrates used in batch culture must undergo processing to achieve an appropriate particle size because to the absence of mastication and rumen contraction in animals. Utilizing a finely pulverized sample diminishes the likelihood of sampling bias, particularly when dealing with forage samples. However, it is possible for tiny particles to escape from the bags before undergoing complete digestion. The lack of standardization of these elements pertaining to batch culture in the laboratory might have a substantial impact on the fermentation outcomes, leading to increased variability and reduced technique dependability. The purpose of this chapter is to examine many important elements that may affect the results of batch culture and to offer valuable insights for optimizing the usage of batch culture technique in assessing feeds or feed additives in ruminant nutrition.

2. BATCH CULTURE

Batch culture is a technique for large-scale production of microbes or microbial products in which, at a given time, the fermenter is stopped and the culture is worked up. The 'batch culture' fermentation is also known as 'closed culture' system. In this system, at the beginning, the nutrients and other additives are added in required amounts. There is no refill of nutrients once the fermentation process has started and the product is recovered at the end of the process. In the beginning, microorganisms grow at a rapid rate due to availability of excess nutrients. As time passes, they increase in number with rapid use of the nutrients and simultaneously produce toxic metabolites. The batch culture that is currently used to evaluate ruminant feeds or feed additives is primarily based on the in vitro technique developed by Tilley and Terry [11] and modified by Goering and Van Soest [12]. The batch culture consists of collection of rumen fluid as inoculum, inoculation of dried, ground feed samples contained in a flask with a buffering and nutritive in vitro medium. Sample digestion is measured following anaerobic degradation by rumen bacteria. The batch culture can measure the kinetics and volume of gas production (mainly CO_2, CH_4), as well as gas profiles, rate, and extent of substrate digestion, which can then be used to evaluate feed values (ranking feed) and feed additive screening. The kinetics of gas production or feed digestion can be a developed model to predict feed intake, microbial protein synthesis, and metabolizable energy.

During the fermentation of feedstuff, the truly digested substrate is partitioned among volatile fatty acids, gas and microbial biomass. Gas production occurs when substrate carbohydrates are fermented to generate acetate or butyrate but no gas is produced with fermentation of carbohydrate to generate propionate. However, gas is also produced when volatile fatty acid causes gas to be released from the bicarbonate buffer [13]. Although gas production is a reflection of the generation of volatile fatty acids and microbial mass as a result of substrate fermentation, gas measurements only account for substrate that is used for volatile fatty acids and gas production and does not consider substrate utilized for microbial growth. Therefore, the volume of gas produced during fermentation is highly associated with the amount of substrate digested. Currently, the gas production technique is commonly used to evaluate and predict feed value and screening feed additives for ruminants. One major advantage of in vitro gas measurement technique is that it focuses on the appearance of fermentation products and non-fermentable substrates do not contribute to gas production [13]. Since gas production does not consider the amount of substrate converted into microbial biomass, the substrate digestibility that is estimated based on gas measurement is considered as apparent digestibility [14]. Feed protein degradation does not contribute to gas but the high ammonia nitrogen concentration in in vitro systems might prevent the release of gas due to its highly basic nature. As one of major measures of batch culture, gas measurement is widely used to predict rate and extent of feed digestion in the rumen as well as feed intake and microbial protein synthesis [14].

3. VARIABLES INFLUENCING BATCH CULTURE FERMENTATION

3.1. Effect of inoculum source

The inoculum is often the major source of variation on the variable measurements in the use of batch culture technique to study fermentation kinetics of ruminant feeds. The effect of inoculum source on in vitro gas production was considerably discussed in a review by Rymer et al. [2]. Considerable animal variation in the quality of rumen fluid inoculum, prepared identically, is known to exist both within and among donor animals [15, 16]. The variation of batch fermentation due to inoculum source is ultimately attributed to the variation of microbial population profiles and microbial activities in the rumen. Therefore, all the factors that potentially affect the ruminal microbial activity would affect inoculum quality, thus varying the batch fermentation. In this section, the effects on batch fermentation of the inoculum from sampling schedule, different species, rumen versus faeces as well as inoculum preparation are discussed.

3.1.1. Effect of donor animals, diet and collection time

The donor animals, type of diet and the inoculum collection time may all have an effect on consistency of fermentation results between cultures. It is well known that there is considerable individual animal variation on rumen pH and rumen fermentation pattern under the same feeding and management conditions. Therefore, it is often recommended to collect rumen inoculum from several animals and then combined to reduce the variation. Recently, we have conducted a batch culture to compare rumen inoculum of cattle with low- and high-feed digestion. It was observed the differences in gas production and dry matter digestibility of barley straw when the low- and high-feed-digesting rumen inocula were used. However, the use of such inoculum did not result overall in the differences in gas production kinetics. The effect of the inoculum sources on the in vitro effective dry matter digestibility agrees with previous reports that a difference in the activity of the inoculum exist among individual donor animals [16]. In another batch culture using always the same inoculum (low- versus high-feed-digesting cattle), we observed that the gas production and dry matter digestibility of barley straw werenot affected by inoculum source. The results suggest that inoculum from high-feed-digesting cattle did not necessary improve in vitro digestion of straw.

The rate and extent of feed degradability in the rumen vary with the type of feeds and feed processing. Therefore, diet is considered as a significant factor influencing the inoculum activity. Cone et al. [17] reported that the degradability of starch from different feed sources was greater for the donor cow fed a diet containing equal concentrate and hay compared with a hay-based diet. However, the composition of the concentrate mixture had only a minor effect on degradability values. It is clear that the ruminal microbial activity was different between cows fed hay versus hay and concentrate mixed diet. However, manipulating concentrate composition would not dramatically change ruminal microbial profiles. Mertens et al. [18] reported that the higher-fibre diets tended to produce more gas than the lower-fibre diets, which may explain by more acetate production with high-fibre diet, since fermentation of substrate fibre generates primarily acetate and gas is produced when substrate is fermented to generate acetate or butyrate rather than propionate. Huntington et al. [19] showed a similar response when dry cows were fed a diet of either straw or grass silage with rolled barley, and no differences in the gas production with a diet of a dried grass. Menke and Steingass [20] indicated that there was little difference in gas production of treated straw when hay in the diet of donor animals was replaced with treated straw. The inconsistent effect of donor animal diet suggests that it is more important to ensure the minimum microbial activity in the rumen fluid, rather than ensuring that donor animals are fed the substrate incubated.

Rumen microbial activity is increasing following feed ingestion, thus different sampling times have been applied to collect inoculum in literature either for obtaining high activity (i.e., 2 h after feeding) or for reducing variation (i.e., before feeding). Cone et al. [21] reported the increased rate of fermentation with rumen fluid that was collected after the morning although the total gas production was not affected. Menke and Steingass [20] stated that sampling rumen contents just before feeding reduced variation in activity of the inoculum. Although differences in microbial activity of inoculum occur at different sampling times, it appears that the most important factor is whether the sampling schedule will allow collection of inoculum with sufficient microbial activity. Payne et al. [22] observed less variation between replicates when rumen fluid was collected either 4 or 8 h after feeding, compared with before or 2 h after feeding.

The rumen fluid preparation procedure had relatively little effect on gas production [23]. However, Bueno et al. [24] reported an increase of in vitro organic matter digestibility by increasing the proportion of the solid phase relative to liquid phase in inoculum preparation and concluded that the contribution of microorganisms from the solid phase of rumen inoculum is important, especially in studies to evaluate high-fibre feeds. Recently, comparing the rumen inocula from low- and high-feed-digesting cattle, we did not find the differences in fibre digestibility of barley straw between the two inoculum sources, which may be explained by the method of inoculum preparation. Although whole ruminal contents were collected, rumen inoculum was obtained by squeezing manually, and it would represent primarily the bacteria associated with liquid or loosely associated with feed particles but not with bacteria tightly associated with particles. The proportion of bacteria associated with rumen feed particulate has been found to range from 50 to 70% and mainly characterized as fibrolytic bacteria [25].

3.1.2. Inoculum from different species

Rumen fluid from sheep is often used as inoculum on the batch culture because housing sheep is easier and less expensive than the cattle, whereas the results obtained with batch culture technique are mainly used to evaluate feeds for beef or dairy cattle. As a result, numbers of studies were conducted in comparison of rumen fluid between cattle and sheep on in vitro gas production and rumen fermentation [24, 26]. Cone et al. [26] compared rumen fluid from cows and sheep fed a similar diet, and they found that the gas production was lower with sheep rumen fluid, but there was a good relationship between volumes of gas produced by the two inocula. They concluded that sheep rumen fluid could replace cow rumen fluid for accurate determination of 24 and 48 h gas production and the gas production profile. However, rumen fluid of cows could not be replaced by that of sheep for the rate of gas production determination. Similarly, Bueno et al. [24] observed the similar gas production and degradability between sheep and cattle under the same feeding and management conditions. However, kinetics of gas production differed between species and so dynamic determinations, such as rate of gas production data, using sheep inoculum cannot be extrapolated to cattle. Bueno et al. [27] found the similar gas production and organic matter degradability of tropic forage between cow and sheep rumen fluid, whereas rumen fluid from sheep resulted in gas production with a longer lag time (6.1 h versus 4.2 h). Differences in microbial composition of rumen fluid from these sheep and cattle appeared to especially affect kinetics of fermentation, but not the end point measures. Few studies were compared between cattle and buffalo on the effects of rumen fluid on rumen fermentation. Calabrò et al. [28] found higher gas volume and earlier maximum rate of substrate degradation with cow than buffalo inoculum. All of these data indicate that species of donor animal will affect rumen fermentation.

3.1.3. Rumen versus faecal inocula

Use of faecal inoculum in batch culture has been paid great attention in scientific community during last two decades as it would overcome the need for surgically modified animals. The comparison between rumen fluid and faecal inoculum on in vitro gas production and extent of feed fermentation were well documented in several review articles [2, 15]. In general, the use of faecal inoculum give lower cumulative gas production and feed digestibility than use of rumen fluid although a good correlation is often determined. It suggests that the microbial activity in faecal inoculum is lower than in rumen inoculum. The difference between rumen and faecal inoculum may vary with feed degradability in the rumen. When the diet of the donor animal is highly fibrous, such that the microbial activity of the rumen is low, then differences between rumen fluid and faecal inoculum would be smaller, but when high-productive animals are used, the faecal inoculum are of limited value. Mauricio et al. [29] stated that the faecal inoculum could replace rumen fluid where incubations were over extended periods and cumulative gas volumes were examined since the gas release kinetics differed up to 48 h of incubation between the two inocula. Cone et al. [26] concluded that cow rumen fluid cannot be replaced by cow faeces for determination of 24 h gas production, but to be a good alternative for cow rumen fluid to accurately determine 48 h gas production. Mould et al. [15] suggested that faeces may replace rumen fluid as an inoculum for end-point measures (i.e., degradability or cumulative gas volume at the end of extended incubation periods); faecal material is likely an unsuitable inoculum for estimating rate of fermentation.

3.2. Manual versus automated methods

The gas generated from batch fermentation is generally measured either manually using the manual pressure transducer developed by Theodorou et al. [5] or automatically with the automated systems as described by Pell and Schofield [30], Cone et al. [21] and Davies et al. [31]. It has been reported that the headspace gas production associated with feed fermentation can be manually measured by inserting a needle attached to a pressure transducer into the vials at fixed time points [1, 8], or measured automatically using a transducer recording system [32]. Theoretically, the automated recording system, which vents gas at regular intervals may be more accurate than the manual system as where headspace gas can reach higher pressures. Accumulation of gas (i.e., the rise of gas pressure) may influence the release of gas from buffered ruminal fluid [3] and reduce the fermentation rate of substrate [5]. In closed systems, where gas is not released and accumulates, the rise of pressure in the headspace may cause a staircase effect in the recorded data. Especially with fast-fermenting substrates, some of the headspace gas may be forced into the liquid phase, and this dissolved gas may not be released instantly in the following reading, thus affecting successive measurements. Several studies were carried out to compare the gas produced using manual pressure transducers and automated pressure systems. The studies by Rymer et al. [2] and Gierus et al. [6] have observed greater gas production with the manual procedure than automated system. Similarly, we previously used two gas production systems, which were differed in gas pressure recording (automated versus manual), headspace and sample size of the bottle. Serum bottles (100 mL) sealed with a rubber stopper were used for manual gas pressure recording and a 500-mL Ankom gas production module (a computerized system with automated pressure transducers, Ankom Technology, Macedon, NY, USA) equipped with an Ankom pressure sensor module including a microchip and a radio transponder was used for automated gas pressure recording. The result also showed that the gas production was different when gas pressure was recorded using the two systems but it was interacted with the type of substrate incubated. The gas production was higher using manual system when the substrates had higher digestibility such as alfalfa hay and wheat distiller grains, whereas no difference in gas production was observed with the incubation of barley straw which had lower digestibility. The similar gas production of barley straw between the two systems may reflect the slower digestion rate of straw generating less gas. In addition, the gas production values from manual and automated recording systems in our study were calculated from different formulas, this may have biased gas production estimates. For the manual system, the gas volume was calculated using the equation described by Mauricio et al. [33]: gas volume, mL $= 0.18 + (3.697 \times \text{gas pressure}) + (0.0824 \times \text{gas pressure}^2)$, whereas for the automated system, the gas volume was estimated according to Avogadro's law (gas volume, mL = gas pressure × [V/RT] × 22.4 × 1000, where V is headspace volume in the bottle in litres, R is the gas constant 8.314472 L kPa/K/mol, and T is the temperature in Kelvin). Rymer et al. (2005) reported the stronger relationships between laboratories with manual system than with automated system and suggested that the increased complexity and cost of automated system may not be repaid by increased value of the results. However, the automated system produced good reproducibility among laboratories [21].

3.3. Effect of material delivery

The feed substrates can be incubated directly by dispersing in the medium or incubated in a filter bag. Incubating feeds in filter bags has been widely applied in batch culture [1, 8] because of its practical convenience. In comparison with dispersing the substrate into the medium, enclosing feed in bags has the advantage of being able to simultaneously determine in vitro digestibility of dry matter and fibre without the need to capture residues after incubation. However, incubating feeds in bags can have concerns on restricting microbial access to the substrates, particle loss from the bags during incubation, and the accumulation of the fermentation products which may inhibit microbial activity [34]. The lower in vitro dry matter and fibre digestibility was reported when feeds were incubated in filter bags as compared to when feeds were dispersed in the medium [35]. Krizsan et al. [36] suggested that this lower feed digestibility may arise from the inability of microbes to readily gain access to substrates within the bags, thus lowering the digestion. Additionally, the possible poor fluid exchange within the bags may result in an accumulation of the fermentation products which could further inhibit the fermentation. Ramin et al. [4] reported lower methane production for feed incubated in filter bags than dispersed in medium because of reduced feed digestion in vitro; however, the proportion of methane to total gas production was greater for feeds incubated in bags than for feeds dispersed in the medium. It was suggested an alteration of microbial population or fermentation pattern of the feeds incubated in bags versus feeds dispersed in the medium. There was also interaction between feed and method (i.e., bag versus dispersing) on ranking of methane output. It was concluded that the bag method should not be used when measuring methane emission during 48 h of incubation. In contrast, our recent study [9] showed that the incubation of feeds in filter bags consistently increased the digestibility of dry matter and fibre as compared to when the feeds were dispersed in the medium. The discrepancy with other studies may be resulted from the low densities of feed substrates being incubated. In this study, Barley straw, alfalfa hay and wheat distiller grain were ground through either 1- or 2-mm sieve and were incubated. It was observed that some feed particles were floating on the top of media and adhered to the sides of the bottles as a result of agitation during incubation. Obviously, this portion of the substrate would not come in direct contact with microbial populations and thus feed digestion would be compromised. Additionally, incubating substrate in bags to measure dry matter digestibility may have potentially resulted in overestimation of digestibility due to possible washout of feed particles from bags. The washout fraction varied with substrate and ranked as wheat distiller grain (18.8%) > alfalfa hay (12.1%) > barley straw (5.9%). However, because the washout fraction could primarily depend on the soluble fraction, and the soluble fraction is considered to be highly fermentable in the rumen, the impact of washout fraction would be minimal for the dry matter digestibility at longer incubation hour, for example, 24 h, whereas it would have a significant impact on the gas production kinetic measurement. He et al. [9] suggested that the method of substrate delivery could be a primary factor to be considered if the dry matter digestibility is a key variable measured. Therefore, lower microbial activity within the bags, altering microbial population or fermentation pattern of the samples incubated in bags compared with those directly dispersed in the medium show negative aspects of the bag method, but such disadvantages may have limited impact when using the batch culture for screening feed additives or ranking feedstuff. The practical convenience of using bags is highly attractive, thus commonly used currently.

3.4. Effect of substrate particle size

Feed digestion in the rumen requires that microorganisms colonize and produce enzymes that hydrolyse feed particles. Increasing the feed surface area increases the accessibility of microbes to substrate, thus potentially increasing feed digestibility. Anele et al. [37] reported that more processed barley grain (i.e., lower processing index at 0.75 which is calculated as ratio of density after rolling to the density before rolling) produced more cumulative gas volume than less processed barley samples with processing index of 0.85 due to less fermentable nutrients. Yang et al. [38] also reported higher in vitro gas production and dry matter digestibility of ground barley (1-mm sieve) as compared to dry-rolled barley (processing index at 0.80), suggesting that grinding increased the surface area available for microbial attachment. However, Rymer et al. [2] suggested that with highly soluble feeds such as some cereal grain, as long as the feed has undergone some abrasion, its particle size does not affect estimates of gas production rate. Lowman et al. [39] reported the similar gas production profiles of incubated naked oats that were cut at half, quarter, coarse and finely ground except the whole naked oats. Similarly, McAllister et al. [40] found the similar in situ dry matter digestibility of halved and quartered grains but significant lower dry matter digestibility of whole grain. Seed grains are generally protected by the pericarp and a processing by rolling or grinding is necessary to make the nutrient-rich endosperm available to the microbes, and to increase the rate and extent of digestion. However, there are many evidences that particle size of processed grain has significant impact on in vitro digestibility. Rymer et al. [2] indicated that maize grain that was steam-flaked, rolled or left intact had the same rate and extent of gas production when it was ground through a 1-mm screen but not when it had been ground through a 4-mm screen. With fibrous and more slowly degraded feeds, gas production rate increases as particle size decreases [39] and it appears that the increased gas production has resulted from an increased surface area as a result of grinding, thereby allowing better microbial access. It seems that there is interaction between feed particle size and type of feeds on in vitro gas production and dry matter digestibility. In the study of He et al. [9], a greater digestibility of dry matter and fibre of alfalfa hay ground 1 mm over 2 mm was observed; however, the digestibility of dry matter of barley straw did not respond to the particle size, and even less gas production and fibre digestibility with barley straw ground through a 1-mm screen as compared to a 2-mm screen was noticed. There was no clear explanation on this unexpected finding and authors speculated that finer straw adhered to the bottle more readily due to its greater buoyancy resulting in a lower digestibility.

The feeds that are incubated in vitro are often finely ground (e.g., ground through 1-mm sieve) since the particle size reduction during in vitro incubation is minimal due to absence of mastication and ruminal contractions. The use of finely ground sample also reduces the risk of sampling bias considering usually only less 1.0 g of sample is included in incubations. Yang et al. [38] concluded, based on a comparison of 60 barley samples either ground or dry rolled, that grinding is likely an appropriate processing method to evaluate digestion characteristics of barley using batch culture technique. In fact, the starch digestibility of ground barley after 24 h of incubation was similar to in vivo values observed in the rumen [41]. There was also less variability in digestibility and better correlation between chemical composition of barley and in vitro digestibility for the ground than the rolled barley. The advantage of using finely ground sample has no concern on processing quality associated with kernel uniformity. However, one can make the argument that barley that is tested in vitro should be processed in a manner similar to the form that it is fed to the animal. Although this approach does not consider the impact of mastication on digestion, it is equally clear that fine grinding also eliminates any sample-mediated differences in the particle sizes that may be generated after dry rolling of barley. Nocek [42] stated that reducing the variability of particle size by grinding through 1-mm sieve may not mimic the in vivo conditions ideally but it does tend to improve the precision of both in vitro and in situ measures. Yang et al. [38] reported the low correlation for digestibility of dry matter between ground and rolled barley ($R^2 = 0.12$), and suggested that the processing associated with kernel uniformity affected at least partly the digestibility of rolled barley. It can be concluded that the impact of particle size on in vitro feed digestibility can be significant but vary with the type of feed incubated. If the gas production technique is to be used as a means of feed evaluation, it may be necessary to require a standardized particle size and sample preparation procedure in order to reduce variation among experiments and laboratories. Adoption of a standardized approach to sample preparation may be possible to enable comparison between independently produced gas production and digestion data of different feeds. Additionally, as substrate particles are continually changing shape, size and composition in the gut, it seems unlikely that gas production or dry matter digestion data will represent kinetics of plant biomass as it is digested in the rumen.

3.5. Ratio of rumen inoculum to buffer

The ratio of rumen inoculum to buffer varies considerably in the various batch culture techniques from 1:9 to 1:4 Cabral Filho et al. [43]. Increasing the proportion of rumen inoculum in the incubation medium reduced lag time of gas production, but increased the volume or the rate of gas production [23, 30]. Navarro-Villa et al. [44] incubated with three different ratios of rumen fluid to buffer (i.e., 1:2, 1:4, and 1:6), and observed the increased gas production per unit of dry matter input, CH_4 to gas production and CH_4 to total volatile fatty acid ratio in all feeds incubated with increasing the proportion of rumen fluid in the mixture. The increase in CH_4 output due to change of rumen inoculum to buffer ratio can be resulted from different fermentation pattern, such as for barley grain appeared to be associated with higher acetate to propionate ratios and for barley straw was due to higher volatile fatty acid production. There was also a quadratic response of dry matter digestibility to increased ratios of rumen fluid to buffer with feed dependent, wherein decreasing the ratio resulted in a decline in digestibility with barley grain, an increase with grass silage and an increase (between 1:2 and 1:4) followed by a larger decrease (between 1:4 and 1:6) with barley straw. The decrease in ratio of rumen fluid to buffer would decrease microbial activity of the mixture media, thereby reduced feed digestibility. Pell and Schofield [30] included rumen fluid at the proportions of 5, 10, 20 and 40% in the total medium mixture, and observed the increase of alfalfa hay digestibility with increasing the proportion of ruminal fluid. It suggested that a 20% inoculum is sufficient to ensure the maximum rate of fibre digestion but lower percentages of inoculum are not sufficient. The increased lag time without altering maximum gas productions by lowering the ratio of rumen fluid to buffer appeared to reflect the time required for the microbial numbers to increase to levels comparable with those in the higher inocula. The microbial activity in rumen fluid can be determined by measuring absorbance of the inoculum following a 50-fold dilution at 600 nm and it is recommended a minimum microbial activity of 94 mg bacterial DM/ml [45].

3.6. Effect of concentrate addition on roughage fermentation

The inclusion of readily digestible carbohydrates in forage-based diets for ruminants can restrict microbial digestion of structural polysaccharides because rumen pH can be below the optimum [46]. The rumen pH below the optimum level is especially unfavourable for microbial fibrolysis. However, when poor quality of roughage such as straw is incubated in batch culture, there may be nutrient deficiency to support microbial growth or lack of fermentable carbohydrate to attract microbes to adhesion on the substrate, consequently reducing digestibility of substrate. Barrios-Urdaneta et al. [47] reported that the low available energy content of the straw cell wall that was incubated in vitro resulted in low fibre digestion even after long hours of incubation (i.e., 72 h). In addition, the low energy was also responsible for low numbers of bacteria associated with the substrate and a low level of polysaccharidase activity, both of which were corrected by the inclusion of energy supplements. Several studies indicated that the source of carbohydrate inclusion could also influence in vitro cell wall fermentation of crop straw. The higher in vitro straw cell wall digestion was observed with addition of pectin versus soluble sugars or starch [47] or when supplemented with sugar beet pulp, a source of highly digestible structural carbohydrates, compared with barley grain as a source of starch [46]. Barrios-Urdaneta et al. [47] suggested that the effect on increased cell wall digestion of straw was mainly attributed to higher bacterial adhesion to cell wall particles at early incubation time. We conducted a batch culture to incubate barley straw alone or barley straw plus a concentrate mix. For the treatment of straw + concentrate, 30% of barley straw was replaced by the equal amount of concentrate mix which consisted of 60% corn distillers grain, 22% canola meal, and 18% mineral and vitamin supplement in dry matter basis. The concentrate was incubated in a second bag within serum bottle. We observed greater rate of gas production and a shorter lag time with adding concentrate than the incubation of barley straw alone. An increased soluble fraction and dry matter degradability as well as increased fibre digestibility of straw by adding concentrate were noticed. The concentrate used in our study consisted of primarily corn distillers' grain which contained very low starch, but high protein and fibre. The fibre in corn distillers' grain has twice hemicellulose compared to original corn and it is highly fermentable in the rumen. Additionally, the protein from concentrate would favour microbial growth compared with straw alone by providing necessary nutrients. It is suggested that adding concentrate would increase microbial colonization on straw and consequently improved dry matter and fibre degradation of poor quality substrate in the rumen. In our study, although rate of gas production was higher, the volume of gas production was lower by adding concentrate, and along with higher digestibility of dry matter, it is suggested that the fermentation efficiency would be improved by adding the concentrate. Doane et al. [48] also noted that gas production of the in vitro fermentation was negatively related to fibre degradation. The lower fibre content of the substrate and the increased fibre degradation by adding concentrate may explain the lower volume of gas production in our study. The positive response of in vitro digestion of poor-quality feed substrates to high fermentable carbohydrate addition suggests necessary consideration when needing to determine the potential digestibility of poor-quality roughage.

4. CONCISE SUMMARY

Various aspects such as the source of inoculum, the gas venting system, the size and delivery of substrate particles, the ratios of inoculum to buffer, and the addition of concentrate to the media might affect the results of fermentation in batch culture. The rumen inoculum has a significant impact on the fermentation process in batch culture. The objective of the inoculum is to introduce a specific group of microorganisms that can break down a feed gradually. This breakdown process is then used to determine the rate or extent of feed digestion. The microbial activity of the inoculum can vary significantly depending on the animal species (such as cattle versus sheep), diets, and the timing of sampling after feeding. However, the most important factor to consider is to ensure adequate microbial activity in the inoculum and minimize the variation in microbial activity among different inocula. It is likely necessary to reduce the variation by increasing the number of donor animals and standardizing the inoculum collection time. Several studies have been carried out to examine the characteristics of rumen fluid and feces, with the goal of finding a substitute for rumen fluid. The primary benefit of employing faecal inoculum is to decrease the need for animals with rumen cannulas. Nevertheless, it is important to acknowledge that faecal and rumen inocula exhibit modest variations. Based on long-term in vitro end-point measurements, it seems that faeces could be a suitable substitute for rumen fluid. However, if short-term or kinetic data is required, rumen fluid should be used instead. The assessment of gas production in batch culture is currently a highly versatile and effective research tool in ruminant nutrition research. The debate surrounding various venting systems and substrate supply methods remains unresolved. It implies that these systems could be influenced by various elements such as bottle size, headspace, and type of feeds incubated. The particle size of the substrate being incubated has a consistent impact on both the rate and amount of feed digestion. The determination of the optimal particle size for feed may not be readily available and can vary depending on the type of feed (such as concentrate or roughage) and the specific goals of the study. Manipulating the proportions of rumen fluid to buffer volume might modify microbial activity in fermentation media, hence potentially affecting the rate of fermentation and the duration of the lag phase. The suggestion is to maintain an adequate level of microbial activity in the fermentation media combination, while avoiding excessive amounts of rumen fluid that could lead to an increased proportion of gas from the inoculum compared to the substrate. Ultimately, including easily fermentable carbohydrates is beneficial in order to optimize the fermentation of low-quality meals.

REFERENCES

1. Eun JS, Beauchemin KA, Schulze H. Use of exogenous fibrolytic enzymes to enhance in vitro fermentation of alfalfa hay and corn silage. J. Dairy Sci. 2007;90:1440–1451.
2. Rymer C, Williams BA, Brooks AE, Davies DR, Givens DI. Inter-laboratory variation of in vitro cumulative gas production profiles of feeds using manual and automated methods. Anim. Feed Sci. Technol. 2005;123–124:225–241.
3. Tagliapietra F, Cattani M, Bailoni L, Schiavon S. In vitro rumen fermentation: effect of headspace pressure on the gas production kinetics of corn meal and meadow hay. Anim. Feed Sci. Technol. 2010;158:197–201.
4. Ramin M, Krizsan SJ, Jancik F, Huhtanen P. Short communication: measurements of methane emissions from feed samples in filter bags or dispersed in the medium in an in vitro gas production system. J. Dairy Sci. 2013;96:4643–4646.
5. Theodorou MK, Williams BA, Dhanoa MS, McAllan AB, France J. A simple gas production method using a pressure transducer to determine the fermentation kinetics of ruminant feeds. Anim. Feed Sci. Technol. 1994;48:185–197.
6. Gierus M, Schiborra A, Südekum KH, Rave G, Taube F. Comparison of gas accumulation profiles of several feeds using manual or automated gas production methods. Anim. Feed Sci. Technol. 2008;147:310–325.
7. He ZX, Ding S, Xu L, Beauchemin KA, Yang WZ. Using exogenous enzymes to increase the rumen degradability of wheat dried distillers grains with solubles. Arch. Anim. Nutr. 2013;67:381–392.
8. Elghandour MMY, Salem AZM, Gonzalez-Ronquillo M, Bórquez JL, Gado HM, Odongo NE, Peñuelas CG. Effects of exogenous enzymes on in vitro gas production kinetics and ruminal fermentation of four fibrous feeds. Anim. Feed Sci. Technol. 2013;179:46–53.
9. He ZX, Zhao YL, McAllister TA, Yang WZ. Short communication, effect of in vitro techniques and exogenous feed enzymes on feed digestion. Anim. Feed Sci. Technol. 2016;213:148–152.
10. Mendoza GD, Loera-Corral O, Plata-Pérez FX, Hernández-García PA, Ramírez-Mella M. Considerations on the use of exogenous fibrolytic enzymes to improve forage utilization. Sci. World J. 2014;247:437.

11. Tilley JMA, Terry RA. A two-stage technique for the digestion of forage crops. J. Br. Grassl. Soc. 1963;18:104–111.

12. Goering HK, Van Soest PJ. Forage fiber analysis (apparatus, reagents, procedures and some applications). 1970; Agric. Handbook No. 379. Washington, DC, ARS-USDA.

13. Getachew G, Blümmel M, Makkar HPS, Becker K. In vitro gas measuring techniques for assessment of nutritional quality of feeds: a review. Anim. Feed Sci. Technol. 1998;72:261–281.

14. Blümmel M, Ørskov ER. Comparison of in vitro gas production and nylon bag degradability of roughages in predicting feed intake in cattle. Anim. Feed Sci. Technol. 1993;40:109–119.

15. Mould FL, Morgan R, Kliem KE, Krystallidou E. A review and simplification of the in vitro incubation medium. Anim. Feed Sci. Technol. 2005;123–124:155–172.

16. Soder KJ. Technical note: influence of rumen inoculum source on in vitro dry matter digestibility of pasture. Prof. Anim. Sci. 2005;21:45–49.

17. Cone W J, Cliné-Theil W, Malestein A, Vant Klooster AT. Degradation of starch by inoculum with rumen fluid. A comparison of different starch sources. J. Sci. Agric. 1989;49:173–183.

18. Mertens DR, Weimer PJ, Waghorn GM. Inocula differences affect in vitro gas production kinetics. In: Deaville, E.R., Owen, E., Adesogen, A.T., Rymer, C., Huntington, J.A., Lawrence, T.L.J. (Eds.), In Vitro Techniques for measuring nutrient supply to ruminants. BSAS, Edinburgh, UK, 1998;95–98, BSAS Occ. Publ. No. 22.

19. Huntington JA, Rymer C, Givens DI. The effect of host diet on the gas production profile of hay and high-temperature dried grass. Anim. Sci. 1998;67:59–64.

20. Menke KH, Steingass H. Estimation of the energetic feed value obtained from chemical analysis and in vitro gas production using rumen fluid. Anim. Res. Dev. 1988;28:7–55.

21. Cone JW, van Gelder AH, Visscher GJW, Oudshoorn L. Influence of rumen fluid and substrate concentration on fermentation kinetics measured with a fully automated time related gas production apparatus. Anim. Feed Sci. Technol. 1996;61:113–128.

22. Payne JS, Hamersley AR, Milligan JC, Huntington JA. The affect of rumen fluid collection time on its fermentative capacity and the stability of rumen conditions in sheep fed a constant diet. Proc. Br. Soc. Anim. Sci. 2002;165.

23. Rymer C, Huntington JA, Givens DI. Effects of inoculum preparation method and concentration, method of inoculation and pre-soaking the substrate on

the gas production profile of high temperature dried grass. Anim. Feed Sci. Technol. 1999;78:199–213.

24. Bueno Ives CS, Sergio LS, Cabral Filho SP, Gobbo, Helder Louvandini, Dorinha MSS Vitti, Adibe L Abdall. Influence of inoculum source in a gas production method. Anim. Feed Sci. Technol. 2005;123–124:95–105.
25. Cheng K-J, Akin DE, Costerton JW. Rumen bacteria: interaction with particulate dietary components and response to dietary variations. Fed. Proc. 1977;36:193–197.
26. Cone JW, van Gelder AH, Bachmann H. Influence of inoculum source on gas production profiles. Anim. Feed Sci. Technol. 2002;99:221–231.
27. Bueno ICS, Abdalla AL, Cabral Filho SLS, Vitta DMSS, Owen E, Mauricio RM, Givens I, Sutton JD, Mould FL. Comparison of inocula from sheep and cattle for the in vitro gas production technique under tropical conditions. Proc. Br. Soc. Anim. Sci. 1999;151.
28. Calabrò S, Williams BA, PiccoloV, Infascelli F, Tamminga S. A comparison between buffalo (Bubalus bubalis) and cow (Bos taurus) rumen fluids in terms of the in vitro fermentation characteristics of three fibrous feedstuffs. J. Sci. Food Agric. 2004;84:645–652.
29. Mauricio RM, Owen E, Dhanoa MS, Theodorou MK. Comparison of rumen liquor and faeces from cows as source of microorganisms for the in vitro gas production technique. In: In vitro techniques for measuring nutrient supply to ruminants, Proceedings of Occasional Meeting of the British Society of Animal Science, 8–10 July 1997, University of Reading, UK.
30. Pell AN, Schofield P. Computerised monitoring of gas production to measure forage digestion in vitro. J. Dairy Sci. 1993;76:1063–1073.
31. Davies ZS, Mason D, Brooks AE, Griffith GW, Merry RJ, Theodorou MK. An automated system for measuring gas production from forages inoculated with rumen fluid and its use in determining the effect of enzymes on grass silage. Anim. Feed Sci. Technol. 2000;83:205–221.
32. Cornou C, Storm IMLD, Hindrichsen IK, Worgan H, Bakewell E, Ruiz DRY, Abecia L, Tagliapietra F, Cattani M, Ritz C, Hansen HH. A ring test of a wireless in vitro gas production system. Anim. Prod. Sci. 2013;53:585–592.
33. Mauricio RM, Mould FL, Dhanoa MS, Owen E, Channa KS, Theodorou MK. A semi-automated in vitro gas production technique for ruminant feedstuff evaluation. Anim. Feed Sci. Technol. 1999;79:321–330.
34. Nozière P, Michalet-Doreau B. 2000 In sacco methods. In: D'Mello, J.P.F. (Ed.), Farm animal metabolism and nutrition. CABI Publishing, Oxon, UK, 2000;233–254.

35. Tagliapietra F, Cattani M, Hindrichsen IK, Hansen HH, Colombini S, Bailoni L, Schiavon S. True dry matter digestibility of feeds evaluated in situ with different bags in vitro using rumen fluid collected from intact donor cows. Anim. Prod. Sci. 2012;52:338–346.

36. Krizsan SJ, Jancik F, Ramin M, Huhtanen P. Comparison of some aspects of the in situ and in vitro methods in evaluation of neutral detergent fiber digestion. J. Anim. Sci. 2013;91:838–847.

37. Anele UY, Refat B, Swift ML, He ZX, Zhao YL, McAllister TA, Yang WZ. Effects of bulk density, precision processing and processing index on in vitro ruminal fermentation of dry-rolled barley grain. Anim. Feed Sci. Technol. 2014;195:28–37.

38. Yang WZ, Oba M, Swift ML, McAllister TA. Variation in response to processing, in vitro gas production and fermentation of western Canadian feed barley. Can J. Anim. Sci. 2014;94:725–729.

39. Lowman RS, Theodorou MK, Cuddeford D. The effect of sample processing on gas production profiles obtained using the pressure transducer technique. Anim. Feed Sci. Technol. 2002;97:221–237.

40. McAllister TA, Rode LM, Major DJ, Cheng KJ, Buchanan-Smith JG. Effect of ruminal microbial colonization on cereal grain digestion. Can. J. Anim. Sci. 1990;70:571–579.

41. Yang WZ, Beauchemin KA, Rode LM. Effects of barley grain processing on extent of digestion and milk production of lactating cows. J. Dairy Sci. 2000;83:554–568.

42. Nocek JE. In situ and other methods to estimate ruminal protein and energy digestibility: a review. J. Dairy Sci. 1988;71:2051–2069.

43. Cabral Filho SLS, Abdalla AL, Bueno ICS, Nozella EF, Rodrigues JAS. Ruminal fermentation and degradability of sorghum cultivar whole crop, and grains, using an in vitro gas production technique. Anim. Feed Sci. Technol. 2005;123–124:329–339.

44. Navarro-Villa A, O'Brien M, López S, Boland TM, O'Kiely P. Modifications of a gas production technique for assessing in vitro rumen methane production from feedstuffs. Anim. Feed Sci. Technol. 2011;166–167:163–174.

45. Nagadi S, Herrero M, Jessop NS. The influence of diet of the donor animal on the initial bacterial concentration of ruminal fluid and in vitro gas production degradability parameters. Anim. Feed Sci. Technol. 2000;87:231–239.

46. Fondevila M, Castrillo C, Guada JA, Balcells J. Effect of ammonia treatment and carbohydrate supplementation of barley straw on rumen liquid

characteristics and substrate degradation by sheep. Anim. Feed Sci. Technol. 1994;50:137–155.

47. Barrios-Urdaneta A, Fondevila M, Balcells J, Dapoza C, Castrillo C. In vitro microbial digestion of straw cell wall polysaccharides in response to supplementation with different sources of carbohydrates. Aust. J. Agric. Res. 2000;51:393–399.

48. Doane PH, Schofield P, Pell AN. Neutral detergent fiber disappearance and gas and volatile fatty acid production during the in vitro fermentation of six forages. J. Anim. Sci. 1997;75:3342–3352.

06 MODELING THE KINETICS OF ETHANOL FERMENTATIONS UNDER 1-G CONDITIONS

1. INTRODUCTION: AN OVERVIEW

The motivation for creating industrial bioethanol mostly stems from economic and environmental concerns. Bioethanol can be synthesized by batch, fed-batch, and continuous processes, and in certain instances, by employing flocculating yeasts [1-6]. Accurate dynamic models are necessary for the creation of efficient control techniques for the primary operating variables in ethanol fermentations, including pH, temperature, residual sugars concentration, agitation speed, foam level, and others. Furthermore, mathematical models play a crucial role in the development, enhancement, and regulation of bioreactors. Bioreactor models aim to accurately depict the overall functioning of the bioreactor. These models are composed of two submodels: a balance/transport submodel, which explains the transfer of mass and heat within and between different phases of the bioreactor, and a kinetic submodel, which explains how the growth rate of microorganisms, consumption of substrate, and formation of products are influenced by important local environmental factors [7].

In ethanol fermentation, the main bioreactions can be summarized by the reductive pathway $S \rightarrow X + P + CO_2$. According to this reaction, substrates S (glucose and fructose, which result from hydrolysis of sucrose as the limiting substrate), in anaerobic conditions, are metabolized to produce a yeast population X, ethanol P (mainly produced by yeast through the Embden-Meyerhof-Parnas metabolic pathway), and carbon dioxide (CO_2). The hydrolysis of sucrose promoted by the invertase present in the yeast is not the limiting step of ethanol production in industrial processes. The stoichiometry of ethanol-formation reaction from glucose is given by the classical Gay-Lussac equation: $C_6H_{12}O_6 \rightarrow 2CH_3CH_2OH + 2CO_2$

According to Doran [8], both *Saccharomyces cerevisiae* yeast and *Zymomonas mobilis* bacteria produce ethanol from glucose under anaerobic conditions without external electron acceptors. The biomass yield from glucose is 0.11 g/g for yeast and 0.05 g/g for *Z. mobilis*. In both cases, the nitrogen source is NH_3, and the cell compositions are represented by the formula $C_{1.8}O_{0.5}N_{0.2}$. From these data, Doran [9] proposed the following stoichiometric equation for ethanol fermentation (the values of stoichiometric coefficients *a*, *b*, *c*, *d*, *e*, and *f* are presented in Table 1):

$$aC_6H_{12}O_6 + bNH_3 \rightarrow cCH_{1.8}O_{0.5}N_{0.2} + dCO_2 + eH_2O + fC_2H_6O(\text{molar basis})$$

Table 1. Stoichiometric coefficients for ethanol fermentation.

Microorganism	Stoichiometric coefficients					
	a	*b*	*c*	*d*	*e*	*f*
Yeast	1	0.16	0.81	1.75	0.35	1.72
Bacteria	1	0.074	0.37	1.89	0.17	1.87

Kinetic modeling of growth, ethanol production, and substrate consumption by yeasts has been traditionally conducted using an unsegregated and unstructured approach for the biomass. This approach ignores the presence of individual cells and structural, functional, and compositional aspects of the cell, describing the complex processes of growth, ethanol production, and substrate consumption through simple kinetic equations [10–15]. Figure 1 shows a simplified scheme of this approach for the ethanol fermentation process by yeasts and bacteria.

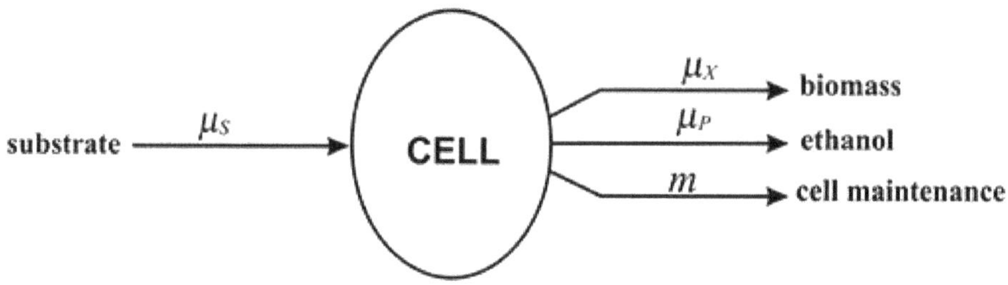

Figure 1. Kinetic modeling of ethanol fermentation based on an unsegregated and unstructured approach for cells (yeasts or bacteria).

2. KINETICS OF CELL GROWTH AND ETHANOL FORMATION

In ethanol fermentation, the kinetics of growth and ethanol production are generally the following:

$$\mu_X = f_1(S)g_1(P) \tag{1}$$

$$\mu_p = f_2(S)g_2(P) \tag{2}$$

where μ_X and μ_p, are, respectively, the specific rate of yeast growth and ethanol production, whereas *S* and *P* represent the limiting substrate and ethanol concentrations.

2.1. Effect of substrate concentration

The functions $f_1(S)$ *and* $f_2(S)$ are generally of the Monod type [11], except when an inhibition caused by high concentrations of substrate or diffusional limitations occurs due to high cell concentrations.

$$f(S) = \frac{\mu_{max} S}{K_S + S} \text{(Monod equation)} \quad (3)$$

The inhibition caused by the excess of substrate has generally been modeled by applying the Andrews equation [16–19], though there are other types of equations that are less commonly used [20].

$$f(S) = \frac{\mu_{max} S}{K_S + S + S^2/K_I} \text{(Andrews equation)} \quad (4)$$

In the case of continuous processes operated near to the steady state, the inhibition concentrations of the substrate are rarely identified. However, inhibitory concentrations can occur during the start-up of these processes or in situations resulting from changes in the substrate feed load.

Atala et al. [21], modeling the effect of temperature upon the kinetics of ethanol fermentation with a high concentration of biomass in a continuous system with total cell retention, used an inhibitory factor (IF) of the exponential type to describe the inhibitory effect of the substrate upon the kinetics of cell growth, which was inserted in the expression of $f(S)$, being $f(S)$, in this case, given by the Monod equation.

$$IF = (e^{-K_i S}) \quad (5)$$

Tsuji et al. [22] evaluating the performance of different ethanol fermentation systems (conventional chemostat, multiple bioreactors, cell recycle bioreactor, extractive bioreactor, and immobilized cell bioreactor) expressed the specific growth rate by an equation analogous to Eq. (1):

$$\mu_X = \mu_1(S)\mu_2(P) \quad (6)$$

One of the analyzed cases considered growth inhibition by substrate, represented by a hyperbolic equation:

$$\mu_1(S) = \frac{\mu_{max} S}{K_S + S}\left(\frac{K_i}{K_i + S}\right) \quad (7)$$

Sousa and Teixeira [23] reported that one of the main disadvantages of the systems that use flocculating cells (bacteria or yeast) is the reduced reaction rates caused by diffusional limitations of the substrate within the flocs and that, in most cases, the diffusion rate is lower than the reaction rate, which means that the process is controlled by diffusion. Sousa and Teixeira [23] reported that it is generally accepted that yeast flocs are formed by a mediator cation (usually $Ca^{2}+$) from the interaction between protein and mannans on adjacent cell walls. According to Sousa and Teixeira [23], one means through which to avoid diffusional limitations within the flocs is by using polymeric additives, which act by widening the bridges formed between adjacent cells.

Fontana et al. [24] reported that when the yeast flocs are suspended in a sucrose solution, various phenomena occur simultaneously: the sugar penetrates by diffusion in the aggregates and is hydrolyzed into glucose and fructose by an invertase that is primarily located on the yeast's cell wall. These two sugars diffuse inside and outside of the particle and are fermented in ethanol and CO_2, which in turn diffuse back in the liquid medium.

Fontana et al. [24] assumed that the Fick's law was valid for the aggregate and that the temporal variation of the concentration of each component involved in the transformation is represented by the following equation:

$$\frac{\partial C_i}{\partial t} = D_{ef,i} \frac{\partial^2 C_i}{\partial x^2} + \sum r_i \quad (8)$$

where C_i is the concentration of the component i in the aggregate in time t and distance x as of the floc surface; $D_{ef,i}$ is the effective diffusion coefficient, while Σr_i is the sum of the consumption or production rates of component i, which are given by Michaelis-Menten-type equations, such as the following:

$$\sum r_i = -r_{S_{max}} \frac{S}{K_S + S} \text{ (Sucrose)} \quad (9)$$

$$\sum r_i = +Y_{G/S} r_{S_{max}} \frac{S}{K_S + S} - r_{G_{max}} \frac{G}{K_G + G} \text{(Glucose)} \quad (10)$$

where S and G represent, respectively, the concentration of sucrose and glucose, while $Y_{G/S}$ is the conversion factor in glucose based on the hydrolyzed sucrose ($Y_{G/S}$ = 0.505g-glucose/g-sucrose). One relation identical to Eq. (10) can be obtained for fructose.

However, the theoretical descriptions of the diffusional resistances in systems that make use of flocculating microorganisms are generally conducted by introducing a factor into the Monod equation that takes into account the reduction in growth rate due to mass transfer limitations. One equation of this type is that proposed by Contois [25].

$$f(S) = \frac{\mu_{max} S}{K_S X + S} \text{(Contois equation)} \quad (11)$$

According to Menezes et al. [26], the Monod model is appropriate at low cell concentrations, while the Contois model is more appropriate at high concentrations, given that the variable saturation term, $K_S X$, can describe the diffusional limitations present in high cell concentrations. Oliveira et al. [27], modeling a continuous process of ethanol fermentation in a tower bioreactor with recycling of flocculating yeasts, obtained a high value for K_S, which was attributed to the diffusional limitations caused by the high cell concentrations reached in the bioreactor.

2.2. Effect of ethanol concentration

The dependence of μ_X and μ_P on the ethanol concentration is due to the fact that this product has been reported in the literature to act as a noncompetitive inhibitor both for growth and its own production [10, 28–32].

Noncompetitive inhibition is characterized by the fact that in the graph of $1/\mu_X$ or $1/\mu_p$ versus $1/S$ (Figure 2), for each ethanol concentration (P), straight lines with different slopes $\left(\frac{K_{S,i}}{\mu_{max,i}g_i(P)}, i = 1,2\right)$ and different intercepts $\left(\frac{1}{\mu_{max,i}g_i(P)}, i = 1,2\right)$ are obtained, but the same intersections with the abscissa are maintained ($-1/K_{S,i}$; $i = 1, 2$).

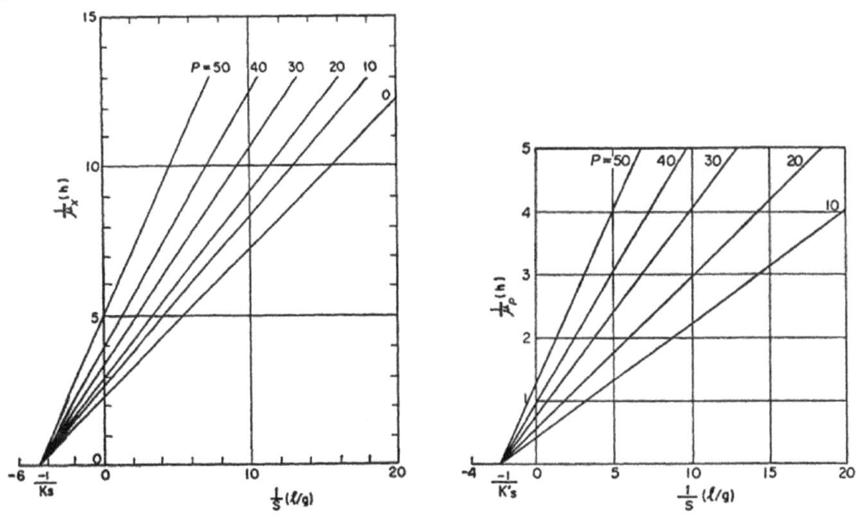

Figure 2. Lineweaver-Burk graph for the specific rates of cell growth (μ_X) and ethanol production (μ_P) (adapted from Aiba et al. [31]).

The molecular base of the mechanism through which the ethanol exerts an inhibitory effect upon fermentation is complex so long as this component, which acts as a denaturing agent, not only acts directly upon the proteins and causes an inactivation or inhibition of the enzymes from the glycolytic pathway but can also act upon the integrity of the lipid membranes, affecting the essential factors, including membrane components, such as transport proteins and the enzymes linked to it [33].

Table 2 presents the main equations proposed for $g_1(P)$ and $g_2(P)$, which are first approximations of much more complicated effects [10, 28–32, 34–42].

Table 2. Types of commonly proposed equations to describe the inhibitory effect of ethanol upon μ_X and μ_P.

Linear (L)	$g(P) = \left(1-\frac{P}{P_m}\right)$ (12)
Generalized nonlinear (GN)	$g(P) = \left(1-\frac{P}{P_m}\right)^n$ (13)
Hyperbolic (H)	$g(P) = \left(\frac{K_P}{K_P+P}\right)$ (14)
Parabolic (P)	$g(P) = \left(1-\frac{P}{P_m}\right)^{0.5}$ (15)
Exponential (E)	$g(P) = \left(e^{-K_P P}\right)$ (16)

The type of inhibition that affects cell growth is not mandatorily the same as that which affects ethanol production, as it is necessary to determine separately each effect, as proposed by Oliveira et al. [27]. According to Bonomi et al. [17], one of the major difficulties during the development of a mathematical model that fits experimental data of ethanol fermentation is the definition of the type of product inhibition exhibited by the yeast's metabolism. Bonomi et al. [17] reported that this inhibition is characterized by the behavior of the specific growth and production rates with an increase in ethanol concentration, while holding constant the substrate concentration. When developing a mathematical model for a batch system of ethanol production, Bonomi et al. [17] set three values for substrate concentration and determined the corresponding pairs of values (μ_X, P) and (μ_P, P) in each fermentation test. These points—(μ_X, P) and (μ_P, P)—were then plotted to define the types of existing relations between the specific rates and ethanol concentration which, in this case, were both exponential. The values of the specific growth and ethanol production rates were calculated based on experimental data, using the geometric approach proposed by Le Duy and Zajic [43].

The conceptual limitation of the hyperbolic and exponential inhibition is that they predict cell growth and production for all of the ethanol concentrations, even though many experimental tests have shown that cell growth and production cease upon reaching a given high concentration of ethanol [44]. The models of linear, generalized nonlinear, and parabolic inhibition consider that there is a determined concentration of ethanol above which growth and production do not occur. In these models, the P_m parameters represent the ethanol concentrations for which the growth and production processes are completely interrupted.

In the linear, generalized nonlinear, and parabolic models, the exponents of the term $(1-P/P_m)$ are called by Levenspiel [40] as "toxic powers." The values of toxic powers are indicative of how the term of inhibition $(1-P/P_m)$ strongly affects the specific growth and ethanol production rates. With the rise in toxic power, the intensity of inhibition increases for a determined ethanol concentration.

In the hyperbolic and exponential inhibition models, the K_p parameters do not admit a physical meaning and can be considered simple empirical constants that apparently depend on the cultivation mode: batch or continuous [31, 32, 37]. Aiba and Shoda [32] argued that the fact that the hyperbolic inhibition constant of the specific growth rate (K_p) has been lower in a batch culture ($K_p = 16.0$ g/L) than in a continuous culture ($K_p = 5$ 5.0 g/L) suggests the possibility that a chemical affinity of ethanol for a key participating enzyme in cell growth appeared in batch experiments. By contrast, the fact that the hyperbolic inhibition constant of the specific ethanol production rate (K'_p) has been lower in continuous cultures ($K'_p = 12.5$ g/L) than in batch cultures ($K'_p = 71.5$ g/L) suggests that the ethanol inhibition upon another key enzyme responsible for the fermentation activity was more expressive in continuous experiments.

Another inhibition model commonly used in the literature is that proposed by Luong [37]:

$$g(P) = 1 - \left(\frac{P}{P_m}\right)^{\beta} \tag{17}$$

where P_m continues to be the ethanol concentration above which no growth or production can occur, and β is an empirical constant.

One different proposal to describe the inhibitory effects of ethanol upon μ_X and μ_P was presented by Wang and Sheu [45] when they applied multiobjective optimization methods to estimate the parameters of kinetic models of batch and fed-batch processes for ethanol production, using one yeast that is highly tolerant to ethanol (*Saccharomyces diastaticus*). In their study, the kinetic models for the specific rate of cell growth and product formation were represented as follows:

$$\mu_X = \left(\frac{\mu_m S}{K_S + S + S^2/K_{IS}}\right)\left(\frac{K_P}{K_P + P + P^2/K_{IP}}\right) \tag{18}$$

$$\mu_P = \left(\frac{v_m S}{K'_S + S + S^2/K'_{IS}}\right)\left(\frac{K'_P}{K'_P + P + P^2/K'_{IP}}\right) \tag{19}$$

Olaoye and Kolawole [46], modeling the kinetics of ethanol fermentation in batch culture of *Kluyveromyces marxianus*, used a semiempirical approach to describe the fermentation process. To model the temporal profile of the biomass concentration, the authors inserted Eq. (20) in the cell mass balance ($dX/dt = \mu_X X$) and analytically integrated the resulting equation to obtain the so-called logistic growth curve (Eq. 21). The ethanol concentration was described, directly applying the modified Gompertz equation (Eq. 22), which represented the empirical part of the proposed mathematical model. The authors did not report the values of the model's parameters.

$$\mu_X = \mu_m\left(1-\frac{X}{X_m}\right) \tag{20}$$

$$X = \frac{X_m}{1+\left(\frac{X_m-X_0}{X_0}\right)e^{-\mu_m t}};\, X_0 = X(0) \tag{21}$$

$$P = P_m \exp\left\{-\exp\left[\frac{\Pr_m \exp(1)}{P_m}(\lambda - t) + 1\right]\right\} \tag{22}$$

where P_m and Pr_m are, respectively, the maximum concentration and maximum productivity of ethanol; λ is the time of duration of the lag phase, anterior to the exponential phase of ethanol production.

The logistic equation has been used to model fermentation kinetics due to its mathematical simplicity. According to Mitchell [7], the logistic equation can, many times in a single equation, offer an adequate approximation of the entire growth curve, including the lag phase and the cessation of growth in the latter stages of fermentation.

2.3. Effect of cell concentration

The inhibitory models that have been developed so far have effectively described a significant number of fermentations. Nevertheless, in continuous processes involving cell recycling, the fermenter achieves large cell densities. To accurately describe the kinetic behavior of the bioprocess, it is important to take into account additional aspects, such as the inhibition induced by an excessive amount of biomass. The suppression of cellular proliferation by cellular densities has been mathematically represented by the follows comprehensive equation [44]:

$$h(X) = \left(1-\frac{X}{X_{max}}\right)^{\delta} \tag{23}$$

where X_{max} is the maximum cell concentration that would be reached if ideal conditions for growth were observed, that is, an adequate supply of nutrients and the absence of inhibitory effects [44]. Analogous to the term of inhibition caused by the product, δ indicates the intensity of the inhibition due to the high cell concentrations.

Jarzebski et al. [47], modeling a continuous ethanol fermentation process with high yeast concentrations in a membrane filtration module system, used the following expressions for μ_X and μ_P, in which other formats can be observed for the terms that describe the inhibitory effects of the biomass itself:

$$\mu_X = \left(\frac{\mu_0 S}{K_S + S}\right)\left[1-\left(\frac{P}{P_m}\right)^{A_1}\right]\left[1-\left(\frac{X}{X_m}\right)^{A_2}\right] \tag{24}$$

$$\mu_P = a\exp(-bX) \tag{25}$$

In cases occurring inhibition of cell growth and product formation by the biomass itself, the expressions of the specific growth and ethanol production rates must be augmented to incorporate these inhibitory effects, that is,

$$\mu_X = f_1(S)g_1(P)h_1(X) \tag{26}$$

$$\mu_P = f_2(S)g_2(P)h_2(X) \tag{27}$$

As regards the procedures of many authors using a kinetic expression for μ_P detached from μ_X, Bu'Lock et al. [10] reported that this does not mean that there is no association between these rates, so long as the ethanol production has been commonly reported in the literature as a process associated with growth. Bu'Lock et al. [10] justified the adoption of such a procedure due to the simplicity and to the better adaptation of such equations to experimental data.

In relating the kinetics of ethanol production to the kinetics of cell growth, the procedure has been to apply the Luedeking-Piret model:

$$\mu_P = \alpha\mu_X + \beta \tag{28}$$

The Luedeking-Piret model is based on the classification of products of the fermentation process as associated ($\alpha > 0$, $\beta = 0$), nonassociated ($\alpha = 0$, $\beta > 0$), and partially associated with cell growth ($\alpha > 0$, $\beta > 0$) [12]. Since ethanol is a product of the primary metabolism of the yeasts, the majority of described cases assumes $\alpha > 0$ and $\beta = 0$, as reported by Oliveira et al. [19] when they modeled the batch ethanol production, using the expression $\mu_P = \alpha\mu_X$, with $\alpha = 4.17$ g/g. However, it is possible to find descriptions in the literature of ethanol production, parts associated and not associated with cell growth, that is, $\alpha > 0$ and $\beta > 0$, as is the case with that reported by Guidini et al. [18] that used the following equation to describe μ_P in a fed-batch ethanol fermentation process with flocculating yeasts (*S. cerevisiae*):

$$\mu_P = \underbrace{\left(\frac{Y_{P/S}}{Y_{X/S}}\right)}_{\alpha} \mu_X + b \tag{29}$$

Rivera et al. [48], modeling a fed-batch ethanol fermentation process with a strain of industrial yeast (*S. cerevisiae*), used a modified version of the Luedeking and Piret model in which the β coefficient is given as a function of the substrate concentration (S):

$$\mu_P = Y_{P/X}\mu_X + \underbrace{\left(\frac{\beta_m S}{K_{\beta S} + S}\right)}_{\beta} \tag{30}$$

Ghosh and Ramachandran [49], analyzing the effect of *in situ* product removal on the stability and performance of a continuous bioreactor with cell separator for ethanol production, emphasized the use of the Luedeking-Piret model to represent the kinetics of product formation.

In Table 3, typical kinetic parameter values for ethanol fermentation are presented [10, 19, 28–32, 34–41].

Table 3. Typical values of kinetics parameters in ethanol fermentation.

	Parameters of $f_1(S)$, $f_2(S)$ $h(X)$	Parameters of $g_1(P)$ $g_2(P)$				
		L	GN	P	H	E
$\mu_{max,1}$ (/h)	0.11–0.56					
$\mu_{max,2}$ (g/g/h)	0.21–1.90					
$K_{S,1}$ (g/L)	0.07–0.57					
$K_{S,2}$ (g/L)	0.33–60.0					
$P_{m,1}$ (g/L)		87.0–95.0	73.0–87.5	93.6		
$P_{m,2}$ (g/L)		114.0–135.0	87.5	99.0		
n_1 (-)			0.41–4.0			
n_2 (-)			0.41			
$K_{p,1}$ (L/g or g/L)					16.0–105.2	0.016–0.029
$K_{p,2}$ (L/g or g/L)					12.5–71.5	0.015–0.094
X_{max} (g/L)	100.0–330.0					

Table 3 shows large variations in the values of kinetic parameters, demonstrating that these parameters are strongly dependent upon the operational conditions for which they were adjusted, from the culture medium and from the microorganisms used in the fermentation.

Oliveira et al. [50] analyzed the scale-up effects on kinetic parameters and predictions of a mathematical model developed for a continuous process, in small scale, of ethanol fermentation in a tower bioreactor with flocculating yeast recycling, and concluded that the scale-up did not affect the parameter values and that the model continued to be valid to describe the process in the newly investigated scale.

Although the great majority of mathematical models reviewed thus far have been developed for free cell systems, these are equally valid for naturally or artificially immobilized cell systems. However, physiological changes in microbial cells caused by immobilization can significantly affect the values of the kinetic parameters of such models. Moreover, internal and external diffusion effects in microbial particles and flocs can affect the fermentation kinetics. Admassu et al. [38], modeling the hydrodynamics and the profile of product concentration in a tower fermenter for the continuous production of ethanol with flocculating yeasts, reported that the growth and reaction rates for these flocculating microorganisms are frequently limited by mass transfer.

Vicente et al. [51], developing a new technique to measure kinetics and mass transfer parameters in flocs of *S. cerevisiae*, modeled the kinetics of oxygen consumption using the following equation:

$$r_C = -\left(\frac{p_1 C}{p_2 + C}\right) X \tag{31}$$

where r_C is the oxygen consumption rate (mg O_2/(L h)) from which the specific rate of respiration q_O can be calculated, C is the dissolved oxygen concentration (mg O_2/L), X is the active biomass concentration (g/L), p_1 corresponds to $q_{O,m}$ (mg O_2/(g h)), $q_{O,m}$ and p_2 corresponds to K_m (mg O_2/L).

According to Vicente et al. [51], although Eq. (31) represents a Monod-type kinetic model, the calculated values of p_1 and p_2 are only apparent and have no direct relationship with the usual kinetic parameters. Vicente et al. [51] argue that the designations $q_{O,m}$ and K_m were not used because they are generally applied to suspended free cell cultures and that, in this case, cell aggregates were studied, which significantly change the overall behavior of the system and, therefore, the meaning of such parameters.

3. KINETICS OF SUBSTRATE CONSUMPTION

The kinetics of substrate consumption can generally be described by the Herbert-Pirt model, according to which the substrate is consumed for cell growth and maintenance and the production of a specific product [26]:

$$\mu_S = \frac{\mu_X}{Y^*_{X/S}} + \frac{\mu_P}{Y^*_{P/S}} + m \tag{32}$$

where $Y^*_{X/S}$ and $Y^*_{P/S}$ are, respectively, the stoichiometric coefficients of substrate conversion in cells and product based on the substrate consumed exclusively for each process.

Substrate consumption for cell maintenance refers to the substrate used in the generation of energy for distinct growth functions, such as the maintenance of the concentration gradients between the interior and exterior environment of the cell (osmotic work), synthesis of the cell components that are being continuously degraded, among others [12].

Equation (32) considers that the specific rate of cell maintenance m is a constant, hypothesis, which Ramkrishna et al. [52] do not adopt. According to these authors, the cells suffer a process of degradation in stages in which, at the first stage, the cells would lose their cell viability and, at a second stage, they would die if their maintenance requirements were not attended. To recover the viability, the nonviable cells would need a substrate that would be the same used for growth (exogenous substrate) or an internally stored substrate (endogenous substrate). From these considerations, the following modification in the mathematical representation of the metabolism of maintenance can be introduced, in turn substituting the constant term m in Eq. (32) by a Monod-type expression [26, 53]:

$$\zeta = \frac{\zeta_{\max} S}{K_{S,m} + S} \tag{33}$$

Equation (33) shows that, at high concentrations of substrate, there is a predominance of exogenous metabolism with $\zeta_{max} \rightarrow m$ when $S >> K_{S,m}$, whereas at low concentrations, the endogenous metabolism predominates with $\zeta \rightarrow 0$ when $S \rightarrow 0$.

Generally, the kinetics of substrate consumption is not described using the Herbert-Pirt model to its full extent. The more common approach is the use of apparent coefficients of substrate conversion in cells ($Y_{X/S}$) and ethanol ($Y_{P/S}$), relating μ_S to μ_X or to μ_p by means of these coefficients (Eqs. (34) and (35)). Another approach to describe μ_S is that represented by Eq. (36).

$$\mu_S = \frac{\mu_X}{Y_{X/S}} \tag{34}$$

$$\mu_S = \frac{\mu_P}{Y_{P/S}} \tag{35}$$

$$\mu_S = \frac{\mu_X}{Y_{X/S}} + m \tag{36}$$

Applications of these approaches can be found in the studies listed in Table 4.

Table 4. Mathematical models used for modeling of substrate-consumption kinetics in 1-G ethanol fermentation processes.

Study	μ_S	Reference
Optimization of an industrial bioprocess of ethanol fermentation with multiple stages and cell recycle, using techniques of factorial design and response surface analysis in combination with phenomenological modeling and simulation	Eq. (34)	[54]
Ethanol fermentation modeling in a tower bioreactor with flocculating yeasts	Eq. (35)	[38]
Analysis of the steady-state stability and modeling of the dynamic behavior of a continuous ethanol fermentation process in a gas-lift tower bioreactor with high cell densities	Eq. (35)	[36]
Bifurcation analysis of two continuous membrane fermentor configurations for ethanol production	Eq. (36)	[55]
Modeling, simulation, and analysis of an ethanol fermentation process with control structure in industrial scale	Eq. (36)	[56]
Modeling of a fed-batch ethanol fermentation process with a strain of industrial yeast (*Saccharomyces cerevisiae*)	Eq. (36)	[48]
Modeling of a fed-batch ethanol fermentation process with flocculating yeasts (*S. cerevisiae*)	Eq. (36)	[18]

Sinclair and Kristiansen [12] emphasize the importance of not confusing the stoichiometric coefficients with apparent coefficients as is normally reported in the literature. The stoichiometric coefficient is a constant that depends on the chemical equation, relating the substrates and the products ($Y^*_{P/S}$ = 0.511g-ethanol/g-glucose in the fermentation of glucose to ethanol). The apparent coefficient is the ratio of the mass of a product formed by the total mass of a consumed substrate, which could be participating in multiple reactions, forming a variety of products, including new cells. In this sense, the following definitions for the stoichiometric and apparent coefficients are convenient:

Table 4. Mathematical models used for modeling of substrate-consumption kinetics in 1-G ethanol fermentation processes.

Study	μS	Reference
Optimization of an industrial bioprocess of ethanol fermentation with multiple stages and cell recycle, using techniques of factorial design and response surface analysis in combination with phenomenological modeling and simulation	Eq. (34)	[54]
Ethanol fermentation modeling in a tower bioreactor with flocculating yeasts	Eq. (35)	[38]
Analysis of the steady-state stability and modeling of the dynamic behavior of a continuous ethanol fermentation process in a gas-lift tower bioreactor with high cell densities	Eq. (35)	[36]
Bifurcation analysis of two continuous membrane fermentor configurations for ethanol production	Eq. (36)	[55]
Modeling, simulation, and analysis of an ethanol fermentation process with control structure in industrial scale	Eq. (36)	[56]
Modeling of a fed-batch ethanol fermentation process with a strain of industrial yeast (Saccharomyces cerevisiae)	Eq. (36)	[48]
Modeling of a fed-batch ethanol fermentation process with flocculating yeasts (S. cerevisiae)	Eq. (36)	[18]

$$\bullet Y^*_{X/S} = \frac{\text{Mass of new cells formed}}{\text{Substrate mass consumed only for the formation of new cells}} \qquad (37)$$

$$\bullet Y_{X/S} = \frac{\text{Mass of new cells formed}}{\text{Total mass of substrate consumed}} \qquad (38)$$

$$\bullet Y^*_{P/S} = \frac{\text{Mass of product formed}}{\text{Substrate mass consumed only for the formation of product}} \qquad (39)$$

$$\bullet Y_{P/S} = \frac{\text{Mass of product formed}}{\text{Total mass of substrate consumed}} \qquad (40)$$

Oliveira et al. [57], modeling a continuous ethanol fermentation process in a two-stage tower bioreactor cascade with flocculating yeast recycle, used simplified

($\mu_S = \mu_P/Y_{P/S}$) and generalized (μ_S $\mu_X/Y^*_X=_S + \mu_P/Y^*_P=_S$ +m) kinetic expressions to describe μS and obtained similar predictions of the state variables by both employed approaches.

Bonomi et al. [17], modeling the ethanol production using cassava hydrolyzate in a batch bioreactor, defined the following equation for the mass balance of substrate:

$$\frac{dS}{dt} = -\frac{1}{2}\left(\frac{1}{Y_{X/S}}\mu_X X + \frac{1}{Y_{P/S}}\mu_P X\right) \quad (41)$$

According to these authors, the definition of the apparent coefficients $Y_{X/S}$ and $Y_{P/S}$ guarantee that the terms $\mu_X X/Y_{X/S}$ and $\mu_P X/Y_{P/S}$ are equal; this equality was also reported by Aiba et al. [31] and Ghose and Tyagi [28]. Bonomi et al. [17] argue that the two terms are not exactly equal due to the fact that the calculated values of $Y_{X/S}$ and $Y_{P/S}$ are affected by different experimental errors and the values of μ_X and μ_P are calculated using estimates of other parameters of the model. These authors justify the introduction of the average among the aforementioned terms in Eq.(41), as a means through which to minimize the propagation of errors discussed above.

By contrast, Jin et al. [42], modeling the kinetics of batch fermentation for ethanol production with *S. cerevisiae* immobilized in calcium alginate gel, presented the following mass balance equation for the substrate, without the introduction of the 1/2 factor in the equation:

$$\frac{dS}{dt} = -\left(\frac{1}{Y_{X/S}}\mu_X X + \frac{1}{Y_{P/S}}\mu_P X\right) \quad (42)$$

One equation like $\mu_S = \mu_X/Y\ \mu_P/Y_{P/S}$ was also employed by Marginean et al. [58] to model, simulate, and develop proportional integral derivative (PID) control strategies for temperature and the pH of an ethanol production process in a continuous stirred tank reactor (CSTR).

A different proposal was presented by Limtong et al. [35] to model a continuous process of ethanol fermentation in a tower bioreactor with recycle of flocculating yeasts. The authors determined linear relations between the product concentration (P (g/L)) and the specific rates of glucose consumption (μ_S) and ethanol production (μ_P). The ratios between the corresponding angular and linear coefficients of the straight lines (1.63/3.74 and 0.020/0.046) provide a reasonable estimate of the value of $Y_{P/S}$ (=0.43 g/g), which demonstrates the consistency of such relations (Eqs. (43) and (44)).

$$\mu_S = -0.046P + 3.74(g/g/h) \quad (43)$$

$$\mu_P = -0.020P + 1.63(g/g/h) \quad (44)$$

Another situation to be analyzed is when there is more than one fermentable sugar present in the medium, as is the case in the production of beer. Ramirez

[59], modeling the dynamic of batch beer fermentation, considered the glucose (G), maltose (M), and maltotriose (N) to be the three majority sugars contained in the fermentative medium. The specific consumption rates of these sugars were described by equations that exhibit a kinetic pattern of preferential use of these substrates, that is, the preferred sugar (G) is first used until its complete exhaustion; next, the second sugar (M), of intermediate preference, is consumed; and lastly, the third sugar (N), the least preferred, is consumed. According to Ramirez [59], this pattern of sequential use is modeled by inserting terms of inhibition of the consumption of a less preferential sugar by one or more preferential sugars in such a way that the specific consumption rates of these sugars μ_i are given by

$$\mu_G = \frac{V_G G}{K_G + G} \tag{45}$$

$$\mu_M = \frac{V_M M}{K_M + M}\left(\frac{K_G'}{K_G' + G}\right) \tag{46}$$

$$\mu_N = \frac{V_N N}{K_N + N}\left(\frac{K_G'}{K_G' + G}\right)\left(\frac{K_M'}{K_M' + M}\right) \tag{47}$$

where V_i is the maximum specific consumption rate of the sugar i (g/g/h), K_i is the saturation constant for the sugar i (g/L), and K_i' is the constant of inhibition caused by the sugar i (g/L).

Additionally, the specific rates of cell growth (μ_X) and ethanol production (μ_E) were given by the following equations [59]:

$$\mu_X = R_{XG}\mu_G + R_{XM}\mu_M + R_{XN}\mu_N \tag{48}$$

$$\mu_E = R_{EG}\mu_G + R_{EM}\mu_M + R_{EN}\mu_N \tag{49}$$

where R_{Xi} and R_{Ei} are, respectively, the stoichiometric yield of the biomass and ethanol per gram of sugar i consumed (g/g).

A similar approach was employed by Lee et al. [60] when they modeled the batch ethanol production by *S. cerevisiae* from a mixture of glucose and maltose. One term ξ was included in the equation of μ_M to represent the glucose repression effect upon the maltose consumption. The final set of the mathematical model equations is presented as follows, highlighting the prediction of diauxic growth in the expression of μ_X and the production of ethanol from the two sugars in the expression of μ_E.

$$\frac{dX}{dt} = \mu_X X = \left(\frac{\mu_{G,\max} G}{K_G + G} + \frac{\mu_{M,\max} M\xi}{K_M + M}\right)\eta X \tag{50}$$

$$\frac{dG}{dt} = -\mu_G X = -\left(\frac{\mu_{G,\max} G}{Y_{X/G}(K_G + G)}\eta\right)X \tag{51}$$

$$\frac{dM}{dt} = -\mu_M = -\left(\frac{\mu_{M,\max} M\xi}{Y_{X/M}(K_M + M)}\eta\right)X \tag{52}$$

$$\frac{dE}{dt} = \mu_E X = \left(\frac{Y_{E/G}\mu_{G,\max}G}{Y_{X/G}(K_G + G)} + \frac{Y_{E/M}\mu_{M,\max}M\xi}{Y_{X/M}(K_M + M)}\right)\eta X \tag{53}$$

$$\eta = \left(1 - \frac{X}{X_{\max}}\right)\left(1 - \frac{E}{E_{\max}}\right) \tag{54}$$

$$\xi = \frac{1}{1 + G/k_i} \tag{55}$$

4. LOSS OF CELL VIABILITY

Many authors have reported a decline in cell viability during continuous ethanol fermentation processes with high cell density. However, only a few studies have taken this phenomena into account in the kinetic modeling of the process. Jarzebski et al. [47] investigated a continuous ethanol fermentation system that included a reactor with a well-mixed solution and a filter equipped with a membrane. The purpose of the membrane was to separate and recycle the cells from the fermenter. The authors conducted a comparison between the predictions of an intrinsic model, which took into account the loss of cell viability, and the predictions of a modified nonintrinsic model that did not address this event. The scientists reached the conclusion that the predictions generated by both models were comparable. Therefore, for the purpose of simulating and conducting further analysis of the process, either model may be employed. The term "intrinsic model" is used to describe a model where the substrate and ethanol concentrations are specified based on a corrected volume that does not take into account the volume occupied by cells in systems with high cell densities. Monbouquette [61] provided a comprehensive mathematical framework for deriving mass balance equations based on intrinsic concentrations. Lafforgue-Delorme et al. [62] conducted a study on a system similar to that of Jarzebski et al. [47]. They emphasized the importance of considering additional factors, beyond the dilution rate and yeast concentration, when modeling processes with high cell densities. This is particularly relevant for continuous ethanol fermentations with cell recycling. The authors constructed a model taking into account the following factors: The factors that can affect the performance of yeast in fermentation include dilution and yeast removal, viscosity of the broth, filter blockage, substrate limitation, physiological state of the yeasts (cell viability), and inhibition caused by ethanol and biomass. Additionally, they proposed the notion of steric "stress," which suggests that when cell densities are high, the specific growth rate decreases because there is insufficient space for cell division. The impact of inhibition, caused by high cell concentrations and steric stress, was characterized by the expressions KX/(KX + XV) and (1-X/Xm), where KX represents an empirical constant, Xv denotes the viable cell concentration, and X represents the total concentration of cells (viable + nonviable). The equations

(56) and (57) provide the ultimate representations of μX and μP in the proposed model. The model's predictions were in good agreement with the experimental data for both the bioreactor's operation with whole recycle and its operation with partial recycle.

$$\mu_X = \left(\frac{\mu_{max} S}{K_S + S}\right) \left(1-\frac{P}{P_m}\right)\left(\frac{K_X}{K_X + X_V}\right)\left(1-\frac{X}{X_m}\right) \tag{56}$$

$$\mu_P = \mu_{pm} \exp\left(-\frac{K_P X_V}{D}\right) \tag{57}$$

Augusto [63], investigating the influence of the specific rate of oxygen consumption in a continuous ethanol fermentation with high cell density, established the range of 0.1–0.8 mmol O_2/(g-cell h) as being that which the oxygen participates in the metabolism as a micronutrient that is essential to the synthesis of the cell membrane compounds, which would in turn increase the tolerance of the membrane to ethanol and to other inhibitors produced in the fermentation. The greatest tolerance resulted in a lower specific rate of cell death and in a greater efficiency of substrate conversion in ethanol due to the reduction in the value of the maintenance coefficient by the activation of the oxidative catabolic pathway. According to this author, this range of oxygen consumption for which the positive effects of this nutrient are observed in the bioconversion of the substrate would be dependent on the microorganism used, on the fermentation medium, and on the mode in which the process is conducted (batch, fed-batch, and continuous). To calculate the many parameters of fermentation, Augusto [63] segregated the microbial population into viable and nonviable cells, this procedure being possible due to the availability of experimental measures of the concentration of each type of cell separately.

Hojo et al. [41], studying the ethanol production with a strain of flocculating yeast in CSTR with and without cell recycle, concluded that the cell viability was of utmost importance in developing the mathematical model of the process with cell recycle and that cell death is a phenomenon that should be considered in the kinetic modeling of prolonged continuous fermentations in cases in which the hydraulic residence time is high. The kinetic expressions for the specific rates of cell growth (μ_X), substrate consumption (μ_S), ethanol production (μ_P), and cell death (μ_d) were represented by

$$\mu_X = \left(\frac{\mu_{max} S}{K_S + S}\right) \left(1-\frac{P}{P^*}\right)^n; \mu_{max} = 0.6/\text{h}^{-1},\ K_S = 0.57\ \text{g/L},\ P^* = 80\ \text{g/L},\ n = 1.8 \tag{58}$$

$$\mu_S = \frac{\mu_X}{Y_g}; Y_g = 0.014\ \text{g/g} \tag{59}$$

$$\mu_P = A + B\mu_S; A = 0.065\ \text{g/g/h}; B = 2.24\ \text{g/g} \tag{60}$$

$$\mu_d = k_d; k_d = 0.0054\ \text{h}^{-1} \tag{61}$$

In the aforementioned works, it was considered that the microbial population consisted solely of viable and nonviable cells, with the latter being incapable of growing and producing the desired product. Although inactive in both processes, it was assumed that the nonviable cells remained intact, which means that cell lysis phenomenon was not considered.

For Borzani [64], when intending to apply such an approach, the segregation of the microbial population must be performed considering the active and inactive cells in the growth process, as well as the active and inactive cells in the production process. According to Borzani [64], this differentiation is justified by the fact that a cell that is considered to be active in a given process may not be active in another, or vice-versa. Though quite realistic, this approach is rarely applied, given the enormous experimental difficulty to quantify the concentration of each group of cells separately.

Using an approach that is quite similar to that suggested by Borzani [64], Ghommidh et al. [65], modeling the oscillatory behavior of *Z. mobilis* in continuous cultures for ethanol production, segregated the microbial population in three distinct groups: viable cells that grow and produce ethanol (X_v), nonviable cells that do not grow but produce ethanol (X_{nv}), and dead cells (X_d). The processes of ethanol production, cell growth, loss of viability, and cell death were represented according to the scheme shown in Figure 3.

$$\mu_v = \left(\frac{\mu_{\max} S}{K_1 + S}\right)\left(1 - \frac{P}{P_c}\frac{S}{K_2 + S}\right) \text{ for } P < P_c(K_2 + S)/S \tag{62}$$

$$\mu_{nv} = \left(\frac{\mu_{\max} S}{K_1 + S}\right)\left(1 - \frac{P}{P_c'}\frac{S}{K_2 + S}\right) - \mu_v \text{ for } P < P_c'(K_2 + S)/S \tag{63}$$

$$S \xrightarrow{\mu_p, (X_v + X_{nv})} P$$

$$X_v \xrightarrow{\mu_v} 2X_v$$

$$X_v \xrightarrow{\mu_{nv}} X_{nv}$$

$$X_v \xrightarrow{\mu_d} X_d$$

$$X_{nv} \xrightarrow{\mu_d} X_d$$

Figure 3. Schematic representation of the cell processes involved in ethanol production by Zymomonas mobilis in continuous cultures, according to the model proposed by Ghommidh et al. [65].

Starting from the scheme proposed by Ghommidh et al. [65], Jarzebski [66] modeled the oscillatory behavior of the state variables X, S, and P in a continuous ethanol fermentation process with *S. cerevisiae*, introducing the concept of combined effect of inhibition by substrate and ethanol simultaneously, since, according to that author, the inhibition by substrate would depend on the ethanol concentration and vice-versa. Taking into account this combined effect of inhibition, Jarzebski [66] proposed the following equations to describe the specific rates of viable cell growth (μ_v), conversion of viable cells into nonviable cells (μ_{nv}), and cell death (μ_d):

$$\mu_d = -\mu_v \text{ for } P < P_c^{'}(K_2 + S)/S \quad (64)$$

Watt et al. [67], using the mathematical model proposed by Jarzebski [66], simulated the continuous ethanol fermentation process for different feed volumetric flow rates and substrate concentrations in the feed stream.

The mathematical modeling of ethanol fermentation processes in which the loss of cell viability occurs is generally conducted by dividing the cell population into two distinct groups: viable cells (X_v) which would be growing and producing ethanol and nonviable or dead cells (X_d), which would be inactive in both processes [68]. The conversion rate from the viable to the nonviable cells is considered to be the first order regarding the concentration of viable cells [12]. The specific rates of cell growth, ethanol production, substrate consumption, and loss of cell viability are defined as regards the viable cell concentration, which refer to the effectively active cells in all of these processes. Mass balance equations for viable and nonviable cells are developed separately. The mass balance equations for ethanol and substrate are similar to those of the conventional model (model that does not incorporate the loss of cell viability) with the previously discussed modifications in the terms involving the specific rates.

Based on these premises, Oliveira et al. [69] developed a mathematical model for a continuous ethanol fermentation process in a tower bioreactor with recycle of flocculating yeasts, in which the loss of cell viability was considered and the predictions of this model were compared with those of the conventional model. Both models provide similar predictions and were equally appropriate for the fermentation process modeling. Later, in another publication, the authors analyzed the scale-up effects on the kinetic parameters and on the predictions of the modified model, and found changes in the values of some of the parameters [70]. In addition, the predictions of the modified model agreed better with the experimental data than did those of the conventional model, especially for the cell concentration variable.

A better description of the fermentation process by the modified model is always the desired result, primarily in those cases in which the levels of cell viability are significantly different than 100%. The cell viability level in ethanol fermentations with high yeast densities has been reported as being strongly dependent on the rate of aeration imposed upon the system [34], varying from 40% to 90% [10, 34, 38,

71]. Under anaerobic conditions, unsaturated fatty acids are not synthesized and the yeasts become more sensitive to ethanol [72]. However, the high levels of cell viability in aerated systems are achieved at the expense of the reduction in ethanol yields [70]. Thus, the rate of aeration is an important variable to be optimized in these systems, seeking to provide an adequate level of oxygen dissolved in the medium [70].

The following studies also employed a segregated technique to assess cell viability and describe the microbial population: Kalil et al. [54], Atala et al. [21], Costa Filho et al. [56], Nelson and Hamzah [53], and Watt et al. [67].

5. CONCISE SUMMARY

The ability to simulate the kinetics of ethanol fermentation processes is possible because the key factors that govern these processes, such as substrate limitation, inhibition, loss of cell viability, and death, are well understood. Additionally, there is a substantial number of mathematical models that have been developed and are readily accessible in the literature. This book consolidates a substantial collection of kinetic models that have the potential to be used for fitting experimental data from ethanol fermentation processes. These models are usable across a wide range of operating circumstances. The models can also be utilized in the production processes of a different generation, such as the extraction of ethanol from lignocellulosic feedstocks (second-generation bioethanol). The literature confirms the application of these models in recent publications:

- Scott et al. [73]: Attainable region analysis for continuous production of second-generation bioethanol.
- Vásquez et al. [74]: Modeling of a simultaneous saccharification and fermentation process for ethanol production from lignocellulosic wastes by *K. marxianus*.
- Liu et al. [75]: Fermentation Process Modeling with Levenberg-Marquardt Algorithm and Runge-Kutta Method on Ethanol Production by *S. cerevisiae*.

Overall, numerous fermentation research have shown that unstructured models are not effective in accurately describing dynamic experiments that involve changes in composition and biomass activity [13, 15]. In contrast, employing a more comprehensive methodology to study cell metabolism, with the goal of accurately characterizing the dynamic nature of the process, might result in the creation of intricate models that incorporate several variables and parameters. In such instances, the process of estimating parameters can be challenging due to the substantial experimental effort involved and the necessity of employing intricate numerical techniques, which may result in getting parameter values that lack physical significance. Rivera et al. [76] employed a structured model to analyze experimental data from a tower bioreactor utilized for ethanol production using immobilized S. cerevisiae. The model comprises 34 kinetic parameters and 9 parameters associated with the glycolytic and respiratory (tricarboxylic acid [TCA]) processes. Therefore, additional experimental and computational endeavors would be necessary to determine the parameters linked to this mathematical model. The class of potentially helpful structured models is created by applying the structured formulation, which involves describing the quantity and qualities of biomass using two or three variables. This results in the formation of two- or three-compartment models. These models integrate a more accurate depiction of the system's behavior with a suitable level of mathematical intricacy and a reduced number of factors [77]. Hence, it is crucial to strike a balance between the intricacy of the model and its identification, and to strive for expressions that are as concise as possible while effectively depicting the process in both dynamic and steady phases [69].

REFERENCES

1. Zanin GM, Santana CC, Bon EP, Giordano RCL, de Moraes FF, Andrietta SR, de Carvalho Neto CC, Macedo IC, Fo DL, Ramos LP, Fontana JD. Brazilian Bioethanol Program. Appl. Biochem. Biotechnol. 2000; 84–86: 1147–1161.
2. Vasconcelos JN, Lopes CE, França FP. Continuous ethanol production using yeast immobilized on sugar-cane stalks. Braz. Chem. Eng. J. 2004; 21: 357–365.
3. Xu TJ, Zhao XQ, Bai FW. Continuous ethanol production using self-flocculating yeast in a cascade of fermenters. Enzyme Microb. Technol. 2005; 37(6): 634-640.
4. Cardona CA, Sánchez OJ. Fuel ethanol production: Process design trends and integration opportunities. Bioresour. Technol. 2007; 98: 2415–2457.
5. Carere CR, Sparling R, Cicek N, Levin DB. Third generation biofuels via direct cellulose fermentation. Int. J. Mol. Sci. 2008; 9: 1342–1360. DOI: 10.3390/ijms9071342
6. Mussatto SI, Dragone G, Guimarães PMR, Silva JPA, Carneiro LM, Roberto IC, Vicente A, Domingues L, Teixeira JA. Technological trends, global market, and challenges of bio-ethanol production. Biotechnol. Adv. 2010; 28: 817–830.
7. Mitchell DA, Von Meien OF, Krieger N, Dalsenter FDH. A review of recent developments in modeling of microbial growth kinetics and intraparticle phenomena in solid-state fermentation. Biochem. Eng. J. 2004; 17: 15–26.
8. Doran PM. Bioprocess Engineering Principles. New York: Academic Press; 1995. 439 p.

9. Doran PM. Solutions Manual: Bioprocess Engineering Principles. Kensington: University of New South Wales; 1997. 164 p.

10. Bu'Lock JD, Comberbach DM, Ghommidh, C. A study of continuous ethanol production using a highly flocculent yeast in the gas lift tower fermenter. Chem. Eng. J. 1984; 29: B9–B24.

11. Bailey JE, Ollis DF. Biochemical Engineering Fundamentals, 2nd ed., New York: McGraw-Hill; 1986. 984 p.

12. Sinclair CG, Kristiansen B. Fermentation Kinetics and Modelling. New York: Taylor & Francis; 1987.

13. Oliveira SC. Mathematical modeling of a continuous process of alcoholic fermentation in a tower-type reactor with flocculating-yeasts recycle [thesis]. São Paulo: University of São Paulo (USP); 2000.

14. Bonomi A, Schmidell W. Mathematical modeling and simulation of fermentative processes. In: Schmidell W, Lima UA, Aquarone E, Borzani W, editors. Industrial Biotechnology: biochemical engineering. S□Paulo: Edgard Blücher; 2001, v.2, p.123–178.

15. Astudillo ICP, Alzate CAC. Importance of stability study of continuous systems for ethanol production. J Biotechnol. 2011; 151: 43–55.

16. Andrews JF. A mathematical model for the continuous culture of microorganisms utilizing inhibitory substrate. Biotechnol. Bioeng. 1968; 10: 707–723.

17. Bonomi A, Aboutboul H, Schmidell W. Simulation of the continuous fermentation of manioc hydrolysate. Biotechnol. Bioeng. Symp. 1981; 11: 333–357.

18. Guidini CZ, Marquez LDS, Silva HA, de Resende MM, Cardoso VL, Ribeiro EJ. Alcoholic fermentation with flocculant Saccharomyces cerevisiae in fed-batch process. Appl. Biochem. Biotechnol. 2014; 172: 1623–1638.

19. Oliveira SC, Oliveira RC, Tacin MV, Gattás EAL. Kinetic modeling and optimization of a batch ethanol fermentation process. J. Bioprocess. Biotech. 2016; 6: 266. DOI:10.4172/2155-9821.1000266

20. Moser A. Kinetics of batch fermentation. In: Rehm HJ, Reed G. Volume Editor: Brauer H. Biotechnology—A comprehensive treatise in 8 volumes. Weinheim: V.C.H. Verlagsgesellschaft; 1985. v.2. pp. 243–283.

21. Atala DIP, Costa AC, Maciel R, Maugeri F. Kinetics of ethanol fermentation with high biomass concentration considering the effect of temperature. Appl. Biochem. Biotechnol. 2001; 91–93: 353–365.

22. Tsuji S, Shimizu K, Matsubara M. Performance evaluation of ethanol fermentor systems using a vector-valued objective function. Biotechnol Bioeng. 1987; 30: 420–426.
23. Sousa ML, Teixeira JA. Reduction of diffusional limitations in yeast flocs. Biotechnol. Lett. 1991; 13(12): 883–888.
24. Fontana A, Bore C, Ghommidh C, Guiraud JP Structure and sucrose hydrolysis activity of Saccharomyces cerevisiae aggregates. Biotechnol. Bioeng. 1992; 40: 475–482.
25. Contois DE. Kinetics of bacterial growth: Relationship between population density and specific growth rate of continuous cultures. J. Gen. Microbiol. 1959; 21: 40–50.
26. Menezes JC, Alves SS, Lemos JM, Azevedo SF. Mathematical modelling of industrial pilot-plant penicillin-G fed-batch fermentations. J. Chem. Tech. Biotechnol. 1994; 61: 123–138.
27. Oliveira SC, Paiva TCB, Visconti AES, Giudici R. Discrimination between ethanol inhibition models in a continuous alcoholic fermentation process using flocculating yeast. Appl. Biochem. Biotechnol. 1998; 74: 161–172.
28. Ghose TK, Tyagi RD. Rapid ethanol fermentation of cellulose hydrolysate. II. Product and substrate inhibition and optimization of fermentor design. Biotechnol. Bioeng. 1979; 21: 1401–1420.
29. Bazua CD, Wilke CR. Ethanol effects on the kinetics of a continuous fermentation with Saccharomyces cerevisiae. Biotechnol. Bioeng. Symp. 1977; 7: 105–118.
30. Novak M, Strehaiano P, Moreno M, Goma G. Alcoholic fermentation: On the inhibitory effect of ethanol. Biotechnol. Bioeng. 1981; 23: 201–211.
31. Aiba S, Shoda M, Nagatani M. Kinetics of product inhibition in alcohol fermentation. Biotechnol. Bioeng. 1968; 10: 845–864.
32. Aiba S, Shoda M. Reassessment of the product inhibition in alcohol fermentation. J. Ferment. Technol. 1969; 47(12): 790–794.
33. Pascual C, Alonso A, García I, Romay C, Kotyk A. Effect of ethanol on glucose transport, key glycolytic enzymes, and proton extrusion in Saccharomyces cerevisiae. Biotechnol. Bioeng. 1988; 32: 374–378.
34. Jones ST, Korus RA, Admassu W, Heimsch RC. Ethanol fermentation in a continuous tower fermentor. Biotechnol. Bioeng. 1984; 26: 742–747.
35. Limtong S, Nakata M, Funahashi H, Yoshida T, Seki T, Kumnuanta J, Taguchi H. Continuous ethanol production by a concentrated culture of flocculating yeast. J. Ferment. Technol. 1984; 62(1): 55–62.

36. Comberbach DM, Ghommidh C, Bu'Lock JD. Steady-state stability and dynamic behavior of continuous ethanol fermentation at high cell densities. Enzyme Microb. Technol. 1987; 9: 676–684.

37. Luong JHT. Kinetics of ethanol inhibition in alcohol fermentation. Biotechnol. Bioeng. 1985; 27: 280–285.

38. Admassu W, Korus RA, Heimsch RC. Ethanol fermentation with a flocculating yeast. Chem. Eng. J. 1985; 31: B1–B8.

39. Gòdia F, Casas C, Sola C. Batch alcoholic fermentation modelling by simultaneous integration of growth and fermentation equations. J. Chem. Tech. Biotechnol.1988; 41: 155–165.

40. Levenspiel O. The Monod equation: A revisit and a generalization to product inhibition situations. Biotechnol. Bioeng. 1980; 22: 1671–1687.

41. Hojo O, Hokka CO, Maior AMS. Ethanol production by a flocculant yeast strain in a CSTR type fermentor with cell recycling. Appl. Biochem. Biotechnol. 1999; 77–79: 535–545.

42. Jin H, Liu R, He Y. Kinetics of batch fermentations for ethanol production with immobilized Saccharomyces cerevisiae growing on sweet sorghum stalk juice. Procedia Environ. Sci. 2012; 12: 137–145.

43. Le Duy A, Zajic JE. A geometrical approach for differentiation of an experimental function at point applied to growth and product formation. Biotechnol. Bioeng. 1973; 15: 805–810.

44. Lee JM. Computer simulation in ethanol fermentation. In: Cheremisinoff PN, Onellette RP, editors. Biotechnology: Applications and Research. Lancaster: Technomic Publishing Company Inc.; 1985. p. 78–87.

45. Wang F-S, Sheu J-W. Multiobjective parameter estimation problems of fermentation processes using a high ethanol tolerance yeast. Chem. Eng. Sci. 2000; 55: 3685–3695.

46. Olaoye OS, Kolawole OS. Modeling of the kinetic of ethanol formation from glucose biomass in batch culture with a non structured model. Int. J. Eng. Res. Appl. 2013; 3: 562–565.

47. Jarzebski AB, Malinowski JJ, Goma, G. Modeling of ethanol fermentation at high yeast concentrations. Biotechnol. Bioeng. 1989; 34: 1225–1230.

48. Rivera EC, Yamakawa CK, Garcia MH, Geraldo VC, Rossell CEV, Maciel Filho R, Bonomi A. A procedure for estimation of fermentation kinetic parameters in fed-batch bioethanol production process with cell recycle. Chem. Eng. Trans. 2013; 32: 1369–1374. DOI: 10.3303/CET1332229

49. Ghosh K, Ramachandran KB. Analysis of the effect of in situ product removal on the stability and performance of a continuous bioreactor with cell separator for ethanol production. Chem. Biochem. Eng. Q. 2007; 21(3): 285–296.

50. Oliveira SC, de Castro HF, Visconti AES, Giudici R. Continuous ethanol fermentation in a tower reactor with flocculating yeast recycle: Scale-up effects on process performance, kinetic parameters and model predictions. Bioprocess Eng. 1999; 20: 525–530.

51. Vicente AA, Dluhý M, Teixeira JA. A new technique for measuring kinetic and mass transfer parameters in flocs of Saccharomyces cerevisiae. Biotechnol. Tech. 1997; 11(2): 113–116.

52. Ramkrishna D, Fredrickson AG, Tsuchiya HM. Dynamics of microbial propagation models considering endogenous metabolism. J. Gen. Appl. Microbiol. 1966; 12(4): 311–327.

53. Nelson MI, Hamzah N. Performance evaluation of bioethanol production trough continuous fermentation with a settling unit. J Energ Power Eng. 2013; 7: 2083–2088.

54. Kalil SJ, Maugeri F, Rodrigues MI. Response surface analysis and simulation as a tool for bioprocess design and optimization. Process Biochem. 2000; 35: 539–550.

55. Garhyan P, Elnashaie SSEH. Bifurcation analysis of two continuous membrane fermentor configurations for producing ethanol. Chem. Eng. Sci. 2004; 59: 3235–3268.

56. Costa Filho MVA, Monteiro JB, Magazoni FC, Colle S. Modeling, simulation and analysis of ethanol fermentation process with control structure in industrial scale. In: Proceedings of the 22nd International Conference on Efficiency, Cost, Optimization, Simulation and Environmental Impact of Energy Systems (ECOS); August 31–September 3 2009; Foz do Iguaçu; 2009.

57. Oliveira SC, De Castro HF, Visconti AES, Giudici R. Mathematical modeling of a continuous alcoholic fermentation process in a two-stage tower reactor cascade with flocculating yeast recycle. Bioprocess Biosyst. Eng. 2015; 38: 469–479.

58. Marginean AM, Trifa V, Marginean C. Simulation of fermentation bioreactor control for ethanol production. In: Proceedings of the 11th International Conference on Development and Application systems; 17–19 May 2012; Suceava; 2012. pp. 17–20.

59. Ramirez WF. Computational Methods for Process Simulation. Stoneham: Butterworth Publishers; 1989.

60. Lee Y-S, Lee WG, Chang YK, Chang HN. Modelling of ethanol production by Saccharomyces cerevisiae from a glucose and maltose mixture. Biotechnol. Lett. 1995; 17(8): 791–796.

61. Monbouquette HG. Modeling high-biomass-density cell recycle fermenters. Biotechnol. Bioeng. 1992; 39: 498–503.

62. Lafforgue-Delorme C, Delorme P, Goma G. Continuous Alcoholic Fermentation with Saccharomyces cerevisiae Recycle by Tangential Filtration: Key points for process modelling. Biotechnol. Lett. 1994; 16(7): 741–746.

63. Augusto EFP. Continuous alcoholic fermentation with high cell density: Influence of the specific oxygen consumption rate [thesis]. São Paulo: University of São Paulo (USP); 1991.

64. Borzani W. Cinética de processos fermentativos. Rev. Brasileira Engenharia.1986; 3(2): 1–51.

65. Ghommidh C, Vaija J, Bolarinwa S, Navarro JM. Oscillatory behaviour of Zymomonas in continuous cultures: A simple stochastic model. Biotechnol. Lett. 1989; 2(9): 659–664.

66. Jarzebski AB. Modelling of oscillatory behaviour in continuous ethanol fermentation. Biotechnol. Lett. 1992; 14(7): 137–142.

67. Watt SD, Sidhu HS, Nelson MI, Ray AK. Analysis of a model for ethanol production through continuous fermentation. Anziam J. 2007; 49: C85–C99.

68. Facciotti MCR, Schmidell W. The new concept of minimum cell viability and its consequences on bioprocess design and operation. Braz. J. Chem. Eng. 1995; 12(1): 22–31.

69. Oliveira SC, Paiva TCB, Visconti AES, Giudici R. Continuous alcoholic fermentation process: Model considering loss of cell viability. Bioprocess Eng. 1999; 20: 157–160.

70. Oliveira SC, De Castro HF, Visconti AES, Giudici R. Scale-up effects on kinetic parameters and on model predictions of a yeast recycle continuous ethanol fermentation model incorporating loss of cell viability. Bioprocess Eng. 2000; 23: 51–55.

71. Comberbach DM, Bu'Lock JD. Continuous ethanol production in the gas-lift tower fermenter. Biotechnol. Lett. 1984; 6(2): 129–134.

72. Kida K, Yamadaki M, Asano S-I, Nakata T, Sonoda Y. The effect of aeration on stability of continuous ethanol fermentation by a flocculating yeast. J. Ferment. Bioeng. 1989; 68(2): 107–111.

73. Scott F, Conejeros R, Aroca G. Attainable region analysis for continuous production of second generation bioethanol. Biotechnol. Biofuels. 2013; 6: 171.

74. Vásquez JE, Quintero JC, Ochoa-Cácere S. Modeling of a simultaneous saccharification and fermentation process for ethanol production from lignocellulosic wastes by Kluyveromyces marxianus. DYNA. 2014; 81(185): 107–115.

75. Liu D, Xu L, Xiong W, Zhang H-T, Lin C-C, Jiang L, Xu B. Fermentation process modeling with Levenberg-Marquardt algorithm and Runge-Kutta method on ethanol production by Saccharomyces cerevisiae. Math Probl. Eng. 2014; 2014: 1–10.

76. Rivera EC, Costa AC, Lunelli BH, Maciel MRW, Maciel Filho R. Kinetic modeling and parameter estimation in a tower bioreactor for bioethanol production. Appl. Biochem. Biotechnol. 2008; 148:163–173.

77. Stremel DP. Development of alternative structured models for the ethanol production process [thesis]. Campinas: State University of Campinas (UNICAMP); 2001.

07 OPTIMIZING MICROBIAL POPULATIONS TO CONTROL AND ENHANCE DARK HYDROGEN FERMENTATION

1. INTRODUCTION: AN OVERVIEW

The growing global recognition of global climate change, urban air pollution, and the need for secure future energy sources has prompted increased research into alternative fuels. Hydrogen is an environmentally friendly and highly promising fuel, particularly when it is obtained exclusively from renewable energy sources. Additionally, it is both efficient and environmentally benign due to its high energy content and the fact that water is the only byproduct [1, 2]. Currently, almost 95% of the hydrogen used for commercial purposes is produced by the process of steam reforming natural gas and coal gasification [3]. Due to their reliance on fossil fuels, these procedures are not ecologically sustainable. One possible solution to reduce reliance on fossil fuels for hydrogen production is to harness the capabilities of hydrogen-producing microbes to generate hydrogen from readily accessible biomass sources. Considering these viewpoints, biological hydrogen production has significant potential as a viable alternative energy source. The process of dark fermentative hydrogen generation from wastewater results in higher rates of hydrogen production compared to other biohydrogen production methods [4]. Additionally, the advantage of this process is that the substrate cost, which is wastewater, is completely free. For example, a fermentative hydrogen-producing process produces hydrogen at a higher rate (0.5–65.0 l $H_2 l^{-1} d^{-1}$) compared to a light-driven process (0.04–4.3 l $H_2 l^{-1} d^{-1}$) [5]. In addition, the major advantages are low energy demands, resulting in minimal pollution, operation without light sources, no oxygen limitation problems, and low capital costs for at least small-scale production facilities (100–1000 m^3 $H_2 \cdot h^{-1}$) [5, 6–9]. Both mesophilic and thermophilic continuous dark fermentative hydrogen production have been investigated. Thermophilic operation is especially suitable for complying with regulations on the treatment of feedstock that contains pathogens or when it is combined with a process that generates waste heat. Alternatively, due to the high energy requirements, thermophilic operation is less likely to be the preferred choice from a technical and economic standpoint. A viable economic method for producing hydrogen is to convert organic wastes, which are frequently considered to have negative value, into gas rich in hydrogen using fermentative bacteria [2, 10]. Hydrogen production has been investigated using different types of organic waste materials and wastewater derived from corn, palm oil, soybean, and meat processing facilities [11, 12]. Due to the presence of non-sterile feedstock, dark-fermentative hydrogen generation systems often utilize mixed microflora obtained from natural sources. In theory, glucose produces 4 moles of hydrogen and 2 moles of acetate together (Eq. 1,3). However, when butyrate is the predominant fermentation product, only 2 moles of hydrogen are created (Eq. 2,4). Based on the aforementioned processes, it can be inferred that acetic acid as the fermentation end product is connected with the maximum theoretical output of hydrogen. Empirically, it has been observed that when the concentrations of acetic acid and butyrate in a mixture are more than those of propionate, the production of hydrogen is higher compared to other scenarios [6, 13]. Usually, around 60-70% of the liquid product produced during sugar fermentation is butyrate, which results in low hydrogen yields (around 2.5-2.9 mol H2/mol glucose). This is lower than the theoretical yield of 4 mol H2/mol glucose that would be obtained if only acetate was the final product of fermentation [14]. Hydrogen yields can be enhanced by augmenting hydrogen generation via reaction (1) and reducing or inhibiting reaction (2). One way to achieve this is by using dark hydrogen fermentation (DHF) using thermophiles or severe thermophiles, which function at temperatures exceeding 60°C [15, 16].

Mesophilic (35°C)

$$C_6H_{12}O_6 + 2H_2O \rightarrow 4H_2 + 2CO_2 + 2C_2H_4O_2 \rightarrow \Delta G^\circ = -184.2\ \text{kJ mol}^{-1} \quad (1)$$

$$C_6H_{12}O_6 \rightarrow 2H_2 + 2CO_2 + C_4H_8O_2 \rightarrow \Delta G^\circ = -244.2\ \text{kJ mol}^{-1} \quad (2)$$

Thermophilic (60 °C)

$$C_6H_{12}O_6 + 2H_2O \rightarrow 4H_2 + 2CO_2 + 2C_2H_4O_2 \rightarrow \Delta G^\circ = -20.1\ \text{kJ mol}^{-1} \quad (3)$$

$$C_6H_{12}O_6 \rightarrow 2H_2 + 2CO_2 + C_4H_8O_2 \rightarrow \Delta G^\circ = -84.2\ \text{kJ mol}^{-1} \quad (4)$$

Elevated temperatures have a thermodynamic preference for the generation of hydrogen. In addition, higher temperatures enhance the death of pathogens and restrict the consumption of hydrogen by hydrogen-consuming microorganisms such as methanogens, homoacetogens, and sulfate reducers. Typically, 67% of the initial organic matter will stay dissolved in the solution, based on the chemical oxygen demand (COD), when the DHF process is carried out under ideal conditions. In order to fully harness the chemical energy stored in biomass, a combined process is necessary. This process combines both the extraction of the remaining organic matter and the generation of methane, electricity, bioplastics, and hydrogen through the photofermentation process. Two-stage procedures have been extensively developed and have the potential to be modified for the production of both hydrogen and methane [17], as well as the generation of hydrogen and power through microbial fuel cells [18]. The effectiveness of DHF (dark hydrogen fermentation) using food waste in anaerobic mesophilic and thermophilic acidogenesis, followed by a two-phase digestion or photo-fermentation, has also been evaluated [19]. In general, numerous technologies have been extensively studied to assess their potential for successful industrial use and long-term viability in generating sustainable bioenergy through improved biohydrogen production. Many different types of microorganisms, including both strict and facultative anaerobic chemoheterotrophs like Clostridia, Enteric bacteria, Caldicellulosiruptor spp., Thermotoga spp., and Thermoanaerobacterium spp., have the ability to efficiently produce hydrogen while breaking down different types of carbohydrates [20]. In order to promote hydrogen production and inhibit the consumption of hydrogen by methanogenic activity, it is essential to create experimental circumstances that utilize mixed microflora. These factors encompass the conditioning of the inoculum, the optimization of working conditions such as hydraulic retention time (HRT), pH, and substrate concentration, as well as the reduction of hydrogen partial pressure [4, 7, 21]. Hawkes et al. [7] have provided a summary of the obstacles involved in optimizing dark hydrogen fermentation systems. In recent years, there has been significant research development in this area. However, a process that is both economically and technically practicable has not yet been established. Typically, control and operation tactics are employed to enhance the efficiency of a process by the manipulation of chemical, physical, and biological components. However, these strategies often only offer temporary answers and might have negative effects on the microbiological properties of the system. The process is often conducted by intricate microbial communities that consist of highly efficient hydrogen generators. In recent times, a multitude of investigations [19, 22–28] have provided molecular data pertaining to these diverse impacts. The majority of Clostridium species are considered beneficial bacteria for mesophilic environments, but Thermoanaerobacterium species, C. thermocellum, C. cellulose, and C. thermoamyloticum are recognized as favorable bacteria for thermophilic circumstances. Understanding the structure and function of microbial communities is crucial for enhancing hydrogen production by optimizing the microbial population. Microbial population optimization is a strategy that utilizes current knowledge of microbial community data to address technical challenges including low hydrogen yields, biomass washout, hydrogen inhibition, unstable hydrogen generation, and short reactor operation duration. Efficient microbial population optimization necessitates a comprehensive understanding of both the microbiology and the physicochemical properties of the process. Understanding microbiological aspects involves studying the structure and function of microbial consortia, as well as the interactions that take place within them. It also involves identifying the essential microbial organisms responsible for hydrogen production and studying their kinetics. Possible strategies after analyzing the population structure and function involve managing the proliferation of unwanted microorganisms (such as methanogens, propionic acid bacteria, and lactic acid bacteria) that consume hydrogen, while simultaneously promoting the growth and stability of hydrogen-producing bacteria.

2. THE DARK HYDROGEN FERMENTATION PROCESS

2.1. Basic principle for dark hydrogen fermentation

Fermentative hydrogen production has a theoretical maximum yield of 4 moles (498 ml-H2/g-1 glucose) of hydrogen from glucose, together with 2 moles of acetate. Additionally, 2 moles (249 ml-H2/g-1 glucose) of hydrogen are created from glucose, along with 1 mole of butyrate. Many types of microorganisms, including both stringent and facultative anaerobic chemoheterotrophs including Clostridia, enteric bacteria, and Thermoanaerobacterium, are highly effective at producing hydrogen. The process of converting glucose into hydrogen, pyruvate, and acetyl CoA through fermentation leads to the production of ATP and the release of acetate. The process of pyruvate oxidation to acetyl CoA necessitates reduction facilitated by ferredoxin (Fd). The process involves the oxidation of Reduced Fd by hydrogenase, resulting in the production of oxidized Fd and the release of electrons in the form of molecular hydrogen (Eq.5–8). The practical yield is further reduced when other metabolic chemicals, such as propionate, ethanol, and lactate, are generated as byproducts of fermentation. Due to thermodynamic restrictions, these metabolic products avoid the main process that produces hydrogen during carbohydrate fermentation [9].

$$C_6H_{12}O_6 + 2H_2O \rightarrow 4H_2 + \rightarrow 2CO_2 + 2C_2H_4O_2 \tag{5}$$

$$C_6H_{12}O_6 \rightarrow 2H_2 + \rightarrow 2CO_2 + C_4H_8O_2 \tag{6}$$

$$\text{Pyruvate} + \text{CoA} + 2\text{Fd}(\text{ox}) \rightarrow \text{Acetyl -CoA} + 2\text{Fd}(\text{red}) + CO_2 \tag{7}$$

$$2\,\text{Fd}(\text{red}) \rightarrow 2\,\text{Fd}(\text{ox}) + H_2 \tag{8}$$

The proton-reducing ability of Fd_{red} and NADH is thermodynamically limited by the maximum hydrogen partial pressures (P_{H2}) of 0.3 and 6x10-^{4} atm (60 Pa), respectively. This confers that as long as the P_{H2} is still less than 0.3 atm, hydrogen production can continue with transferring electrons from Fd_{red} which contains electrons from oxidative decarboxylation of pyruvate by pyruvate:ferredoxin oxidoreductase (PFOR). Meanwhile, the oxidation of NADH by NADH:Fd oxidoreductase (NFOR) can generate Fd_{red} that subsequently generates additional hydrogen when the P_{H2} is maintained less than 60 Pa. However, the P_{H2} limited to hydrogen generation via the oxidation of NADH could be increased to 0.1–0.2 atm at a temperature of 70°C [16]. Therefore, increasing cultivation temperature is necessary to overcome thermodynamic limitation, thereby resulting in a decrease of the Gibbs free energy of conversion according to the second law of thermodynamics ($\Delta G = \Delta H\text{-}T\ \Delta S$) [29]. Thermophilic microorganisms produce generally higher hydrogen yields compared to mesophiles because they are thermodynamically favorable [30]. High hydrogen yields in the range of 314.0–473.0 ml-H_2/g^{-1} sugars have been previously reported using thermophiles such as *C. thermocellum* and *Thermoanaerobacterium thermosaccharolyticum* and extreme thermophiles such as *Thermotoga elfi, Caldicellulosiruptor saccharilyticus,* and *Caldanaerobacter subterraneus* [15, 31–34]. In a practical sense, through controlling the fermentative types of microorganisms, it is possible to maximize the amount of hydrogen produced by fermentation.

2.2. Dark hydrogen fermentation by mixed cultures

The combination of dark fermentation process and environmental biotechnology for the treatment of organic wastes or residues using industrial biotechnology, with the goal of maximizing hydrogen production and utilizing mixed culture fermentation, is becoming increasingly appealing compared to pure culture fermentation. This is because mixed cultures are originally used in waste treatment applications. heterogeneous culture fermentation, in contrast to pure culture fermentation, eliminates the need for media sterilization, provides enhanced adaptability due to its abundant microbial population and the ability to co-ferment heterogeneous substrates, and enables a continuous fermentation process. [35]. Undefined mixed cultures taken from different natural sources need pretreatment or enrichment, by manipulating the operation of the fermentation process and/or by varying the sources of the natural inoculum in order to obtain the required metabolic capacities and the corresponding microbial population for development of the dark fermentation process [36, 37]. To prepare the inoculum for hydrogen production by fermentation of carbohydrates, the original anaerobic sludge is first pretreated to suppress methanogenic archaea, which consume hydrogen generated and subsequently enrich hydrogen-producing bacteria in various reactor configurations [38]. Pretreating anaerobic seed sludge under harsh conditions, spore-forming bacteria involved in anaerobic conversion of carbohydrates to hydrogen could have a better chance to survive compared to the non-spore-forming methanogenic archaea. The spores formed can be activated when the required environmental conditions are provided during subsequent enriching for hydrogen production [39]. Methods, including heat shock, load shock, acid, base, and chemical pretreatments are usually applied to pretreat anaerobic seed sludge for fermentative hydrogen production

2.2.1. Heat shock

The most prevalent and efficient technique for removing methanogenic archaea is heat shock, which involves subjecting the seed sludge to steam heating at temperatures ranging from 75 to 121°C for a duration of 15 to 120 minutes. This process is reasonably simple and cost-effective. The heat shock can inhibit the activity of propionate makers that do not generate spores, but it is not able to efficiently destroy homoacetogens [21, 40]. Homoacetogenic bacteria reduce hydrogen production by using the hydrogen produced during fermentation for the synthesis of acetate [41]. Furthermore, Duangmanee et al. [42] previously noted that an inoculum subjected to heat shock was not consistently effective for hydrogen production in the continuous reactor. They found that repeated heat treatments were necessary on a monthly basis to maintain some level of stability in hydrogen production.

2.2.2. Load shock

During load shock using the pulse load technique in batch and organic fermentation, or hydraulic shock in continuous fermentation, volatile fatty acids (VFAs) tend to accumulate in the fermentative reactor in high concentrations, associated with acidic conditions, and they inhibit methanogens [42, 43]. Applying a load shock with a pulse load of about 40–50 g-sugar/l, the pretreated anaerobic sludge effectively suppressed methanogenic activity [24, 44]. Furthermore, O-Thong et al. [24] have described that load shock-pretreated seed sludge could result in high level of hydrogen production similar to the heat shock-pretreated seed sludge and that load shock would be technically easier to do and more economical than heat shock for implementation on an industrial scale.

2.2.3. Acid and alkali pretreatment

The bioactivity of methanogens during the conventional anaerobic process treatment of organic wastes occurs in neutral to slightly alkaline environments (pH 6.8–8.0) [38]. Limiting methanogenesis can be achieved by adjusting the acidity of the anaerobic sludge substantially away from the preferable range to either pH 3–4 or pH 12. The acid or alkali pretreatment is considered to be technically easier than the heat shock pretreatment for industrial scale implementation [21]; however, the inoculum obtained from an acid or alkali pretreatment requires a much longer acclimatization time of 10 to 30 days to establish hydrogen production [45].

2.2.4. Methanogen inhibitors

2-bromoethanesulfonate acid (BESA), an analog of the coenzyme-M in methanogens, is a chemical that deactivates methanogens. Using BESA at concentrations of 25–100 mM has been found to effectively inhibit the bioactivity of methanogens; however, treating an anaerobic sludge at these levels would not be cost effective for a commercial scale operation [39].

3. MOLECULAR METHODS FOR MICROBIAL COMMUNITY STRUCTURE AND FUNCTION STUDIES

Molecular monitoring techniques such as fluorescence in situ hybridization (FISH) [46], a combination of FISH and microautoradiography (FISH–MAR) [47], stable isotope probing (SIP) [48], denaturing gradient gel electrophoresis (DGGE) [49], ribosomal intergenic spacer analysis (RISA) [23], and clone libraries have been developed for studying microbial community and function. These methods are used intensively in natural and engineered systems for wastewater treatment. Principles of and deeper insights on these molecular tools are available elsewhere (e.g. [50]). Among these techniques, cloning and the creation of a gene library, DGGE, TRFLP, RISA, and FISH stand out. DGGE was one of the first techniques used to describe DHF microflora [51, 52]. DGGE is a rapid and simple method that provides characteristic band patterns for different samples, allowing quick sample profiling, while retaining the possibility of a more thorough genetic analysis by sequencing of particular bands. DGGE provides information about the structure of microbial communities and can relatively quantify species abundance through DNA band intensities. Cloning provides very precise taxonomical information, but is time consuming and requires specialized personnel and hence, its introduction in the DHF process has been slow. FISH helps identify microorganisms at any desired taxonomical level, depending on the specificity of the probe used. It is the only quantitative molecular biology technique, although quantification is either complex or tedious and subjective. Combination with a confocal laser scanning microscope allows the visualization of three-dimensional microbial structures (granules and biofilms). Both DGGE and FISH have been extensively employed. Other techniques such as RISA [23] provide information on microbial diversity and species dominance. The advantages and disadvantages of the molecular techniques frequently applied to microbial ecology research in DHF process are shown in Table 1.

Table 1. Brief description of frequently molecular methods that have been used for microbial community analysis in dark fermentative hydrogen production.

Molecular methods	Nucleic acid extraction	PCR	Advantages	Disadvantages
Fluorescence *in situ* hybridization (FISH)	No	No	- Direct analysis and quantification - Suitable for targeting specific group/species -Easy and fast if required -Allows direct visualization of non-cultured microorganisms -Differential/preferential detection of active microorganisms	- Require genes/RNA with high number of copies - Limit for total diversity mapping - The design of a specific and unambiguously restrictive probe for a certain group of microorganisms is not always possible - The design and optimization of a new probe is a difficult process - Structural analysis of granular requires a confocal microscope

Molecular methods	Nucleic acid extraction	PCR	Advantages	Disadvantages
Denaturing gradient gel electrophoresis (DGGE)	yes	yes	- Permits rapid and simple monitoring of the spatial temporal variability of microbial populations if just band patterns are considered - It is relatively easy to obtain an overview of the dominant species of an ecosystem - It is adequate for analysis of a large number of samples (far more than cloning)	- Bias from PCR - The number of detected bands is usually small, which implies: the number of identified species is also small; the bands correspond, although not necessarily, to the predominant species in the original sample - The sequences of the bands obtained from a gel correspond to short DNA fragments (200–600 bp), and so phylogenetic relations are less reliably established than with cloning of the whole 16S rRNA gene. In addition, short sequences are less useful for designing new specific primers and probes. - GC content of the amplified DNA can modulate the *Taq* polymerase activity
Cloning and sequencing	Yes	Yes	- Contain larger sequence - Complete 16S rRNA sequencing allows: very precise taxonomic studies and phylogenetic trees of high resolution to be obtained - Identification of microorganisms that have not been yet cultured or identified - Covers most microorganisms, including minority groups, which would be hard to detect with genetic fingerprinting methods	- A large number of clones must be sequenced for positive diversity - Sequences need to be compared with each other and libraries - Very time consuming and laborious, making it unpractical for high sample throughput - It is not quantitative. The PCR step can favor certain species due to differences in DNA target site accessibility - Bias from PCR, total universal primers cannot be totally universal bacteria - Exponential amplification of the DNA mixture may result in ratio discrepancies between the amplified 16S rDNA fragments and the original mixture

For decades, a biological reactor has been considered as a black box. The new insights in microbiology have helped to improve the design and performance of new generation reactors [53, 54]. Probably it is true that it is not essential to know the phylogenetic position of the individual microorganisms that dwell inside a system for the design of a biohydrogen facility. But the knowledge on microbial community structure and function is needed. The more recent reports on microbial community structures of DHF processes still interpret the results with reactor performance and metabolic by-products (indirect function) [55, 56]. However, we are still uncertain regarding which microorganisms can function effectively in DHF and whether the whole community takes part. Thus, deeper insight into the function is required, not just community structure. The latter is due to a general shortcoming in all these molecular tools. However, some attempts have been made in this direction, as FISH–MAR and FISH combined with biosensors could be applied to reveal the microbial community structure and function in parallel. Furthermore, other techniques such as DNA microarrays are being developed with the goal of being able to infer the in situ physiology of the microorganisms [57], and these should find application in the hydrogen-producing biosystems.

Post-genomic research and systems biology tools such as metaproteomics will greatly contribute to the development by providing functional performance insights of the microorganisms and their metabolism [55]. Recent work on post-genomics involving microbial ecosystems has expanded to both natural microbial biofilms and activated sludge [56, 58]. These cutting-edge technologies are aimed at using new molecular tools to understand the microbial community structures in relation to functions [55] or metabolic transformations [58]. It is commonly known that 16S rRNA genes approaches have copy numbers and PCR bias problems. Housekeeping genes with a single copy are now the focus for population genomic analysis. Multilocus sequence typing (MLST) of housekeeping genes could provide a deeper insight on how microbial populations evolved [59]. These modern molecular monitoring techniques are vital tools and could also be applied for DHF, as they will particularly break new ground for the quantification and dynamics of microorganisms in complex consortia.

A whole variety of analytical methods for both microbial community structure and function are now available. However, microbiologists and engineers should take efforts to apply these tools for quantitative studies of DHF. With a more thorough understanding of the microbial community and its dynamics, an improvement of expectations and optimization of fermentative processes will be possible. The microbial community structure and microbial function may be further optimized by adding species and specific nutrients for the dominant species.

4. UTILIZING MOLECULAR EVIDENCE TO STUDY DARK HYDROGEN FERMENTATION PROCESSES

4.1. The impact of different types of inoculum and conditioning on the organization of microbial communities

It has been previously reported that the methods for seed preparation can affect both start up and overall efficiency of the hydrogen-producing reactors [7]. Quick recovery from process upsets in full-scale applications may also require large quantities of readily available hydrogen-producing seeds. Therefore, induction of hydrogen accumulation in fermentative consortia is related to the inhibition of hydrogen consumers which is essential for its further scale-up and industrial applications. Several types of inocula have been used for anaerobic hydrogen fermentation, such as anaerobic-digested sludge [60], sewage treatment sludge [61], agricultural soil [62], sludge compost [63], and isolated bacteria [64]. In addition, several methods are used for conditioning inocula such as acid conditioning [65], heat conditioning [60, 62], chemicals conditioning such as 2-bromoethanesulfonic acid (BESA) [66], short hydraulic retention time (HRT) without conditioning [67], and overload conditioning [24]. All conditioning methods aid in inhibiting methane formation, as well as accelerating the enrichment of hydrogen producing bacteria, such as spore-forming *Clostridium* species, as these are highly tolerant to extreme environments [68]. The effect of conditioning on hydrogen production rates is inoculum dependent, with appreciable hydrogen production yields being demonstrated with anaerobic-digested sludge and agricultural soil [69]. Several studies (e.g. [23, 51, 70–73] reveal that heat-conditioned anaerobic-digested sludge guarantees the highest hydrogen production yields. Heat shock treatment of sludge gave highest hydrogen yield (2 mol H_2/mol glucose), while base treatment of sludge gave lowest hydrogen yield (0.48 mol H_2/mol glucose) [74]. Sung et al. [71] illustrated that hydrogen production using heat-treated seeds declined after 1-month operation and repeated heat treatment of sludge to recover from reactor every month is not credible. However, others claim that high yields can be achieved without heat treatment [68]. Zhu and Beland [75] have demonstrated that heat shock and acid treatment methods completely repressed methanogenic activity, while base treatment methods did not completely repress methanogenic activity and also significantly affected hydrogen production. Hwang et al. [76] reported that the acidic conditions (pH 4.5–6) can act as a weak inhibitor, but not complete long-term inhibition of methanogenic activity. Elsewhere, it has been shown that acid pretreatment is particularly effective for enhancing the growth of lactic acid bacteria (LAB) [52, 77]. Five methods for preparation of hydrogen-producing seeds (base, acid, 2-bromoethanesulfonic acid (BESA), and load shock and heat shock treatments) as well as an untreated anaerobic digested sludge were evaluated for their hydrogen production performance and responsible microbial community structures under thermophilic conditions (60°C) by O-Thong et al. [24]. The results showed that the load shock treatment method was the best for enriching thermophilic hydrogen-producing seeds from mixed anaerobic cultures as it completely repressed methanogenic activity and gave a maximum hydrogen production yield of 1.96 mol H_2 mol^{-1} hexose with a hydrogen production rate of 11.2 mmol H_2 l^{-1} h^{-1}.

In general, microbial profiles in fermentative production processes occur as a result of a combination of process conditions, such as feedstock characteristics, environmental conditions (pH, temperature, and H_2 partial pressure), and metabolic pathways existing in the microbes involved [51]. Iyer et al. [51]

investigated hydrogen-producing bacterial communities from a heat-treated soil inoculum by RISA. They found that species of Clostridiaceae, Bacillaceae, and Enterobacteriaceae responded to hydrogen production at 30°C and a 30-h HRT. The gene pool at 30-h HRT, as determined by 16S rRNA gene sequences, was more diverse than at the 10-h HRT, as only Clostridiaceae were detected at this later point. The application of DGGE indicated *Clostridium tyrobutyricum*, *Lactobacillus ferintoshensis*, *Lactobacillus paracasei*, and *Coprothermobacter* species to be dominant in bacterial communities developed from pH-pretreated inocula [68]. *Lactobacillus* species are common coexisting bacteria in hydrogen fermentation processes. However, they have adverse effects on hydrogen production by competing for sugars and producing acidic products [78, 79]. Interference by lactic acid bacteria is often prevented by feedstock heat treatment at 50°C or by thermophilic fermentation at temperatures beyond 50°C [80]. Load shock and heat shock treatments under thermophilic conditions resulted in a dominance of *T. thermosaccharolyticum* while base- and acid-treated seeds were dominated by *Clostridium* and BESA-treated seeds were dominated by *Bacillus* sp. [24]. The comparative experimental results from hydrogen production performance and microbial community analysis showed that the load shock treatment method was better than base- and acid-treated, heat shock, BESA-treated methods for enriching thermophilic hydrogen-producing seeds from anaerobic-digested sludge. Load shock-treated sludge was implemented in palm oil mill effluent (POME) fermentation and was found to give maximum hydrogen production rates of 13.34 mmol H_2 $l^{-1}h^{-1}$ and resulted in a dominance of *Thermoanaerobacterium* spp. Load shock treatment is an easy and practical method for enriching thermophilic hydrogen-producing bacteria from anaerobic-digested sludge. The efficiency of preparation methods could be considered based on hydrogen production yield together with microorganisms revealed in the process. Therefore, the microbiological aspects and hydrogen production performance information are needed to identify effective methods for preparation of hydrogen-producing seeds.

4.2. Influence of reactor design and operation on the composition of microbial communities

Various reactor types seeded with the same inoculum and operating under similar process conditions could develop microbial communities with different properties. For instance, in batch mode under mesophilic conditions with glucose as a substrate, microbial communities became dominated by *Clostridium butyricum*-like species [51], *Clostridium* spp. [52] *C. butyricum* [81], and *Clostridium sp*_T5zd [77]. Conversely, a continuous stirred tank reactor (CSTR) was dominated by *Clostridium sporogenes*-like and *Clostridium celerecrescens*-like species [82]. Yet, in an anaerobic membrane reactor (MBR), the main population consisted of Clostridiaceae, Flexibacteraceae,

Clostridium acidisoli, Linmingia china, and *Cytophaga* [23]. *Clostridium* spp. were also dominant in a CSTR used to produce hydrogen from sucrose at 35°C, pH 5.5, and HRT 12 h, as analyzed by DGGE [51, 71]. Xing et al. [83] followed communities in a CSTR operating on molasses at a low pH with acidophilic bacteria from sewage, which established an ethanol–acetate hydrogen-producing community after 28 days. This was also consistent with other studies, i.e., the hydrogen production rate increased with the increase of *Ethanologenbacterium* sp., *Clostridium* sp., and *Spirochaetes.* Some types of *Clostridium* sp., *Acidovorax* sp., *Kluyvera* sp., and *Bacteriodes* were found throughout all periods of reactor operation [83]. It appeared that hydrogen production depended not only on hydrogen producers but also on cometabolism in the whole community.

In common with many other systems, in batch fermentations without pH control, it has been found that microbial communities change with pH [77], and their biodiversity decreased considerably as the pH decreased from 6.5 to 4.5. Kim et al. [81] reported the effect of substrate concentration on dark hydrogen fermentation using a CSTR. At the peak of hydrogen production yields, all bacterial species detected by DGGE analysis were *Clostridium* spp. and at inlet sucrose concentrations below 20 gCOD l^{-1}, the hydrogen yield per hexose consumed decreased, while *Clostridium scatologenes* (an H_2-consuming acetogen) was found in the sludge. Moreover, it has been shown that short HRT operation without anaerobic sludge preparation allowed for more microbial diversity and increasing the system robustness [22]. Species that differ in optimal growth conditions but are metabolically similar are then present, sharing the same function. Such advantages allow for flexibility in performance when perturbations in process conditions occur. Overall, under mesophilic conditions, hydrogen may be produced by a large group of bacteria such as the three main groups belonging to the low-GC(guanine-cytosine)gram positive bacteria, i.e., Clostridaceae, Enterobacteriacea, and Bacillaceae. A number of studies have focused on the analysis of the 16S rRNA gene to understand the species richness of microbial communities in lab-scale reactors under mesophilic conditions, as shown in Table 2.

Table 2. Microbial community structure, operational conditions and reactor performance of fermentative hydrogen production process from various organic wastes under mesophilic condition.

Substrate	Processes and operation condition	Dominating microorganisms	H_2 yield (mol H_2 mol^{-1} hexose)	H_2 rates (l H_2 $l^{-1}d^{-1}$)	References
Glucose	Batch experiment, pH 5.5 and 36°C	Clostridiaceae Enterobacteriaceae *Streptococcus bovis*	0.47	4.6	[52]

Substrate	Processes and operation condition	Dominating microorganisms	H_2 yield (mol H_2 mol-1 hexose)	H_2 rates (l H_2 l-1d-1)	References
Carbohydrate-containing wastewater	Two-step process using CSTR, pH 5.5, HRT 6 h, 36°C and complete-mix cylindrical photoreactor, HRT 25 h, pH 8.0, 32°C	*Clostridia* *Rhodobacter capsulatus*	2.1 and 2.5	4.5 and 0.3	[84]
Glucose	Anaerobic membrane reactor, HRT 3.3 h, pH 5.5, mixed at 200 rpm and 35°C	Clostridiaceae Flexibacteraceae *Clostridium acidisoli* *Linmingia chinaCytophaga*	1.1	15.36	[23]
Food waste	CSTR, HRT 5 d, pH 5.6 and 35°C	*Thermotogales* *Bacillus* spp. *Prevotella species*	0.03–0.1	0.22	[19]
Sucrose	CSTR, HRT 24 h, 37°C and pH 5.5	*Clostridium* sp. *Bacillus* sp.	2.3	0.1	[71]
Glucose	CSTR, pH 5.5 and 30°C at 30 h and 10 h HRT	Bacillaceae Clostridiaceae Enterobacteriaceae Only Clostridiaceae at HRT 10 h	1.61	10.4	[51]
Rice slurry	Batch experiment, pH 4.5 and 45°C	*Clostridium* sp. 44a-T5zd	2.5	2.1	[77]
Sucrose	CSTR, gas sparging at 300 ml/min, pH 5.3 and 35°C	*Clostridium tyrobutyricum* *Clostridium ptoteolyticum* *Clostridium acidisoli*	1.68	6.45	[81]
Sucrose	CSTR, HRT 12 h, pH 6.8 and 35°C	*Clostridium ramosum*	0.9–3.5	9.1	[85]
Glucose	CSTR; glucose to peptone ratio (5:3) 35°C, pH 7 and HRT 12 h	*Clostridium sporogenes* *Clostridium celerecrescens*	0.6	6.8	[82]
Glucose	Batch; glucose to peptone ratio (5:3) 35°C, pH 7 and HRT 12 h	*Clostridium butyricum*	1.11	5.2	[83]

Different microbial community structures develop within different temperature regimes. For instance, a comparative study on hydrogen production from food waste between mesophilic and thermophilic acidogenic conditions revealed that biogas produced in thermophilic conditions was methane free, whereas methane was still detected under mesophilic conditions [19]. Species such as *Thermoanaerobacterium thermosaccharolytium* and *Desulfotomaculum geothermicum* were detected in the thermophilic acidogenic culture, while *Clostridium* and *Bacillus* species were detected in the mesophilic acidogenic culture with DGGE. The composition of microbial communities in thermophilic dark hydrogen fermentation production was investigated in more detail using quantitative FISH and DGGE [8, 22, 86]. This demonstrated that *Thermoanaerobacterium* made up almost half of the total community in thermophilic dark hydrogen fermentation production.

In thermophilic fermentative hydrogen production, a number of microbial species are known, including *C. thermoamylolyticum* [84], *C. cellulose, C. thermocellum, T. thermosaccharolytium* [22, 84], *D. geothermicum* [19], *Saccharococcus* sp. clone ETV-T2 [8], *Mitsuokella jalaludinii* [84]. *Thermoanaerobacterium* sp. and the related genotypes are found to be dominant in many thermophilic fermentations operating at 55°C and neutral pH with feedstocks, including starch, organic waste [22], and cellulose-rich materials [84]. Thus, many studies on microbial consortia of thermophilic fermentations resulted in the detection of the same dominant species. This is in contrast to observations from mesophilic fermentations, and it might therefore indicate that thermophilic conditions lead to a convergence of microbial populations. In this way, thermophilic reactors can provide an additional benefit for the application in sludge population optimization. One of the problems of bioreactor operation is washout of microorganisms. Trickling biofilter reactors (TBR) have been proposed as a solution to this problem, with continuous hydrogen production under thermophilic conditions being successfully demonstrated [22, 23]. In those studies, the TBR was dominated by *T. thermosaccharolytium* and *Clostridia* and *Bacilli* in the phylum *Firmicutes.*

Microbial community structure dynamics in the ASBR for biohydrogen production from palm oil mill effluent during changing of hydraulic retention time (HRT) and organic loading rate (OLR) was studied by denaturing gradient gel electrophoresis (DGGE) aiming at improved insight into the hydrogen fermentation microorganisms. The microbial community structure was strongly dependent on the HRT and OLR. DGGE profiling illustrated that *Thermoanaerobacterium* spp., such as *T. thermosaccharolyticum,* and *T. bryantii,* were dominant and probably played an important role in hydrogen production under thermophilic conditions. The shift in the microbial community from a dominance of *T. thermosaccharolyticum* to a community where *Caloramator proteoclasticus* also constituted a major component occurred at suboptimal HRT (1 d) and OLR (80 gCOD $l^{-1}d^{-1}$) conditions [25]. The information showed that the hydrogen production performance was closely correlated with the bacterial community structure. A number of studies have focused on the analysis of the 16S rRNA gene to understand the species richness of microbial communities in lab-scale reactors under thermophilic conditions, as shown in Table 3.

Table 3. Microbial community structure, operational conditions and reactor performance of fermentative hydrogen production process from various organic wastes under thermophilic condition.

Substrate	Processes and operation condition	Dominating microorganisms	H_2 yield (mol H_2 mol^{-1} hexose)	H_2 rates (l H_2 l^{-1}d^{-1})	References
Glucose	Fed batch experiment, HRT 0.5 d, pH 6.6, and 60°C	*Thermoanaerobacterium thermosaccharolyticum* KU-001	2.4	3.5	[70]
Cellulose	Batch experiment, stirring at 200 rpm, pH 6.4, and 60°C	*Clostridium* and *Bacillus* *T. thermosaccharolyticum* *Clostridium thermocellum* *Clostridium cellulosi*	2.0	1.35	[87]
Starch in wastewater	Batch experiment, pH 6.0 and 55°C	*Thermoanaerobacteriaceae* *Saccharococcus* sp. clone ETV-T2	0.68	2.8	[8]
Cellulose	Batch experiment, pH 6.5 and 55°C	*Thermoanaerobacterium* *Clostridium thermoamylolyticum*	0.2	0.82	[84]
Food waste	CSTR, HRT 5 d, pH 5.6, and 55°C	*Thermoanaerobacterium thermosaccharolytium* *Desulfotomaculum geothermicum*	0.9–1.8	4.56	[19]
Glucose	Trickling biofilter reactor (TBR), HRT 2 h and 55–64°C	*T. thermosaccharolyticum*	1.1	23.25	[22]
Food waste	CSTR; HRT 5 d, pH 5.5, and 55°C	*T. thermosaccharolyticum*	2.2	1.4	[86]
Artificial garbage slurry	Jar fermentor; HRT 1d, pH 6.0, and 60°C	*T. thermosaccharolyticum*	1.99	4.46	[88]

In addition to volatile fatty acids (VFAs), anaerobic fermentations may also lead to the production of reduced end products such as ethanol, butanol, and lactate [5], thus reducing H_2 yield potential. Therefore, bacterial metabolism must avoid

VFAs by efficient product removal [7, 8] or metabolic engineering. Stripping gas may be used to remove H_2 from the liquid phase to prevent product inhibition. N_2 is used often, but it increases the costs of H_2 purification. For economical reasons, CO_2 might be a better choice, as it is relatively easy to separate from the gas phase. Using CO_2 rather than N_2 for stripping H_2 resulted in a higher production of H_2 and butyrate [79, 89]. High CO_2 partial pressures had little effect on hydrogen-producing bacteria but were inhibitory to other competitive microorganisms such as acetogens and lactic acid bacteria. The microbial community structure under CO_2 sparging conditions was dominated by *C. tyrobutylicum*, *C. proteolyticum*, and *C. acidisoli*. CO_2 sparging has another beneficial effect on reactor performance by improving mixing and contact between substrate and microorganisms and also decreased the effects of hydrogen partial pressure [89, 90].

4.3. Microbial key players in dark hydrogen fermentation

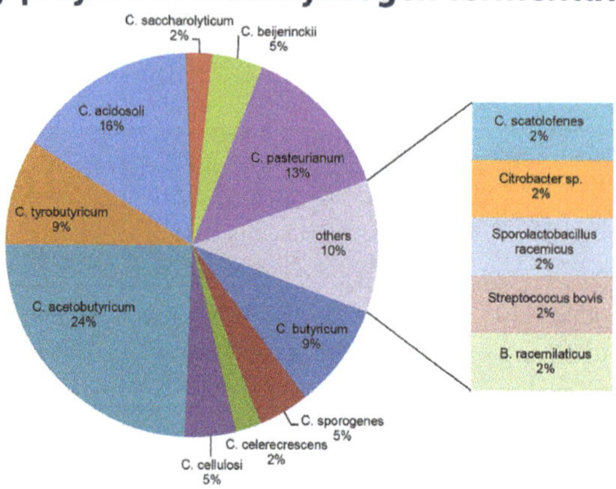

Figure 1. Summary of all fermentative hydrogen-producing bacteria frequently observed based on molecular tools studied under mesophilic conditions.

Figure 1 summarizes the richness of the microbial key players of mesophiles. Fermentative hydrogen production has been studied for a large group of pure cultures, including species of *Enterobacter*, *Bacillus*, and *Clostridium*. However, hydrogen-producing microflora obtained from natural sources, which are able to survive on non-sterile substrates, contain mostly *Clostridium* spp., such as *C. butyricum*, *C. acidosoli*, *C. tyrobutylicum*, and *C. acetobutylicum*. Although the numbers of case studies are still low to infer solid conclusions, they indicate that the *Clostridium* genus represents the major group in dark mesophilic fermentation under mesophilic conditions. Various *Clostridium* species are found in mesophilic environments, but only four species are highly frequently observed (*C. acetobutylicum* (24%), *C. tyrobutyricum* (9%), *C. acidisoli* (16%), and *C. pasteurianum* (13%)) and related with high hydrogen yield [52, 81]. However, *C. saccharolyticum, C. butyricum, C. sporogenes, C. celerecrescens, C. cellulosi*, and *C. beijerinkii* were also found to be strong hydrogen producers [83]. Others species (*Citrobacter sp., Sporolactobacillus racemicus, Streptococcus bovis,* and *B. racemilaticus*) that differ in optimal growth conditions from *Clostridium* but are metabolically similar are allow for flexibility in performance when perturbations in process conditions occur.

The *Thermoanaerobacterium* genus represents the major group in dark thermophilic fermentation. Figure 2 summarizes the richness of the microbial key players of thermophiles. Thermophilic conditions clearly show that *T. thermosaccharolyticum* is a key player in fermentative hydrogen production. *Thermoanaerobacterium* spp. have also been found to dominate in a long-term hydrogen production reactor. Bacteria species are highly frequently observed under thermophilic conditions and they are *Thermoanaerobacterium* sp. (47%) and *T. thermosaccharolyticum* (30%). The microbial community structure of thermophilic mixed culture sludge used for biohydrogen production from palm oil mill effluent was analyzed by fluorescence in situ hybridization (FISH) and 16S rRNA gene clone library techniques. The microbial community was dominated by *Thermoanaerobacterium* species (~66%). The remaining microorganisms belonged to *Clostridium* and *Desulfotomaculum* spp. (~28% and ~6%, respectively). The hydrogen-producing bacteria were isolated and their ability to produce hydrogen was confirmed. Three hydrogen-producing strains, namely HPB-1, HPB-2, and HPB-3, were isolated. The 16S rRNA gene sequence analysis of HPB-1 and HPB-2 revealed a high similarity to *T. thermosaccharolyticum* (98.6% and 99.0%, respectively). The *Thermoanaerobacterium* sp. HPB-2 strain was a promising candidate for thermophilic fermentative hydrogen production with a hydrogen yield of 2.53 mol H_2 mol^{-1} hexose from organic waste and wastewater containing a mixture of hexose and pentose sugars. *Thermoanaerobacterium* species play a major role in thermophilic hydrogen production as confirmed by both molecular and cultivation-based analyses [91]. Various Clostridium species (*C. cellulose, C. thermoamyloticum, and C. thermocellum*) that differ in optimal growth conditions from *Thermoanaerobacterium* but are metabolically similar are allow for flexibility in performance when perturbations in process conditions occur. Other species (*Saccharococcus spp., D. geothermicum*, and *Bacillus* spp.) could allowed for more microbial diversity and increasing the system robustness.

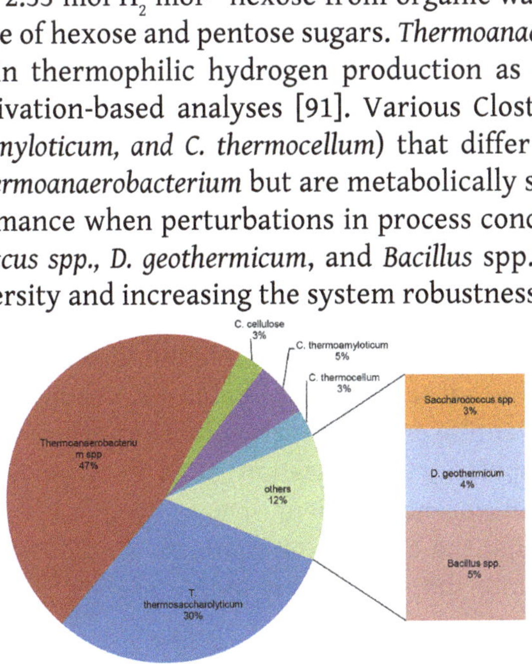

Figure 2. Summary of all fermentative hydrogen producing bacteria frequently observed based on molecular tools studied under thermophilic conditions.

5. OPTIMIZING THE MICROBIAL POPULATION FOR DARK HYDROGEN FERMENTATION

Different species likely possess different growth properties (growth rates, affinity constants with substrates, and yields), and perhaps different capacities in coping with stress arising from variations in growth conditions. Obviously, the species with the most desirable properties would be selected to perform a required function. The possibility of selecting species with better properties has huge potential for improving the performance (efficiency and reliability) of a DHF system. Unfortunately, we still lack knowledge concerning the species to be selected and how they may be selected. Furthermore, 16S rRNA sequence-based identification does not allow inference of functional properties. The correlation between microbial community composition and reactor performance would provide a rationale to further improve the efficiency of fermentative hydrogen production. The characterization of the microbial community as a whole contributes to meaningful data regarding structure and function of such communities and their activities.

The interest in hydrogen as a clean energy carrier has strongly increased recently. Cost-effective generation of hydrogen through fermentation will have an important role in making this idea a reality. Future dark hydrogen fermentation from organic wastes depends on a thorough understanding of the microbiological community structure and function for enhanced or controllable hydrogen production and reactor. Sludge population optimization aims to obtain the best performance of a system through maximizing the properties of the sludge such as kinetics, yields, and robustness to environmental disturbance. A systematic investigation on the effects of a number of operational conditions on fermentative hydrogen production community and their properties is essential for sludge population optimization. The operational parameters to be studied include pH, temperature, hydraulic retention time, sludge retention time, organic loading rate, and nutrient concentration.

Additional improvements of microbial communities should be considered such as creating conditions that select for the stable and productive growth of desired microbes, while preventing or limiting growth of organisms that reduce hydrogen yields. Microbial population optimization could be achieved by biostimulation using the additive of various nutrient species specifically for the community, bioaugmentation using the additive of dominant species or efficient hydrogen-producing bacteria into the system, and online process control for maintaining their community.

A successful selection of such organisms, in particular those responsible for hydrogen production, will be used for recovery from off-set reactors by bioaugmentation strategy. To achieve high and stable hydrogen yield and long-term operation, it is necessary to control the growth of undesirable microorganisms such as hydrogen-consuming bacteria, propionic acid bacteria, and lactic acid bacteria via pH adjustment and reducing of H_2 partial pressure. The absence of hydrogen-consuming bacteria leads to relatively high hydrogen concentrations in the biogas and would significantly reduce costs for gas purification. Enhancement of hydrogen-producing bacteria via specific nutrient supplements will improve the reliability and performance of the process. Sludge population optimization strategies under thermophilic conditions shown in Figure 3.

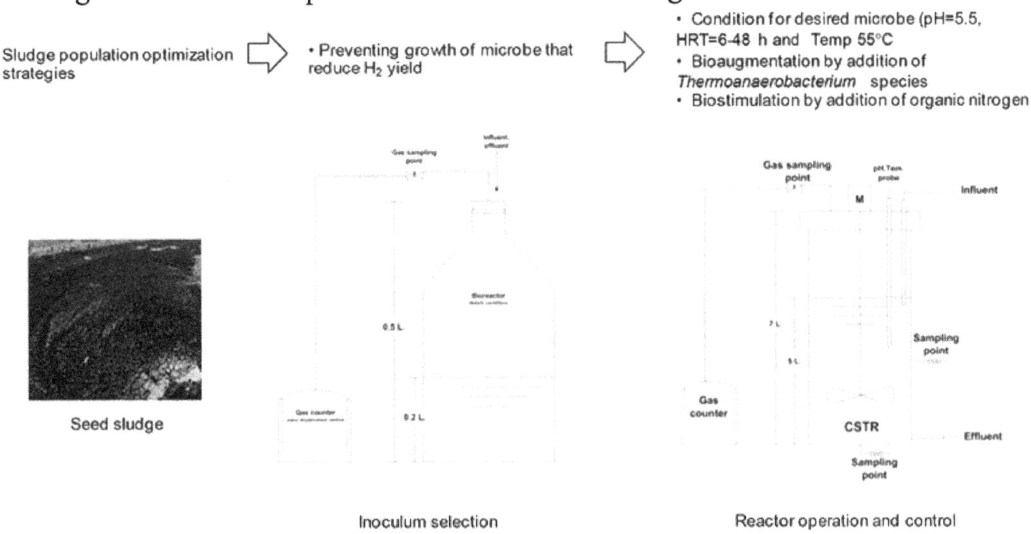

Figure 3. Summary of sludge population optimization strategies under thermophilic conditions.

6. POTENTIAL AVENUES FOR FUTURE DEVELOPMENT

Hydrogen's utilization as an environmentally friendly energy transporter has recently garnered significant attention. The economical production of hydrogen by fermentation will play a significant part in this undertaking. The future development of dark fermentation hydrogen (DHF) from organic wastes relies on the specific composition and activity of the microbial population to achieve improved and manageable hydrogen production in the reactor. The objective of sludge population optimization is to achieve optimal system performance by maximizing the characteristics of the sludge, such as kinetics, yields, and resilience to environmental disturbances. Conducting a methodical study on the impact of various operational factors on the community and characteristics of fermentative hydrogen generation is crucial for optimizing the sludge population. The factors that affect the function of microbial species being researched are pH, temperature, hydraulic retention time, sludge retention time, organic loading rate, and nutrient content. Further enhancements to microbial communities should be taken into account, namely by establishing conditions that favor the consistent and efficient development of desired microorganisms, while inhibiting or restricting the growth of organisms that would decrease hydrogen yields. Microbial population optimization can be achieved through biostimulation, which involves adding specific nutrients to support their community, bioaugmentation by introducing dominant species or efficient hydrogen-producing bacteria, and implementing online process control to sustain their community. The successful identification and inclusion of these species, namely those that are responsible for hydrogen production, will be beneficial for the restoration of the reactor through the employment of a bioaugmentation method. In order to attain a substantial amount of hydrogen production and ensure continuous operation, it is crucial to regulate the proliferation of unwanted microorganisms, including hydrogen-consuming bacteria, propionic acid bacteria, and lactic acid bacteria, by adjusting the pH and reducing the pH2 levels. The lack of hydrogen-consuming bacteria results in elevated hydrogen concentrations in the biogas, which would greatly decrease expenses for gas purification. The dependability and efficacy of the hydrogen production process can be enhanced by providing particular nutrient supplements to the microorganisms involved.

REFERENCES

1. Nielsen, A. T., Amandusson, H., Bjorklund, R., Dannetum, H., Ejlertsson, J., Ekedahl, L. G., Lundstrom, I. and Svensson, B. H. 2001. Hydrogen production from organic waste. Int. J. Hydrogen. Energy. 26: 547–550.
2. Das, D. and Veziroglu, T. N. 2001. Hydrogen production by biological processes: a survey of literature. Int. J. Hydrogen. Energy. 26: 13–28.
3. Elam, C.C., Padro, C.E.G., Sandrock, G., Luzzi, A., Lindbland, P. and Hagen, E.F.2003. Realizing the hydrogen future: the International Energy Agency's efforts to advance hydrogen energy technologies. Int. J. Hydrogen. Energy. 28: 601–607.
4. Nath, K. and Das, D. 2004. Improvement of fermentative hydrogen production: various approaches. Appl. Microbiol. Biotechnol. 65: 520–529.
5. Levin, D. B., Pitt, L. and Love, M. 2004. Biohydrogen production: prospects and limitations to practical application. Int. J. Hydrogen. Energy. 29: 173–185.
6. Ren, N., Li, Y., Wang, A., Li, J., Ding, J. and Zadsar, M. 2006. Hydrogen production by fermentation: Review of a new approach to environmentally safe energy production. Aquat. Eco. Health. Manage. 9: 39–42.
7. Hawkes, F. R., Dinsdale, R., Hawkes, D. L. and Hussy, I. 2002. Sustainable fermentative hydrogen production: challenges for process optimization. Int. J. Hydrogen Energy. 27: 1339–1347.

8. Zhang, T., Liu, H. and Fang, H. H. P. 2003. Biohydrogen production from starch in wastewater under thermophilic condition. J. Environ. Manage. 69: 149–156.

9. Angenent, L., Karim, L., Al-Dahhan, M., Wrenn, B. and Domiguez-Espinosa, R. 2004. Production of bioenergy and biochemicals from industrial and agricultural wastewater. TRENDS Biotechnol. 22: 477–485.

10. Montgomery, R. 2004. Development of biobased products. Bioresour. Technol. 91: 1–29.

11. Claassen, P. A. M., Van Lier, J. B., Lopez Contraras, A. M., Van Niel, E. W. J., Sijtsma, L., Stams, A. J. M., de Vries, S. S. and Weusthuis, R. A. 1999. Utilization of biomass for the supply of energy carriers. Appl. Microbiol. Biotechnol. 52: 741–755.

12. Kapdan, I.K. and Kargi, F. 2006. Bio-hydrogen production from waste materials. Enzyme Microb. Technol. 38: 569–582.

13. Ren, N., Qin, Z. and Li, J. 2003. Comparison and analysis of hydrogen production capacity with different acidogenic fermentative microflora. Huan Jing Ke Xue. 24 (1): 70–74.

14. Liu, W., Chan, O. And Fang, H. 2002. Microbial community dynamics during start-up of acidogenic anaerobic reactors. Water Res. 36: 3203–3210.

15. Van Niel, E. W. J., Budde, M. A. W., de Haas, G. G., van de Wal, F. J., Claassen, P. A. M. and Stams, A. J. M. 2002. Distinctive properties of high hydrogen producing extreme thermopiles, Cadicellulosiruptor saccharolyticus and Thermotoga elfii. Int. J. Hydrogen. Energy. 27: 1391–1398.

16. Van Niel, E. W. J., Claassen, P. A. M. and Stams, A. J. M. 2003. Substrate and product inhibition of hydrogen production by the extreme thermophile, Caldicellulosiruptor saccharolyticus. Biotechnol. Bioeng. 81: 254–262.

17. Hallenbeck, P.C., Ghosh, D., 2009. Advances in fermentative biohydrogen production: the way forward? Trends Biotechnol. 27 (5), 287–297.

18. Logan, B. 2005. Production of electricity from acetate or butyrate using a single-chamber microbial fuel cell. Environ. Sci. Technol. 39: 658–662.

19. Shin, H. S., Youn, J. H. and Kim, H. S. 2004. Hydrogen production from food waste in anaerobic mesophilic and thermophilic acidogenesis. Int. J. Hydrogen Energy. 29: 1355–1363.

20. Nandi, R. and Sengupta, S. 1998. Microbial production of hydrogen: an overview. Crit. Rev. Microbiol. 24 (1): 61–84.

21. Hawkes, F. R., Hussy, I., Kyazze, G., Dinsdale, R. and Hawkes, D. L. 2007. Continuous dark fermentative hydrogen production by mesophilic microflora: Principles and progress. Int. J. Hydrogen Energy. 32 (2): 172–184.

22. Ahn, Y., Park, E. J., Oh, Y. K., Park, S., Webster, G. and Weightman, A. J. 2005. Biofilm microbial community of a thermophilic trickling biofilter used for continuous biohydrogen production. FEMS Microbiol. Lett. 249: 31–38.

23. Oh, S. E., Lyer, P., Bruns, M. A. and Logan, B. E. 2004. Biological hydrogen production using a membrane bioreactor. Biotechnol. Bioeng. 87(1): 119–127.

24. O-Thong, S., Prasertsan, P. and Birkeland, N.K. 2009. Evaluation of methods for preparing thermophilic hydrogen producing seeds from anaerobic digested sludge with microbial communities aspects. Bioresour. Technol. 100: 909–918.

25. Prasertsan, P., O-Thong, S. and Birkeland, N.K. 2009. Optimization and microbial community analysis for production of biohydrogen from palm oil mill effluent by thermophilic fermentative process. Int. J. Hydrogen Energy. 34: 7448–7459.

26. O-Thong, S., Prasertsan, P., Intrasungkha, N., Dhamwichukorn, S. and Birkeland, N.K. 2007. Improvement of biohydrogen production and pollution reduction from palm oil mill effluent with nutrient supplementation at thermophilic condition using an anaerobic sequencing batch reactor. Enzyme Microb Technol. 41: 583–590.

27. Mamimin, C., Singkhala, A., Kongjan, P., Suraraksa, B., Prasertsan, P., Imai, T. and O-Thong, S. 2015. Two-stage thermophilic fermentation and mesophilic methanogen process for biohythane production from palm oil mill effluent. Int. J. Hydrogen Energy. 40(19): 6319–6328.

28. Mamimin, C., Thongdumyu, P., Hniman, A., Prasertsan, P., Imai, T., O-Thong, S. (2012) Simultaneous thermophilic hydrogen production and phenol removal from palm oil mill effluent by Thermoanaerobacterium-rich sludge. Int. J. Hydrogen Energy. 37(20):15598–15606.

29. Stams A. 1994. Metabolic interactions between anaerobic bacteria in methanogenic environments. Antonie van Leeuwenhoek. 66(1), 271–294.

30. Kengen S.W.M., Goorissen H.P., Verhaart, M.; Stams, A.J.M.; van Niel, E.W.J.; Claassen, P.A.M. 2009. Biological hydrogen production by anaerobic microorganisms. In Biofuels; John Wiley & Sons, Ltd.: Chichester, UK: 197–221.

31. Bothun, G., Knutson, B., Berberich, J., Strobel, H., Nokes, S. 2004. Metabolic selectivity and growth of Clostridium thermocellum in continuous culture under elevated hydrostatic pressure. Appl. Microbiol. Biotechnol. 65(2), 149–157.

32. Ivanova G, Rákhely G, Kovács K. 2009. Thermophilic biohydrogen production from energy plants by Caldicellulosiruptor saccharolyticus and comparison with related studies. Int. J. Hydrogen Energy. 34(9): 3659–3670.

33. O-Thong S, Prasertsan P, Karakashev D, Angelidaki I. 2008. Thermophilic fermentative hydrogen production by the newly isolated Thermoanaerobacterium thermosaccharolyticum PSU-2. Int. J. Hydrogen Energy. 33(4), 1204–1214.

34. Yokoyama H, Moriya N, Ohmori H, Waki M, Ogino A, Tanaka Y. 2007. Community analysis of hydrogen-producing extreme thermophilic anaerobic microflora enriched from cow manure with five substrates. Appl. Microbiol. Biotechnol. 77(1), 213–222.

35. Kleerebezem R, van Loosdrecht M. 2007. Mixed culture biotechnology for bioenergy production. Curr. Opin. Biotechnol. 18(3), 207–212.

36. Temudo M, Muyzer G, Kleerebezem R, van Loosdrecht M. 2008. Diversity of microbial communities in open mixed culture fermentations: impact of the pH and carbon source. Appl. Microbiol. Biotechnol. 80(6), 1121–1130.

37. Ozmihci S, Kargi F. 2010. Comparison of different mixed cultures for biohydrogen production from ground wheat starch by combined dark and light fermentation. J. Ind. Microbiol. Biotechnol. 37:341–347.

38. Demirel B, Scherer P, Yenigun O, Onay T. 2010. Production of Methane and Hydrogen from Biomass through Conventional and High-Rate Anaerobic Digestion Processes. Crit. Rev. Environ. Sci. Technol. 40(2), 116–146.

39. Li C, Fang H. 2007. Fermentative hydrogen production from wastewater and solid wastes by mixed cultures. Crit. Rev. Environ. Sci. Technol. 37(1), 1–39.

40. Arooj M, Han S, Kim S, Kim D, Shin H. 2008. Continuous biohydrogen production in a CSTR using starch as a substrate. Int. J. Hydrogen Energy. 33(13), 3289–3294.

41. Gavala H, Skiadas I, Ahring B. 2006. Biological hydrogen production in suspended and attached growth anaerobic reactor systems. Int. J. Hydrogen Energy. 31(9), 1164–1175.

42. Duangmanee T, Padmasiri S, Simmons J, Raskin L, Sung S. 2007. Hydrogen production by anaerobic microbial communities exposed to repeated heat treatments. Water Environ. Res. 79, 975–983.

43. Kaparaju P, Serrano M, Angelidaki I. 2009. Effect of reactor configuration on biogas production from wheat straw hydrolysate. Bioresour. Technol. 100, 6317–6323.

44. Luo G, Xie L, Zou Z, Wang W, Zhou Q. 2010. Evaluation of pretreatment methods on mixed inoculum for both batch and continuous thermophilic biohydrogen production from cassava stillage. Bioresour. Technol. 101,959-964.

45. Valdez-Vazquez, I., Rios-Leal, E., Esparza-Garcia, F., Cecchi, F. and Poggi-Varaldo, H. M. 2005. Hydrogen production from the organic fraction of

municipal solid waste using semi-continuous solid substrate anaerobic reactors: influence of the temperature. Int. J. Hydrogen Energy. 30 (13–14): 1383–1391.

46. Amann, R.I., Fuchs, B.M., Behrens, S. 2001. The identification of microorganisms by fluorescence in situ hybridization. Curr. Opin. Microbiol. 12: 231–236.
47. Lee, N., Nielsen, P.H., Andreasen, K.H., Juretschko, S., Nielsen, J.L., Schleifer, K.H., Wagner, M. 1999. Combination of fluorescent in situ hybridization and microautora diography-a new tool for structure-function analyses in microbial ecology. Appl. Environ. Microbiol. 65: 1289–1297.
48. Radajewski, S., McDonald, I.R., Murrell, J.C. 2003. Stable-isotope probing of nucleic acids: a window to the function of uncultured microorganisms. Curr. Opin. Biotechnol. 14: 296–302.
49. Muyzer, G., de Waal, E.C., Uitterlinden, A.G. 1993. Profiling of complex microbial populations by denaturing gradient gel electrophoresis analysis of polymerase chain reaction amplified genes coding for 16S rRNA. Appl. Environ. Microbiol. 59: 695–700.
50. Daims, H., Taylor, M.W., Wagner, M. 2006. Wastewater treatment: a model system for microbial ecology. Trends Biotechnol. 24(11): 483–489.
51. Iyer, P., Bruns, M.A., Zhang, H., Van Ginkel, S., Logan, B.E. 2004. Hydrogen producing bacterial communities from a heat-treated soil inoculum. Appl. Microbiol. Biotechnol. 66:166–173.
52. Fang, H., Zhang, T. and Liu, H. 2002. Microbial diversity of a mesophilic hydrogen-producing sludge. Appl. Microbiol. Biotechnol. 58: 112–118.
53. Jenssen, T.K., Laegreid, A., Komorowski, J., Hovig, E. 2001. A literature network of human genes for high-throughput analysis of gene expression. Nat. Genet. 28: 21–28
54. van Lier, J.B., Tilche, A., Ahring, B.K., Macarie, H., Moletta, R., Dohanyos, M., Pol, L.W., Lens, P., Verstraete, W. 2001. New perspectives in anaerobic digestion. Water Sci. Technol. 43 (1): 1–18.
55. Wilmes, P. and Bond, P. 2006. Metaproteomics studying functional gene expression in microbial ecosystems. Trends Microbiol. 14: 92–97.
56. Wilmes, P. and Bond, P. 2006. Towards exposure of elusive metabolic mixed-culture processes: the application of metaproteomic analyses to activated sludge. Water Sci. Technol. 54:217–226.
57. Burja, A.M., Dhamwichukorn, S., Wright, P.C. 2003. Cyanobacterial postgenomic research and systems biology. Trends Biotechnol. 21: 504–511.

58. Ram, R.J., VerBerkmoes, N.C., Thelen, M.P., Tyson, G.W., Baker, B.J., Blake, R.C., Shah, M., Hettich, R.L., Banfield, J.F. 2005. Community proteomics of a natural microbial biofilm. Science. 308: 1915–1920.

59. Banfield, J.F., Verberkmoes, N.C., Hettich, R.L., Thelen, M.P. 2005. Proteogenomic approaches for the molecular characterization of natural microbial communities. OMICS. 9:301–333.

60. Okamoto, M., Miyahara, T., Mizuno, O. and Noike, T. 2000. Biological hydrogen potential of materials characteristic of the organic fraction of municipal solid wastes. Water. Sci. Technol. 41: 25–32.

61. Yu, H., Z. Zhu, W. Hu, and H. Zhang. 2002. Hydrogen production from rice winery wastewater in an up-flow anaerobic reactor by using mixed anaerobic cultures. Int. J. Hydrogen Energy. 27: 1359–1365.

62. Van Ginkel, S.W. and Logan, B. 2005. Increased biological hydrogen production with reduced organic loading. Water Res. 39: 3819–3826.

63. Kataoka, N., Miya, A. and Kiriyama, K. 1997. Studies on hydrogen production by continuous cultures system of hydrogen producing anaerobic bacteria. Water. Sci.Techol. 36: 41–47.

64. Wang, C. C., Chang, C. W., Chu, C. P. and Lee, D. J. 2003. Sequential production of hydrogen and methane from wastewater sludge using anaerobic fermentation. J. Chin. Inst. Chem. Engrs. 34(6): 683–687.

65. Chang, J. S., Lee, K. S. and Lin, P. J. 2002. Biohydrogen production with fixed bed bioreactors. Int. J. Hydrogen Energy. 27: 1167–1174.

66. Kotsopoulos, T., Zeng, R. and Angelidaki, I. 2006. Biohydrogen production in granular up-flow anaerobic sludge blanket (UASB) reactors with mixed cultures under hyper-thermophilic temperature (70 degrees C). Biotechnol. Bioeng. 94(2): 296–302.

67. Fan, K. S., Kan, N. R. and Lay J. J. 2006. Effect of hydraulic retention time on anaerobic hydrogenesis in CSTR. Bioresour. Technol. 97: 84–89.

68. Kawagoshi, Y., Hino, N., Fujimoto, A., Nakao, M., Fujita, Y., Sugimura, S. And Furukawa, K. 2005. Effect of inoculum conditioning on hydrogen fermentation and pH effect on bacterial community relevant to hydrogen production. J. Biosci. Bioeng. 100: 524–530.

69. Hussy, I., F. Hawkes, R. Dinsdale, and D. Hawkes. 2003. Continuous fermentative hydrogen production from a wheat starch co-product by mixed microflora. Biotechnol. Bioeng. 84: 619–626.

70. Ueno, Y., Haruta, S., Ishii, M. and Igarashi, Y. 2001. Characterization of a microorganism isolated from the effluent of hydrogen fermentation by microflora. J. Biosci. Bioeng. 92(4): 397–400.

71. Sung, S., Raskin, L., Duangmanee, T., Padmasiri, S. and Simmons, J. J. 2002. Hydrogen production by anaerobic microbial communities exposed to repeated heat treatments. Proc. of the 2002 U. S. DOE Hydrogen Program Review.

72. Hasyim, R., Imai, T., Reungsang, A., O-Thong, S. (2011) Extreme-thermophilic biohydrogen production by an anaerobic heat treated digested sewage sludge culture. Int. J. Hydrogen Energy. 36 (14): 8727–8734.

73. Liu, D., R. Zeng, and I. Angelidaki. 2006. Hydrogen and methane production from household solid waste in the two-stage fermentation process. Water Res. 40(11): 2230–2236.

74. Mu, Y., Wang, G. and Yu, H. Q. 2005. Kinetic modeling of batch hydrogen production process by mixed anaerobic cultures. Bioresour. Technol. 97(11): 1302–1307.

75. Zhu, H., and M. Beland. 2006. Evaluation of alternative methods of preparing hydrogen producing seeds from digested wastewater sludge. Int. J. Hydrogen Energy. 31: 1980–1988.

76. Hwang, M.H., N.J. Jang, S.H. Hyun, and I.S. Kim. 2004. Anaerobic bio-hydrogen production from ethanol fermentation: the role of pH. J. Biotechnol. 111: 297–309.

77. Fang, H. H. P., Li, C. and Zhang, T. 2005. Acidophilic biohydrogen production from rice slurry. Int. J. Hydrogen Energy. 31: 683–692.

78. Lay, J. J., Fan, K. S., Chang, J. I. and Ku, C. H. 2003. Influence of chemical nature of organic wastes on their conversion to hydrogen by heat-shock digested sludge. Int. J. Hydrogen Energy. 28: 1361–1367.

79. Kawano, T., Wada, K., Li, Y.Y. and Noike, T. 2004. Effect of substrate concentration and pH on hydrogen fermentation of mixed substrate by microflora. J. Soc. Water. Environ. 27: 473–479.

80. Noike, T., Takabatake, H., Mizuno, O. and Onhba M. 2002. Inhibition of hydrogen fermentation of organic wastes by lactic acid bacteria. Int. J. Hydrogen Energy. 27: 1367–1371.

81. Kim, D. H., Han, S. K., Kim, S. H., Shin, H. S. 2006. Effect of gas sparging on continuous fermentative hydrogen production. Int. J. Hydrogen Energy. 31: 2158–69.

82. Wang, X., Hoefel, D., Saint, C.P., Monis, P.T. and Jin, B. 2007. The isolation and microbial community analysis of hydrogen producing bacteria from activated sludge. Appl. Microbiol. 103(5):1415–1423.

83. Xing, D., Ren, N., Li, Q., Lin, M., Wang, A., Zhao, L. 2006. Ethanoligenens harbinense gen nov., sp. Nov., isolated from molasses wastewater. Int. J. Syst. Evol. Microbiol. 56(4):755–760.

84. Liu, H., Zhang, T. and Fang, H. H. P. 2003. Thermophilic H_2 production from a cellulose-containing wastewater. Biotechnol. Lett. 25: 365–369.

85. Lin, C., Lee, C., Tseng, I. and Shiao I. 2006. Biohydrogen production from sucrose using base-enriched anaerobic mixed microflora. Process Biochem. 41: 915–919.

86. Shin, H. S. and Youn, J. H. 2005. Conversion of food waste into hydrogen by thermophilic acidogenesis. Biodegradation. 16: 33–44.

87. Ueno, Y., Haruta, S. and Igarashi, Y. 2001. Microbial community in anaerobic hydrogen producing microflora enriched from sludge compost. Appl. Microbiol. Biotechnol. 57: 555–562.

88. Ueno, Y., Sasaki, D., Fukui, H., Haruta, S., Ishii, M., Igarashi, Y. 2006. Changes in bacterial community during fermentative hydrogen and acid production from organic waste by thermophilic anaerobic microflora. Appl. Microbiol. 101: 331–343.

89. Mizuno, O., Dinsdale, R., Hawkes, D. L. and Noike, T. 2000. Enhancement of hydrogen production from glucose by nitrogen gas sparging. Bioresour. Technol. 73: 59–65.

90. Mandal, B., Nath, K. and Das, D. 2006. Improvement of biohydrogen production under decreased partial pressure of H_2 by Enterobacter cloacae. Biotechnol. Lett. 28: 831–835.

91 Nitipan, S., Mamimin, C., Intrasungkha, N., Birkeland, N.K. and O-Thong, S. 2014. Microbial community analysis of thermophilic mixed culture sludge for biohydroge.

08 APPLICATION OF BIOSENSORS IN FERMENTATION

1. INTRODUCTION: AN OVERVIEW

Biosensor is an interdisciplinary field that draws on theoretical and technical foundations from several disciplines such as biological science, physics, analytical chemistry, and information technology. Biosensors are a field within analytical biology. The primary objective is to develop efficient, reliable, and user-friendly analytical instruments and associated analytical technologies. Biosensors are a revolutionary analytical technology that possess tiny dimensions, high sensitivity, high analytical specificity, and quick accessibility, making them suitable for reagentless studies. This technology has made significant progress and garnered attention since its inception in the 1960s. In 1962, Professor Clark and his colleagues successfully created and utilized a biosensor, namely a biological electrode, to accurately measure medical information. Figure 1 depicts a schematic of it. The electrode is created by attaching a layer of immobilized glucose oxidase (glucose oxidase, EC 1.1.3.4) membrane to an ion selective electrode that can measure dissolved oxygen concentrations. During the process of work, glucose oxidase acts as a catalyst to facilitate the transformation of the substrates β-D-glucose (which is present in the solution) and molecular oxygen into gluconic acid and hydrogen peroxide (H2O2), which are the resulting products. The electrode can detect changes in oxygen levels in the environment as it is consumed by the enzymatic reaction. It then converts this information into voltage signals, which can be used to calculate the concentration of glucose. This is possible because the glucose concentration is directly proportional to the concentration of dissolved oxygen within a specific range. This technology combines the precise specificity of enzymatic reactions with the strong detection capabilities of electrochemical electrodes, allowing for the monitoring of biological reactions using a physicochemical detector. Following this demonstration, biosensors have become highly discussed subjects among researchers worldwide. Biosensors have now become prevalent in various forms of commercially available goods, serving both practical uses and scientific research.

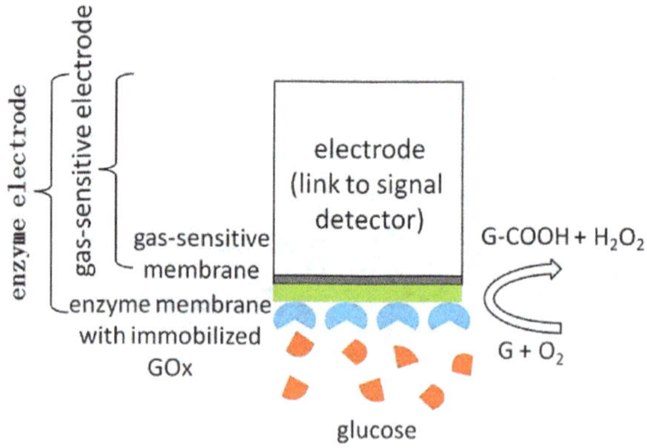

Figure 1. A schematic representation of the classic Clark enzyme electrode with glucose oxidase (Gox) as its biological element.

1.1. Classification of biosensors

As is recommended by IUPAC in 1999, a biosensor is an independently integrated receptor transducer device, which is capable of providing selective quantitative or semiquantitative analytical information using a biological recognition element [1].

A typical biosensor is made up of two main parts: (a) a biological element that can give any form of detectable signals, enzymatically catalyzed reactions, and biomolecular recognitions are the most referred, among others. Enzymes, antibodies, and nucleic acids are often exploited as the biological element; (b) a transducer by which the signal produced by the biological element can be detected and converted into measurable electrical signal.

To construct a biosensing system, three main elements are often required, they are as follows: a biological element, a transducer, and a signal processing system [2]. A schematic of a typical biosensor system can be seen in Figure 2.

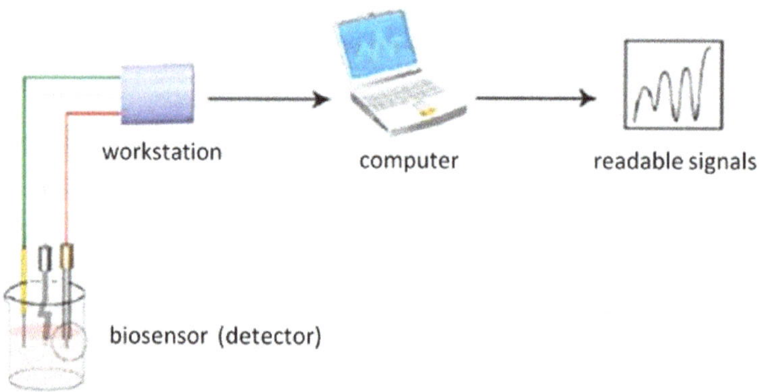

Figure 2. Basic elements to construct a biosensor system.

According to the difference of the biological element and transducer utilized, biosensors can be divided into several categories as what shall we introduced in the following listed in Figure 3.

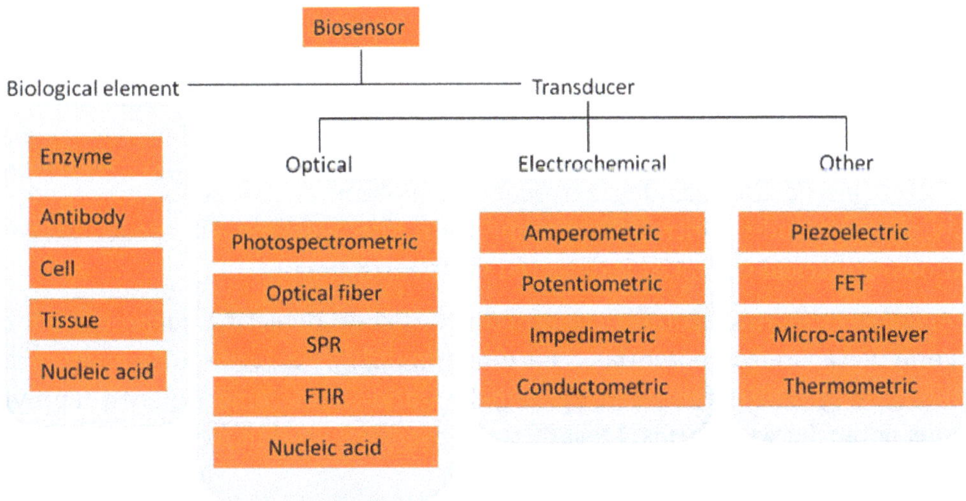

Figure 3. Classifications of common biosensors.

As is shown in the figure, electricity, light, sound, heat, and force, almost all physical dimensions have been available to be used as the target property for constructing a biosensor. Biosensors based on different physical principles have both their advantages and disadvantages and befit various kinds of analyzing targets. Electrochemical biosensors are inexpensive, easy to be prepared, and available to meet various ranges of analyte concentration [3]; this strategy is most often used in both fermentation and other practical applications. Optical ones can detect biological parameters with UV-visible [4], infrared [5], fluorescent [6], and chemiluminescent [7] lights. Thermal ones detect heat released in physicochemical processes. For fermentation uses, heat is released by both cellular and non-cellular processes to facilitate monitoring the fermentation progress [8]. Other strategies are often seen in comparatively highly specific uses, for example, SPR and microcantilever sensors are good choices to facilitate biomolecular researches [9], although they are often cost non-saving and high in instrumental and operative requirements.

Biosensors, other than those fit into the definition given above, and some sensing technologies that are aimed at detecting biologically related parameters yet contain not any biologically active elements are also counted, and they are usually called "generalized biosensors." Examples of generalized biosensors include for instance: mass spectrometric measurements used in off-gas analyses in fermentation processes [10, 11] and cytometry for fermentation process controls [12].

Out of its prevalence, robust analyzing capability, cost-saving, and facile instrumenting, in this text, we put the most focus on the most studied and relatively mature in practical applications such as electrochemical biosensors.

1.2. Basic principles

Electrochemical biosensors are based on various kinds of electrodes by which electrical signals can be produced and sensed. There are three main techniques widely used for electrochemical biosensors [13]:

Potentiometric. It includes zero-current potentiometry and techniques of applying amplitude controlled current onto the working electrode. In these methods, electrical potential is detected for measurements.

Amperometric. Amperometric method is based on detecting current produced by applying a known potential. There are mainly two forms of amperometric methods: constant-potential amperometry and amperometry with applying various potential waveforms. Most of the biosensors used in fermentations are of amperometric type.

Impedimetric. It is also called impedance spectrometry, which is based on detections of impedance, conductance, and capacitance of a certain electrochemical system. It is more often used in theoretical analyses of the sensing surface.

2. APPLICATIONS

2.1. Amino acid detection

Amino acid is one of the most important biomolecules for life, and they act as building blocks of numerous proteins, precursors in the synthesis of many biologically functional molecules and energy resource, in some cases. For human beings, counting in the eight essential and two semi-essential ones, in total about 22 natural amino acids are required to maintain a healthy human body. In food and medicine applications, amino acid is a vital material.

Biosensing tactics for amino acids are mainly realized by using enzyme biosensors. The enzymes are often amino acid oxidases as L-amino acid oxidase [14], glutamate oxidase [15–17], leucine dehydrogenase [18], tyrosinase [19, 20], and L-phenylalanine dehydrogenase [21].

Glutamate is very important in medical uses [22, 23], and sodium salt is a widely used seasoning additive, which is now mainly produced by large-scale cellular fermentations. Amperometric enzyme electrodes that detect glutamate is one of the most widely and maturely used biosensors. In majority of the now commercially available glutamate biosensors, a typical strategy is to integrate a glutamate oxidase onto a platinum electrode. When use, preset potential is applied upon the electrode to electrochemically catalyze the oxidation of the enzymatic reaction's product—hydrogen peroxide. Computer records the electrical current produced in this process and translates it into the corresponding concentration as the readout [24]. In a research, glutamate oxidase (EC 1.4.3.11) and NADP+-

dependent glutamate dehydrogenase (EC 1.4.1.3) were co-immobilized onto an oxygen electrode to fabricate an enzymatic MSG detector for food uses. By exploiting those two enzymes, monosodium salt of glutamate and glutamic acid can be distinguished to perform more accurate measurements [25]. Rita et al. [26] constructed a glutamate enzyme sensor by immobilize L-glutamate oxidase (GLOD, EC 1.4.3.11) and Gox on glass carbon electrode. To minimize the interference, the enzyme electrode was then modified with the polymer Nafion, a very widely used material to improve the sensing selectivity of amperometric enzyme sensors. The electrode can perform simultaneous measurement of L-glutamate and glucose without any obvious interference. Tang et al. [27] constructed glutamate enzyme electrode with NAD+-dependent glutamate dehydrogenase (EC 1.4.1.3). To improve its sensitivity, the electrode was modified with nanocomposite. The electrode has a very rapid response and good stability (remains 85% sensing intensity after 4 weeks). For quantifications of total L-amino acid, enzyme electrode based on immobilized L-amino acid oxidase can be a very good choice [28].

Stasyuk et al. [29] used recombinant yeast cells as the arginine activity source to establish an amperometric biosensor along with immobilized urease. The cell-enzyme coupled sensor reportedly exhibited a linear range cross 3 orders of magnitude up to 0.6 mM and can give the result within no more than 1 min. Another strategy is by coupling arginase (EC 3.5.3.1) and urease (EC 3.5.1.5). An arginine biosensor was constructed on ion-selective field effect transistors (ISFETs) surface via co-immobilizing arginase and urease. When working, arginine catalyzes the conversion of arginine into ornithine with release of urea, which, in turn, is degraded by urease to produce ammonium ions. Production of ammonium is accompanied by the subtle change of pH and thus can be detected by the transmitter [30].

All amino acids exist in humans are of L-form, for enzyme dealing with D-amino acids are in lack in human body. Therefore, were there D-amino acids in food or medical products, it is of danger to cause problems of safety and monitoring of their presence is often one part of many fermentation products' quality control. To meet this need, an enzyme electrode was developed by co-immobilizing D-amino acid oxidase (DAAO, EC 1.4.3.3) and peroxidase onto polymer electron mediator modified electrode [31]. Zain et al. devised a D-serine-sensitive electrochemical detector via immobilizing DAAO onto polymer functionalized metal electrode. The detector is reported to have ideal interference resistance toward most of the neurochemicals and can be used for *in vivo* D-serine detections [32].

Other than enzyme-biosensing methods, amino acid measurement can also be accomplished by enzyme-free techniques. Dai et al. [33] prepared nanoporous nickel-modified boron-doped diamond electrode with electron-assisted hot filament chemical vapor deposition method, and the electrode can capture redox processes of L-alanine and can perform anti-interference, sensitive detections of it. Seki et al. developed a tryptophan sensitive, potentiometric detector with microbial cells as the biological element. In the scenario, an auxotrophic bacterial strain *Escherichia coli* WP2—a mutant requiring tryptophan for its growth was monitored with a light-addressable potentiometric sensor. When L-tryptophan is in present, the bacterial metabolic result of it will cause pH changes, which can be detected and used for quantification by the sensor [34].

2.2. Sugar detection

In fermentation processes, sugars can serve as essential substrates for cellular fermentations or, in certain instances, they can be the desired end product. For example, oligosaccharides in isomalto-oligosaccharide and chito-oligosaccharide preparations, as well as glucose obtained from enzymatically destroyed starch, can be targeted products. The current focus of commercially available and insufficiently studied biosensors is the detection of monosaccharides. There is a scarcity of highly developedsensingplatformsforoligo-andpolysaccharides,mostlyduetothechallenge of obtaining biological elements that have strong biorecognition towards them. Currently, glucose is the primary analyte of interest in the majority of biosensor research, with a particular focus on its medical applications. There are various types of commercial biosensors available for fermentation purposes, most of which are operated in an offline manner. A commonly used glucose biosensor is similar to the previously mentioned glutamate enzyme electrode, with the exception that the enzyme used is typically glucose oxidase (Gox). Glucose biosensors, which are made using rare metal-based enzymes, are simple to create, reasonably priced, and can be reused. In a common method involving a Pt-based Gox electrode, the enzyme is attached to a membrane that allows certain substances to pass through. The biosensor is then formed by tightly covering the side of the membrane that contains the enzyme onto the surface of the Pt electrode. When the detected signals noticeably decrease after a period of usage, operators only need to replace the enzyme membrane with a new one from the same manufacturer to renew the sensor. Platinum (Pt) is electrochemically inert, meaning it does not react chemically during usage. As a result, it is resistant to repeated use, as seen in Figure 4. Challenges in modifying the basic electrode to make it useful, as well as the expensive nature of the electrode that prevents it from being transformed into disposable forms, often serve as motivations for researchers to create more advanced alternatives. The commercial blood glucose biosensor was employed to measure the glucose concentration in the fermentation broth. The results demonstrate that the method is effective in meeting the requirements of formation and serves as a viable alternative to HPLC analysis and reducing sugar test [37]. White et al.

employed a screen-printed glucose oxidase (Gox) electrode as a glucose detector in order to carry out real-time fermentation control [38]. The electrode is a conventional type of electrochemical detector that is cost-effective, highly consistent, and capable of large-scale manufacture. In order to address the issue of real-time sensing, the researchers implemented a flow injection analysis system to bridge the gap between the sampling process and the sensor. High-temperature processing is frequently necessary in fermentations to prevent biological contaminations throughout the fermentation operations. Nevertheless, biosensors generally lack the ability to withstand high temperatures. In order to address the issue, Phelps et al. introduced a glucose-sensing device that is capable of withstanding autoclaving and can be utilized multiple times [39]. The electrode differs from traditional immobilized enzyme electrodes in that it primarily consists of a semipermeable membrane. This membrane serves to protect the sensing surface from fouling by the fermentation broth. Additionally, a Pt electrode functions in the same way as conventional versions. The electrode also includes a chamber with conduits that allow for the filling or discarding of electrolyte and enzyme solutions. The design avoided the contact between the harsh autoclaving conditions and the temperature-sensitive enzyme.

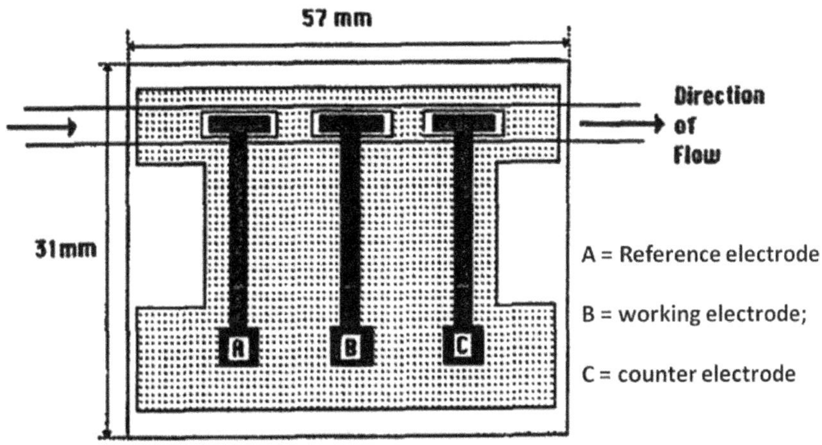

Figure 4. A schematic representation of screen-printed electrode. The base electrode is derived from a printing technology called "screen printing," which uses a mesh with predrawn patterns hollowed out and let the printing ink to through it so that the pattern is printed onto the base material. Enzyme electrodes prepared with this method are near two dimensional so they are potable. The high reproducibility of its preparing progress grants high reproducibility to electrodes made in the same batch.

Another approach to on-line biosensor design involves the use of non-immobilized, liquid Gox as the biological component. The base electrode does not have direct physical contact with the enzyme. Its function is solely to detect the catalytic product present in the enzyme solution [40]. In addition to its online applications, this design introduces a novel approach to developing biosensors specifically designed to measure high-concentration glucose, which is particularly valuable in the context of fermentations. Many fermentation processes frequently have excessively high glucose concentrations at the beginning and early stages, beyond the upper detection limit of most enzyme-immobilized sensors. Thus, achieving continuous monitoring is challenging without the use of gradient dilutions, which are frequently the primary cause of sample inaccuracy. The enzyme-injected mode enables the electrode to directly detect large concentrations of glucose in broths. The development of enzyme electrodes has progressed through three primary stages, with the Gox electrode serving as a notable illustration.

Stage I: Gas-sensitive electrodes represented by Clark enzyme electrode. The measurement is performed in potentiometric method by detecting changes of dissolved oxygen or any acid or base produced in the enzymatic reaction. The problem of this strategy is as dissolved oxygen consumed, measuring results are liable to become awry of the electrode's linear range and as a result brings huge errors to the quantification.

Stage II: Electron mediator-functionalized electrodes overcome the shortcut of gas-sensitive electrodes and established new strategy for enzyme electrode designs. Electron mediators were found by researcher that when they were integrated into the biosensing interfaces, electrons produced by the enzymatic reaction can be relied by the mediator, which, in turn, is oxidized by the base electrode. By this way, electrochemical detection can become oxygen consumption independent. Electron mediators can be either natural substances, for example cytochromes and co-enzymes, or artificial ones as some organic dyes, ferrocene and its derivatives, metal complexes, and some conductive polymers. By using electron mediators, detecting potentials can be effectively lowered so that much interference would thus be eliminated.

Stage III: Direct enzyme electrode was proposed after phenomenon that some proteins can make direct electrochemical communications with the electrode, which is represented by the finding of reversible cyclic voltammetric response on the electrode. Principle of the phenomenon is suggested to be when the redox center of the protein molecule is in close adjacent to the electrode surface, electrons can transport between them directly without the aid of any mediators. By using this mechanism, an obviously improved electron transfer efficiency can be obtained and the sensing capability is enhanced. However, it is often difficult to achieve the required conditions for enzymes. Taking Gox for example, its electrochemical active center—FAD+ cofactor—is wrapped into the space formed by the dimmer subunits and therefore is difficult to build direct electrochemical communication between the electrode and the enzyme. When Gox is tightly adsorbed onto some electrochemical active materials such as carbon nanotubes and graphenes, the direct electron transfer can be observed. Researchers have to pay more attention and efforts to develop direct enzyme electrodes, as it is a promising strategy to construct more rapid, sensitive, and reagentless biosensors.

2.3. Alcohol detection

The alcohol level in fermentation broths can be determined using many standard methods, such as hydrometry and gas chromatography. Due to their high error limit, expensive nature, and time-consuming methods, biosensors are considered a favorable alternative.

Kuswandi et al. suggested a colorimetric biosensor. The sensor was fabricated using a polyaniline film that immobilized alcohol oxidase. The introduction of ethanol causes a noticeable color change from green to blue as a result of the oxidation of polyaniline by the enzyme reaction product H2O2. The computer processing software can quantitatively determine the range of alcohol between 0.01 and 0.8% [41]. Gotoh et al. developed an amperometric alcohol sensor that utilizes co-immobilized alcohol dehydrogenase and coenzyme NAD+. The enzyme electrode exhibited a linear response to solutions with ethanol concentrations ranging from 0.05 to 10v/v%. As a reagentless enzyme sensor, it can endure weeks of continuous detections without the need for adding the coenzyme [42].

3. PROSPECTS FOR THE FUTURE

Despite significant progress in the use of fermentation processes and the availability of numerous commercially viable biosensors, there remain several unanswered concerns that require more investigation. Currently, the variety of biologically active species that may be used as biological elements in biosensors for fermentation processes is limited. As a result, the detectable parameters for these biosensors are constrained to only a few types of target constituents. On one hand, this can be expanded throughout time by discovering and separating new appropriate biological components from the natural environment. Dehydrogenase is currently the most commonly employed enzyme in biosensors. More than 400 dehydrogenases have been identified or separated; dehydrogenase sensors could potentially detect several biological components. Dehydrogenases frequently possess isozymes and rely on NAD+, NADP+, or quinones as cofactors. Unlike oxidases, dehydrogenases have a redox core that is not securely surrounded by the protein components. This makes them more susceptible to establishing direct electronic communication with electrodes. One of the most challenging issues is to develop techniques for immobilizing cofactors in order to create reagentless biosensors. Conversely, new generations may present prospects. A potential approach is the creation of molecular imprinting sensors, which utilize synthetic polymers that replicate the structure of natural enzymes, antibodies, or antigens to achieve a similar level of specificity. Although potent polymers can be created in certain cases, there is still a significant amount of progress needed before the technology can be applied widely. An alternative approach is to create aptamers, which are typically short chains of nucleic acids or peptides that are carefully designed and manufactured. Aptamers are viable substitutes that can perform equivalent biorecognition tasks as their natural counterparts. A multitude of functional aptamers have been developed, and it is of utmost need to establish a high-throughput selection method in order to expedite the process of discovering novel aptamers.

Furthermore, the pursuit of enhancing the foundational electrode is an ongoing undertaking that remains a central focus in the field of electrochemistry. Through alteration, electrodes can acquire additional functionalities or have their sensor capabilities improved. Promising materials for electrode functionalization encompass nanocomposites, conductive polymers, new electron mediators, liquid ions, and similar substances. Researchers are particularly interested in nanocomposites like carbon nanotubes, graphene, and metal nanoparticles due to their strong physical and chemical properties. These properties can be utilized to enhance sensing capabilities, such as the limit of detection, sensitivity, selectivity, anti-interference, and electrochemical stability. Additionally, these nanocomposites have great potential for immobilizing biological elements. The nanocomposite itself serves as a framework for creating intricate composites by combining the aforementioned elements.

Furthermore, the current state of downsizing, integration, and automation of biosensors in fermentation applications is still in its first phase. While commercial biosensors have greatly aided in the detection of various components, it remains challenging to implement autonomous control of many parameters for the entire fermentation process. In addition to the development of increasingly varied biosensors suitable for different targets and testing situations, microfabrication technology and the Internet of Things offer promising means to achieve this objective.

REFERENCES

1. Thevenot DR, Toth K, Durst RA, et al. Electrochemical biosensors: recommended definitions and classification. Pure and Applied Chemistry. Chimie pure et appliqué, 1999, 71 (12): 2333-2348. DOI: 10.1351/pac199971122333

2. Morrison D W G, Dokmeci M R, Utkan D, et al. Biomedical Nanostructures: Clinical Applications of Micro- and Nanoscale Biosensors. 2008, John Wiley & Sons, Inc., New Jersey. DOI: 10.1002/9780470185834.ch17

3. Kauffmann J M, Pravda M, Kauffmann J M. The electrochemical biosensor era. Analytical Chemistry, 2012, 84(2): 685-707.

4. Nidhi N, Ashutosh C. A colorimetric gold nanoparticle sensor to interrogate biomolecular interactions in real time on a surface. Analytical Chemistry, 2002, 74(3): 504-509. DOI: 10.1021/ac015657x

5. Kleiren E, Ruysschaert J M, Goormaghtigh E, et al. Development of a quantitative and conformation-sensitive ATR-FTIR biosensor for Alzheimer's disease: the effect of deuteration on the detection of the Aβ peptide. Spectroscopy, 2010, 119(15): 5055-5061. DOI: 10.3233/spe-2010-0405

6. Hung Y P, Albeck J, Tantama M, et al. Imaging cytosolic NADH-NAD(+) redox state with a genetically encoded fluorescent biosensor. Cell Metabolism, 2011, 14(14): 545-554. DOI:10.1016/j.cmet.2011.08.012

7. Zhang Q, Xu G, Gong L, et al. An enzyme-assisted electrochemiluminescent biosensor developed on order mesoporous carbons substrate for ultrasensitive glyphosate sensing. Electrochimica Acta, 2015, 186: 624-630. DOI:10.1016/j.electacta.2015.10.081

8. Hundeck H G, Weiss M, Scheper T, et al. Calorimetric biosensor for the detection and determination of enantiomeric excesses in aqueous and organic phases. Biosensors & Bioelectronics, 1993, 8: 205–208. DOI: 10.1016/0956-5663(93)85034-L

9. Wu L, Zhang Q, Su L, et al. Effects of small molecular inhibitors on the binding between HIV-1 reverse transcriptase and DNA as revealed by SPR biosensor. Sensors & Actuators B Chemical, 2007, 122(1):243-252. DOI: 10.1016/j.snb.2006.05.025

10. Heinzle E, Oeggerli A, Dettwiler B. On-line fermentation gas analysis: error analysis and application of mass spectrometry. Analytica Chimica Acta, 1990, 238(00): 101-115. DOI: 10.1016/S0003-2670(00)80528-0

11. Yanqiang D, Jiemin L, Jianguo L, et al. Characteristic analysis for odor gas emitted from food waste anaerobic fermentation in the pretreatment workshop. Journal of the Air & Waste Management Association, 2013, 63(10): 1173-1181. DOI: 10.1080/10962247.2013.807318

12. Steudler S, Böhmer U, Weber J, et al. Biomass measurement by flow cytometry during solid-state fermentation of basidiomycetes. Cytometry Part A, 2015, 87(2): 1-13. DOI: 10.1002/cyto.a.22592.

13. Mikkelsen S R. Bioelectrochemistry, 2002, Wiley-VCH, Weinheim.

14. Nieh C H, Yuki K, Osamu S, et al. Sensitive D-amino acid biosensor based on oxidase/peroxidase system mediated by pentacyanoferrate-bound

polymer. Biosensors & Bioelectronics, 2013, 47(28): 350-355. DOI: 10.1016/j.bios.2013.03.042.

15. Batra B, Pundir CS. An amperometric glutamate biosensor based on immobilization of glutamate oxidase onto carboxylated multiwalled carbon nanotubes/gold nanoparticles/chitosan composite film modified Au electrode. Biosensors & Bioelectronics, 2013, 47(18): 496-501. DOI: 10.1016/j.bios.2013.03.063.
16. Rita M, Hanna C, Youssef S. Amperometric and impedimetric characterization of a glutamate biosensor based on Nafion and a methyl viologen modified glassy carbon electrode. Biosensors & Bioelectronics, 2007, 22(11): 2682-2688. DOI: 10.1016/j.bios.2006.11.003.
17. Zhang M, Mullens C, Gorski W. Amperometric glutamate biosensor based on chitosan enzyme film. Electrochimica Acta, 2006, 51(21): 4528-4532. DOI: 10.1016/j.electacta.2006.01.010.
18. Labroo P, Cui Y. Amperometric bienzyme screen-printed biosensor for the determination of leucine. Analytical & Bioanalytical Chemistry, 2014, 406(1): 367-372. DOI: 10.1007/s00216-013-7443-7.
19. Mangombo Z A, Key D, Iwuoha E I, et al. Development of L-phenylalanine biosensor and its application in the real samples. In Science Journal, 2013, 03(01): 1-23. DOI: 10.5640/insc.030101.
20. Kanchana P, Lavanya N, Sekar C. Development of amperometric L-tyrosine sensor based on Fe-doped hydroxyapatite nanoparticles. Materials Science & Engineering C, 2014, 35(2): 85-91. DOI: 10.1016/j.msec.2013.10.013.
21. Villalonga R, Fujii A, Shinohara H, et al. Supramolecular-mediated immobilization of L-phenylalanine dehydrogenase on cyclodextrin-coated Au electrodes for biosensor applications. Biotechnology Letters, 2007, 29(3): 447-452. DOI: 10.1007/s10529-006-9259-4.
22. Villarta R L, Cunningham D D, Guilbault G G. Amperometric enzyme electrodes for the determination of L-glutamate. Talanta, 1991, 38(1): 49-55. DOI: 10.1016/0039-9140(91)80008-N
23. Wolf M E. The role of excitatory amino acids in behavioral sensitization to psychostimulants. Progress in Neurobiology, 1998, 54(6): 679-720. DOI: 10.1016/S0301-0082(97)00090-7.
24. Chen Y, Feng D, Bi C Y, et al. Recent progress of commercially available biosensors in China and their applications in fermentation processes. Journal of Northeast Agricultural University, 2014, 21(4): 73-85. DOI: 10.1016/S1006-8104(15)30023-4.

25. Basu A K, Chattopadhyay P, Roychudhuri U, et al. A biosensor based on co-immobilized L-glutamate oxidase and L-glutamate dehydrogenase for analysis of monosodium glutamate in food. Biosensors & Bioelectronics, 2006, 21(10): 1968-1972. DOI: 10.1016/j.bios.2005.09.011.

26. Rita M, Hann C a, Youssef S. Amperometric and impedimetric characterization of a glutamate biosensor based on Nafion and a methyl viologen modified glassy carbon electrode. Biosensors & Bioelectronics, 2007, 22(11): 2682-2688. DOI: 10.1016/j.bios.2006.11.003.

27. Tang L H, Zhu Y H, Xu L H, et al. Amperometric glutamate biosensor based on self-assembling glutamate dehydrogenase and dendrimer-encapsulated platinum nanoparticles onto carbon nanotubes. Talanta, 2007, 73(3): 438-443. DOI: 10.1016/j.talanta.2007.04.008.

28. Lata S, Pundir C S. L-amino acid biosensor based on L-amino acid oxidase immobilized onto NiHCNFe/c-MWCNT/PPy/GC electrode. International Journal of Biological Macromolecules, 2013, 54(3): 250–257. DOI: 10.1016/j.ijbiomac.2012.12.004.

29. Stasyuka N Ye, Gaydaa G Z, Gonchar M V. L-Arginine-selective microbial amperometric sensor based on recombinant yeast cells over-producing human liver arginase I. Sensors and Actuators B: Chemical, 2014(204): 515–521. DOI: 10.1016/j.snb.2014.06.112.

30. Sheliakina M, Arkhypova V, Soldatkin O, et al. Urease-based ISFET biosensor for arginine determination. Talanta, 2014(121): 18–23. DOI: 10.1016/j.talanta.2013.12.042.

31. Nieh C H, Yuki K, Osamu S, et al. Sensitive D-amino acid biosensor based on oxidase/peroxidase system mediated by pentacyanoferrate-bound polymer. Biosensors & Bioelectronics, 2013, 47(28): 350-355. DOI: 10.1016/j.bios.2013.03.042.

32. Zain Z M, O'Neill R D, Lowry J P. Development of an implantable D-serine biosensor for in vivo monitoring using mammalian D-amino acid oxidase on a poly (o-phenylenediamine) and Nafion-modified platinum–iridium disk electrode. Biosensors & Bioelectronics, 2010, 25(25): 1454-1459. DOI: 10.1016/j.bios.2009.10.049.

33. Dai W, Li M, Li H. Amperometric biosensor based on nanoporous nickel/boron-doped diamond film for electroanalysis of L-alanine. Sensors & Actuators B Chemical, 2014, 201(4): 31-36. DOI: 10.1016/j.snb.2014.05.005.

34. Seki A, Kawakubo K, Iga M, et al. Microbial assay for tryptophan using silicon-based transducer. Sensors & Actuators B Chemical, 2003, 94(3): 253-256. DOI: 10.1016/S0925-4005(03)00381-2.

35. Ding W U,Zou Y H,Wang Y, et al. Technique of isomaltooligosaccharide preparation with immobilizing enzymes. Food Science, 2005, 26(3): 125-127.

36. Cabrera J C, Cutsem P V. Preparation of chitooligosaccharides with degree of polymerization higher than 6 by acid or enzymatic degradation of chitosan. Biochemical Engineering Journal, 2005, 25(2):165-172. DOI: 10.1016/j.bej.2005.04.025.

37. Thornton A J, Brown D E. Fermentation glucose assay using the Exactech blood glucose biosensor. Biotechnology Techniques, 1991, 5(5): 363-366. DOI: 10.1007/BF00185015.

38. White S F, Tothill I E, Newman J D, et al. Development of a mass-producible glucose biosensor and flow-injection analysis system suitable for on-line monitoring during fermentations. Analytica Chimica Acta, 1996, 321(2–3): 165-172. DOI: 10.1016/0003-2670(95)00514-5.

39. Phelps M R, Hobbs J B, Kilburn D G, et al. An autoclavable glucose biosensor for microbial fermentation monitoring and control. Biotechnology & Bioengineering, 1995, 46(6): 514-24. DOI: DOI: 10.1002/bit.260460604.

40. Liu G S, Gao, X J, Jia Z Y. Research and fabrication of enzyme-injected glucose biosensor. Transducer & Microsystem Technologies, 2012, 31(7): 95-101.

41. Kuswandi B, Irmawati T, Hidayat M A. A simple visual ethanol biosensor based on alcohol oxidase immobilized onto polyaniline film for Halal verification of fermented beverage samples. Sensors, 2014, 14(2): 2135-2149. DOI: 10.3390/s140202135.

42. Gotoh M, Karube I. Ethanol biosensor using immobilized coenzyme. Analytical Letters, 1994, 27(2): 273-284. DOI:10.1080/00032719408001072.

09 BIOGAS IS A PROCESS THAT CONVERTS WASTE INTO ENERGY THAT IS ENVIRONMENT FRIENDLY

1. INTRODUCTION: AN OVERVIEW

The push to decrease reliance on petroleum has spurred the advancement of innovative technologies that harness other energy sources to produce various chemicals, including new types of fuels. Biogas, one of the biofuels currently in use, has been created for sanitary purposes for over 2000-3000 years [1]. However, the first recorded production of biogas was in 1895 in England, using a meticulously planned infrastructure. The interest in the utilization of biogas increased during World War II when France and Germany commenced the construction of biogas facilities and employed them as a source of fuel for automobiles and tractors. Following the war, there was a decline in the enthusiasm for biogas. However, it regained popularity during the oil crisis of 1973 due to advancements in technology. Currently, Germany holds the dominant position as the global frontrunner in the production of biogas.

Biogas is produced through the process of anaerobic digestion (AD) in a bioreactor, which is also known as a digester unit. The manufacture of this substance can be achieved using either a batch or continuous method, involving one, two, or several phases. It mostly utilizes organic matter derived from trash as the substrate. The biofuel is classified as carbon-neutral since it utilizes carbon dioxide that was recently absorbed by plants from the atmosphere and is capable of releasing it back into the environment through the fermentation of waste leftovers [2]. This biofuel also safeguards the environment from diseases by diminishing the trash that would decompose in the atmosphere, so minimizing the likelihood of attracting disease-carrying vectors. Furthermore, it significantly decreases air and water pollution, aids in the preservation of trees, and substitutes inorganic fertilizer with its digested leftovers [3]. As per the European Union, biogas has the capacity to provide 25% of total renewable energy. It has the capability to generate electricity, heat, and fuel for vehicles, thereby replacing traditional energy sources that emit greenhouse gases.

Biogas production has shown a significant boost in recent years. This is demonstrated by the swift establishment of biogas facilities, which have been constructed solely in Europe. In 2012, global biogas production amounted to 17.2 ktoe/year, which is comparable to millions of tons of oil per year. Europe accounted for 60% of total production, producing approximately 10.5 ktoe/year. European Union output experienced a significant growth of 27.6% in 2013, reaching 13.4 ktoe/year. It is projected to further increase to 33.0 ktoe/year by 2022. Numerous European nations confront significant challenges concerning the surplus generation of organic waste originating from industry, agriculture, and homes. AD can also aid in waste reduction by preventing the buildup of hazardous and long-lasting wastes, while also reducing the costs associated with trash disposal.

This chapter will provide a comprehensive overview of the generation of biogas, considering its significant importance as a clean energy source. Henceforth, the subsequent subjects will be discussed: The text covers the following topics: (1) the composition of biogas, (2) the types of substrate utilized in biogas production, (3) an overview of biogas production, (4) the physical and chemical process of anaerobic digestion, and (5) anaerobic bioreactors. The focus will be on key elements of fermentation, including: (1) the microorganisms and trophic groups responsible for each stage (hydrolytic bacteria, acidogenic bacteria, acetogenic bacteria, methanogenic groups); (2) factors that impact the efficiency of biogas production (temperature, pH, and chemical properties of biomass); (3) the biochemical substrates utilized by the microorganism population. This text will provide a brief overview of the many types of bioreactors and their tactics for producing biogas. However, the focus will primarily be on key elements of fermentation, including: (1) the specific microorganisms and trophic groups involved in each stage (hydrolytic bacteria, acidogenic bacteria, acetogenic bacteria, and methanogenic groups); (2) the factors that impact the effective generation of biogas (temperature, pH, and chemical properties of biomass); (3) the biochemical alterations in substrates caused by the microorganism population.

2. BIOGAS COMPOSITION

The typical composition of biogas is methane (CH_4), carbon dioxide (CO_2), and sulfuric elements (H_2S). The approximate percentage of biogas components is shown in the Table 1 [4].

The main cause of the high variation in percentages of biogas composition (Table 1) is due to the substrate utilized. The fact that methane is present at high concentration makes biogas a very attractive source of energy considering that methane has a heating value of 8500 kcal/m^3 and that CO_2 has no energy associated with it. The heating value of biogas is on average 5000–7000 kcal/m^3, approaching nearly 12,000 kcal/m^3 when in a high degree of purity (65% CH_4). Comparatively, a cubic meter of biogas has the same calorific power as 0.613 L of gasoline, 0.579 L of kerosene, 0.553 L of diesel, 0.454 L of cooking gas, 1.536 kg of wood, and 0.790 L of ethanol and produces the equivalent power of 1.4208 kW.

Table 1. Approximate percentage of biogas components [4].

Biogas composition	Typical analysis (%/volume)
Methane	55–65
Carbon dioxide (CO_2)	35–45
Nitrogen (N_2)	0–3
Hydrogen (H_2)	0–1
Hydrogen sulfide (H_2S)	0–1
Oxygen (O_2)	0–2
Ammonia (NH_3)	0–1

Typically, 0.2–3% of biogas is composed of gases that enter the digester with air included in the substrate (N_2 and O_2). Among these, nitrogen and CO_2 (produced during the digestion process) are included in the inert gases that compose the total biogas mix. On the other hand, the remaining NH_3, O_2 and H_2S gases are unwanted gases due to their toxicity to strict anaerobes that are essential for the process. Both O_2 and H_2S can be removed from biogas through chemical processes such as iron based processes, for example, with the addition of iron chloride, while NH_3 can be degassed through an H_2SO_4 absorber.

Another component, hydrogen sulfide (H_2S), is normally present in biogas as a by-product from anaerobic digestion. It is considered a major cause of corrosion of metal parts and degradation of engine oil, and during the fermentation process, it can precipitate metal elements. This gas is prevenient to the degradation of sulfur-containing proteins (i.e., cysteine and methionine), and besides being prevenient to normal metabolism of fermentation organisms, it has to be removed from the biogas before utilization.

3. TYPES OF SUBSTRATES

The most utilized residues for biogas production are found in animal manure, agriculture residues, and general organic wastes from food (both vegetable and animal in origin), organic fractions of municipal waste and from catering, sewage sludge and residues from crops dedicated to energy (i.e., biofuels), such as sugar cane and sorghum. These can be classified into various criteria: its origin, organic content, methane yield and dry matter content (Table 2). These substrates usually have a high content of sugar, starch, proteins, or fats, which are decomposed through AD. Table 2 shows several substrates and their classifications according to organic content, carbon:nitrogen ratio, percentage of dry matter, percentage of volatile solids in dry matter, and its biogas yield [5]. It is noticeable how the utilization of different biomasses has a consequence in the biogas yield, for example, it can vary from 0.15 m^3/kg VS (volatile solids) (utilizing straw) to 0.9 m^3/kg VS. When the utilized substrate is concentrated whey, a 500% increase in growth can be observed (Table 2). Generally, the C:N ratio also affects the production of biogas. As can be seen in Table 2, low C:N ratios (between 3 and 20) produce a yield ranging between 0.25 and 0.78 m^3/kg VS. Higher C:N ratios (above 20, reaching up to 150) do not produce greater yields, since the greater yield obtained is 0.56 m^3/kg VS, approximately 30% lower than that obtained at lower C:N ratios.

In spite of the numerous advantages of utilizing biogas digesters, there are still challenges that need to be overcome in order to maximize fuel production. Methanogenic archaea, microorganisms that produces methane, have specific requirements such as temperature and pH, and they must be maintained within specific ranges for optimal production, which increases the production cost of biogas [6]. Another challenge is hydraulic retention time (HRT), which is the normal time that the input substrate spends in the digester before it is removed. At tropical temperatures, the HRT is 30–50 days, although in colder atmospheres, it may go up to 100 days without heating, which requires a bigger digester volume and raises costs. While digesters can save energy at small-scale production on farms, finding the right economic balance for large-scale production is yet another challenge.

Table 2. Substrates commonly utilized for biogas production, its composition, and average biogas yield [5].

Biomass type	Organic content	C:N ratio	DMa (%)	VS[table2] (% of DM)	Biogas (yield m^3/kg VS)
Pig slurry	Carbohydrates, proteins, lipids	3–10	3–8	70–80	0.25–0.50
Cattle slurry	Carbohydrates, proteins, lipids	6–20	5–12	80	0.20–0.30
Poultry slurry	Carbohydrates, proteins, lipids	3–10	10–30	80	0.35–0.60
Stomach/intestine content	Carbohydrates, proteins, lipids	3–5	15	80	0.40–0.68
Whey	75–80% lactose, 20–25% protein	NR	8–12	90	0.35–0.80

Biomass type	Organic content	C:N ratio	DMa (%)	VS[table2] (% of DM)	Biogas (yield m^3/ kg VS)
Concentrated whey	75–80% lactose, 20–25% protein	NR	20–25	90	0.80–0.90
Flotation sludge	65–70% proteins, 30–35% lipids	NR	NR	NR	NR
Fermented slops	Carbohydrates	4–10	1–5	80–95	0.35–0.78
Straw	Carbohydrates, lipids	80–100	70–90	80–90	0.15–0.35
Garden wastes	NR	100–150	60–70	90	0.20–0.50
Grass	NR	12–25	20–25	90	0.55
Grass silage	NR	10–25	15–25	90	0.56
Fruit wastes	NR	35	15–20	75	0.25–0.50
Fish oil	30–50% lipids	NR	NR	NR	NR
Soya oil/margarine	90% vegetable oil	NR	NR	NR	NR
Alcohol	40% alcohol	NR	NR	NR	NR
Food remains	NR	NR	10	80	0.50–0.60

[a]Dry matter.

[b]Volatile solids. NR, not reported.

4. BIOGAS PRODUCTION: AN OVERVIEW

Biogas generation is a well-established method, however there is less knowledge regarding the microorganisms that are involved when processing various types of garbage. Therefore, it is essential to comprehend the activity of microorganisms and the variables that can impact the composition of biogas in order to optimize fermentation efficiency and minimize operational expenses. Thus, to determine the microorganisms participating in anaerobic digestion, researchers have conducted sequencing of 16SrRNA and metagenomics [7]. Additionally, they have analyzed the methyl-coenzyme M reductase encoding gene, which serves as a marker for identifying archaea that are specifically involved in methanogenesis [8]. The DNA extracted from various bioreactors utilizing different substrates revealed a clear and direct correlation between the kind of reactor and the taxonomic groups present. For instance, a study reported that in a stirred digester that was fed with fodder beet silage, the predominant microbial groups were Bacilli, Clostridiales, Deltaproteobacteria, and Bacteroidetes [9]. Another study characterized a thermophilic digester with a high abundance of Clostridia in its microbial community [10]. Another crucial correlation exists between the microorganisms and the physical location of the digester [11]. Multiple studies have concluded that during the first and intermediate stages of Alzheimer's disease (AD), a minimum of 58 species belonging to 18 genera are implicated in the process of biogas production, namely through mixed fermentation.

4.1. Microorganisms and the biochemistry of AD

Biogas production is carried out by a group of microorganisms through four primary reactions: hydrolysis, acidogenesis, acetogenesis, and methanogenesis. These reactions involve organisms from the bacteria and archaea domains, which work together in consortia to convert the substrate into gases such as CH4 and CO2, among others. Figure 1 provides a depiction of the many types of microorganisms involved and a summary of the substrate process.

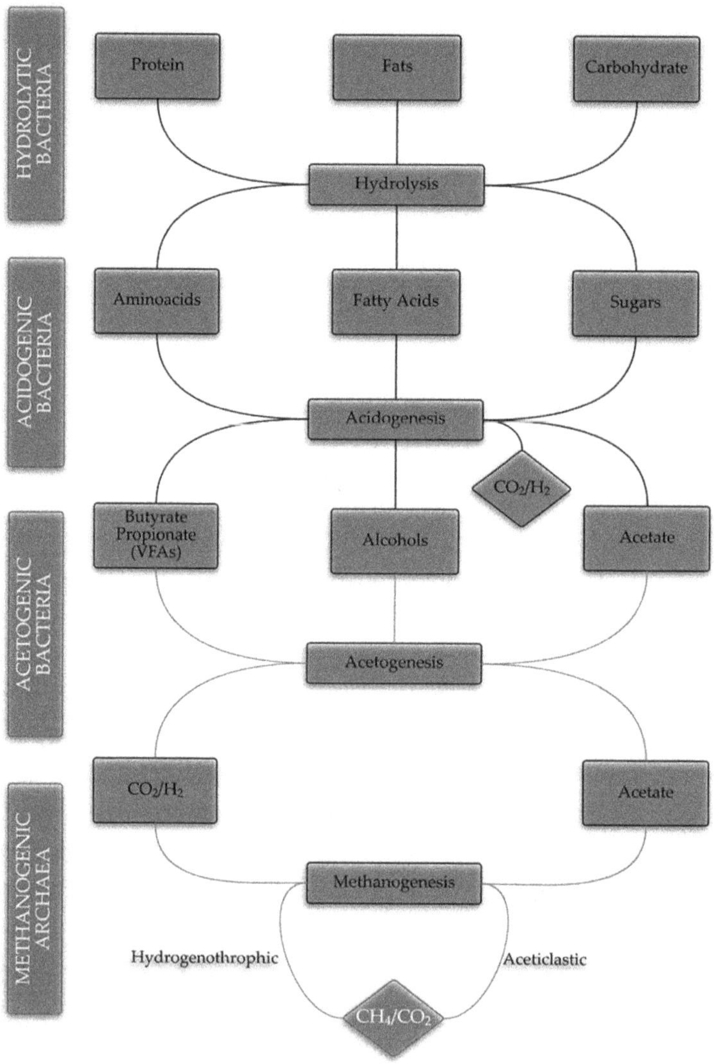

Figure 1. Microorganisms involved in each catabolic step during biogas biosynthesis.

4.1.1. Hydrolytic bacteria

Anaerobic digestion starts with the polymer hydrolysis of fats, proteins, and carbohydrates into monomers that are suitable for further digestion. Hydrolytic bacteria, which can be either facultative or strict anaerobes, are capable of hydrolyzing the bonds of these compounds, converting them into oligomers, monomers, amino acids, and unsaturated fats. For example, cellulose $[(C_6H_{12}O_6)_n]$, an insoluble substrate commonly found in sludge, is hydrolyzed by bacteria from the genus *Cellulomonas*, resulting in glucose monomers. The hydrolysis of polymers that are difficult to decompose restrains the rate of waste processing, and just half of these compounds experience hydrolysis in a one-stage digester. In some cases, pretreatment involving an aerobic step can be added. The concept of aerobic treatment consists in the knowledge that some aerobic microorganisms can produce hydrolytic enzymes that are able to generate monomers from the polymers present in the biomass. Moreover, inhibitory macromolecules such as lignin may also be transformed, resulting in a less toxic substrate to the microorganisms that start the AD process [12].

Anaerobic digesters that utilize substrates derived from wastewater treatment from industry, such as dairy and agro industries, are usually composed of soluble organic compounds and therefore do not experience this kind of hydrolysis. However, different sugars such as sucrose and lactose must be hydrolyzed despite being soluble, since they are larger than most cells can absorb [13].

4.1.2. Acidogenic bacteria

In regard to the second reaction stage, acidogenic bacteria will then convert these molecules into volatile fatty acids (VFAs) with high carbon numbers such as butyrate, propionate, and alcohols in addition to CO_2, H_2, and acetate [14]. These biochemical steps depend on various factors, like pH, enzyme production by bacteria, diffusion, and adsorption of enzymes by the biomass undergoing the process of digestion. This is executed by microorganisms from the group of anaerobic bacteria of genera such as *Streptococcus* and *Enterobacteria*.

However, VFAs produced during this stage may negatively affect the AD process depending on its concentration in the bioreactor. When unstable, the AD process accumulates VFAs inside the reactor, which results in a drop of pH-value and consequently a decrease in methane yield. This is explained by the low tolerance of methanogenic archaea in an acidic environment. It is demonstrated that different digesters can react differently in response to the same amount of VFA, where, in one digester, the concentration may be optimal and, in another, it is a considerable inhibitor to methane production. One conceivable explanation is the microorganism population, which varies from digester to digester. It can also be explained by the buffering capacity of the substrate.

4.1.3. Acetogenic bacteria

For the third reaction stage, acetogenic bacteria convert VFAs into acetate. Acetogenic bacteria are obligate proton-reducing bacteria (OPR) and are known for the production of H_2 during acetate production. Some VFA conversions are displayed below in Eq. (1):

$$\begin{aligned} &\text{propionate} + 3H_2O \rightarrow \text{acetate} + HCO_{3^-} + H^+ + 3H_2 \Delta G = +76.1\text{kJ/mole} \\ &\text{butyrate} + 2H_2O \rightarrow 2\,\text{acetate} + H^+ + 2H_2 \Delta G = +48.1\text{kJ/mole} \\ &\text{ethanol} + 2H_2O \rightarrow \text{acetate}^- + H^+ + 2H_2 \Delta G = +9.6\text{ kJ/mole} \end{aligned} \quad (1)$$

In accordance with the examples above, it is important to note that all of them require energy input. However, in the presence of low hydrogen concentrations provided by the digester, the reaction moves to the product side to maintain equilibrium. To this end, they only live in coexistence with a $H_{2,}$ utilizing species, which are the methanogenic archaea. A genus such as *Desulfovibrio* oxidizes alcohols and organic acids into acetate and transfers the electrons released to sulfate. Genera such as *Aminobacterium* and *Acidaminococcus* ferment amino acids, trans-aconitate and citrate into acetate, $CO_{2,}$ and H_2. Sulfate-reducer organisms such as the acetogenic *Desulfovibrio*, which oxidizes organic acids and alcohols to acetate and transfers the released electrons to sulfate resulting in a higher energy yield than fermentation, are deeply involved in compound decomposition by AD. These bacteria form cultures from obligated and facultative anaerobes to ferment available substrates such as lactate and alcohol from the acidogenic step.

4.1.4. Methanogenic group

The last phase of anaerobic digestion is catalyzed by a group of microorganisms from the archaea group. This group is subdivided into two groups: a hydrogenotrophic methanogenic group and aceticlastic methanogenic group. The first group utilizes the H_2 produced by the OPR group. Their affinity to uptake hydrogen is on the order of parts per million, making them very efficient in maintaining the substrate with a very low hydrogen partial pressure. The aceticlastic methanogenic group consists of only two genera: *Methanosarcina* and *Methanothrix*. These microorganisms can produce methane from acetic acid, and approximately 70% of all methane produced in biogas reactors originates from this conversion. The reactions of the processes are displayed below (Eqs. (2) and (3)).

$$CO_2 + 4H_2 \rightarrow CH_4 + 2H_2O \quad (2)$$

$$CH_3COOH \rightarrow CH_4 + CO_2 \quad (3)$$

Methanogenic archaea have, in their metabolism, the enzyme methyl-CoM reductase. This hexamer is a large complex composed by two copies of three different subunits (α, β, and γ) containing a unique coenzyme, the nickel phorphinoid factor

F_{430} and with activity deep inside the complex for protection from the surrounding water. This complex catalyzes the release of the CH_4 from methyl-CoM [15]. The F_{430} ring needs a nickel atom that is stabilized in the reactive state, which is an important property of this enzyme because the substrate methyl-coenzyme M is rather inert, which makes the reaction easier.

Acetoclastic archaea are well known for their slow doubling time (1–12 days in thermophilic conditions) because of their relative inefficiency in taking up acetate, but on the other hand, hydrogenotrophic methanogenic bacteria are extremely productive and have moderately quick doubling times (0.5–2 days in thermophilic conditions) [16].

5. PHYSICAL AND CHEMICAL AD PARAMETERS

The growth and metabolism of anaerobic microorganisms are essentially impacted by physical and chemical conditions such as temperature, pH value, nutrient supply, mixing intensity, and the additional presence of inhibitors.

5.1. Temperature

A large portion of reactor cost comes from the energy spent to maintain its temperature stable. Thus, an optimum temperature setting is the most critical factor in temperate countries since more energy is needed to maintain the temperature of AD and consequently methane production. Temperature parameters for AD can take place at different levels: cryophilic (below 25 °C) mesophilic (25–45 °C), and thermophilic (45–70°C). There is an inverse relationship between the temperature range and the HRT, meaning that thermophilic digesters have a shorter retention time than mesophilic and cryophilic ones.

Many facilities operate their biodigesters at the optimum temperature of thermophilic microorganisms because this reduces number of pathogens, favors methanogenic bacteria growth, improves the separation of liquid and solid fractions, and improves degradation of the substrate since there is more metabolic activity. Moreover, the methane production in thermophilic digesters is 25% greater than in mesophilic digesters. Nevertheless, the utilization of thermophilic temperatures also has disadvantages such as a higher degree of imbalance due to an increased production of volatile fatty acids. When dealing with manure, for example, reactors had optimal production in mesophilic reactors with the temperature between 30 and 35 °C, with only a 3% difference in the methane yield between these two temperatures. The same substrate at 25 °C had a decrease in methane yield of 17.4% [17]. In another study, two reactors, a one-stage reactor operated at mesophilic temperatures and a two-stage reactor operated at thermophilic (first stage) and mesophilic (second stage) temperatures, had their volatile solid consumption compared. The results demonstrated that a thermophilic (60 °C) stage was especially effective in degrading sludge waste substrates, with a 35% reduction in VFAs compared to the one-stage mesophilic digester.

5.2. pH

The pH value of utilized substrates affects AD by influencing the methanogenic-organisms' doubling time. Moreover, pH also influences the dissociation of some important compounds, such as ammonia, sulfide, and some organic acids. Methane generation takes place in the range of 5.5–8.5 pH, with optimal production in the 7.0–8.5 pH range. Most of the problems in AD can be attributed to acid accumulations and a consequent drop in the pH value. Considering that CO_2 solubility decreases when the temperature increases, the pH of thermophilic reactors is higher than mesophilic ones and therefore has less carbon dioxide dissolved in carbonic acid form, making it more endurable for methanogenic groups. In a two-phase digester, the hydrolytic-acidogenic and acetogenic phases are separated from methanogenesis, and with this, the pH can be controlled to the optimum range for the first phase (4.0–6.0) and second phase (7.0–8.5). In a single-phase reactor, the pH is usually maintained around the tolerance of the methanogenic group (6.6–8.0) since the other population groups of organisms can tolerate these conditions [18].

5.3. Ammonia

Nitrogen in the form of ammonia (NH_4) is present in the environment of the digester as a gas. It originates from protein degradation and from animal slurry, due to its high ammonia concentration. The precise concentration of free ammonia at which it starts to be toxic remains uncertain, but when dealing with a non-adapted digester (i.e., a digester that has not had enough time to acclimate its methanogenic population to a high ammonia concentration), its inhibition starts at 0.08–0.15 gN/L of free ammonia and 2.5 gN/L of total ammonia. In an adapted digester, it is 0.7–1.1 gN/L of free ammonia and 4–6.5 gN/L of total ammonia [19]. Methanogenic bacteria are very sensitive to the presence of ammonia as its presence can disturb the process in two forms, (1) inhibiting methanogenic enzymes in archaea and (2) entering the archaea cell and causing an unbalance in the electrons and disrupting the process [20].

5.4. Micronutrients (trace elements)

The impact of trace elements and changes in its concentration in bioreactors depends on various factors, such as the microbial community structure; population dynamics; individual trophic group metabolism; and meta-community (e.g., the microbial community as a group, incorporating compounds as well as cells). With that in mind, it is hard to fix micronutrient concentrations that are fully satisfactory for the microorganisms' community present in the reactor.

Although nutritional demand for each microorganism species varies, this topic will explore general guidelines of micronutrients, which are limiting for methane-forming archaea. These microorganisms have specific methanogenic enzyme systems with different requirements when compared to other microorganisms. These systems need specific micronutrients that must be incorporated or added to the substrate for its proper degradation and efficiency of CH_4 production.

Cobalt, iron, nickel, and sulfide are obligatory micronutrients, because they are cofactors of the methane pathway enzymes that convert acetate into methane. In some cases, molybdenum, tungsten, and selenium can be obligatory micronutrients as well as barium, calcium, magnesium, and sodium [21].

These micronutrients are usually present in municipal wastewater, although the digester effluent, in some cases, must be analyzed to ensure their presence in enough quantities and guarantee that these nutrients are in a soluble form since micronutrient deficiency can be mistaken with toxicity from the accumulation of volatile fatty acids.

Simple variations in the amounts of elements can disturb the environment inside the digester by unbalancing the substrate process and then causing inhibition of the whole process. For example, under co-limiting conditions, methanogenic activity was lost within ten days by acidification of a methylotrophic digester. In other study, Zn deprivation affected methane production significantly, which could not be later restored by a continuous supply of Zn [22].

6. ANAEROBIC BIOREACTORS

The biodigester (or anaerobic bioreactors) must guarantee optimal conditions for feedstock transformation to occur, such as the retention of the active biomass and favorable environmental conditions for biomass degradation of organic matter [23]. A report, dating from the 1880s, presents a biodigester, named by its inventor, Donald Cameron (Exeter, England), as a "Septic Tank," which was much more efficient than previous, and more rudimentary tanks since its design promoted microbial growth by adopting an organic material entry and exit system below water level in order to minimize the entry of air and turning of the upper part of the tank [24]. The precursor tank, called the "automatic scavenger," was built by Jean Louis M. Mouras, author of the first reference to the liquefaction of organic matter of wastewater under anaerobic conditions (patented in 1881) [24]. However, it is worth noting that this is not the first AD bioreactor, but one of the first reports in the literature.

The increase in demand of organic matter degradation has allowed for further development of these bioreactors, such as the addition of a heating system [25] and mechanical agitation- -Patent US2605220 [26]. Additionally, there are many studies regarding bioreactor design and the way the digestion is conducted, as described in the next Section (6.1).

6.1. Bioreactors types

The digestion unit is the most important part of a biogas plant; after all, it is where organic matter is reduced into biogas by microorganisms. An anaerobic digester design should allow for a continuously high load rate of organic matter, short hydraulic retention time (to reduce bioreactor volume), and a maximization of methane production. The shape of the bioreactor should take important considerations into account, such as the exchange of heat and the mixture, which is not observed in underground reactors (Figure 2). In general, these bioreactors are built from concrete blocks in a rectangular or square shape format that does not benefit the mixture. Furthermore, they have accumulated points (edges) of raw materials that lead to reduction in plant efficiency and require more frequent maintenance and thus idle time [27].

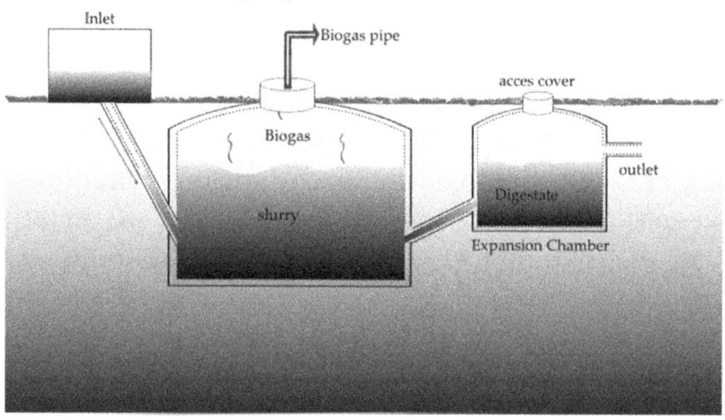

Figure 2. Underground reactor.

The choice of bioreactor for biogas production will depend directly on the characteristics of the raw materials utilized such as dry matter content, rate of degradation, and risk of inhibition. Among the main processing technology options available, there are feeding systems, reactor type, temperature reactor, number of phases, and agitation system (Figure 3) [28]. Nevertheless, only the most frequently used options of reactor type and number of phases will be described in more detail in this chapter.

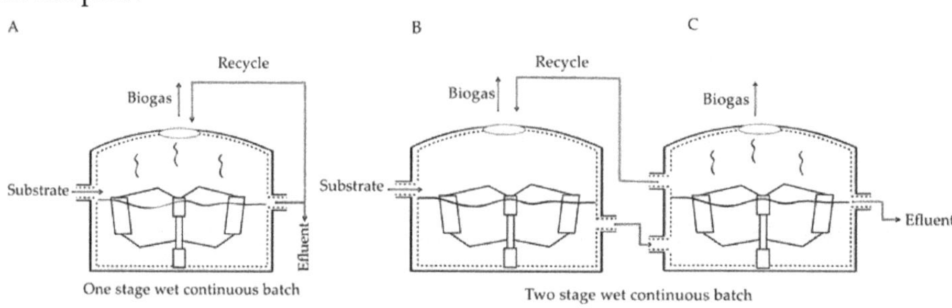

Figure 3. Fermentation modes utilized for biogas production batch digester—one-stage continuously fed system (A); two- or multistage continuously fed system: first stage (B) and second stage (C).

They may be dry or wet, batch or continuous, one step or multistep, and one phase or multiphase and may operate under different temperature conditions (mesophilic or thermophilic). However, the main bioreactor groups commonly employed are as follows: (1) batch bioreactors (Figure 3A); (2) continuous fed system: (a) one stage (Figure 3B); and (b) two stage or multistage [29] (Figure 3C).

6.1.1. Batch bioreactors

In this type of system (Figure 3A), a digestion vessel is loaded once with the feedstock then sealed off and left to ferment until gas production decreases. Then, the bioreactor is emptied and filled again with a new batch of feedstock. It is worth noting that part of the digestate should be left in the vessel, which will serve as inoculum for the next batch [30]. This type of bioreactor is generally utilized for feedstock that has a high solid content (between 30 and 40%) and with a high fiber content [31], and it requires little daily attention and it is notable for its simplicity. Moreover, batch reactors may be more suitable when using small amounts of substrate [32].

However, batch bioreactors have some limitations, for example (1) high variation in gas quality and production; thus, a series of batch digesters are employed, which are fed sequentially to generate a reasonably homogenous production of biogas; (2) a considerable time requirement to empty and load the batch digesters; (3) biogas losses during discharging the bioreactors; and (4) limited bioreactor heights [29]. The production of methane may vary from 44.6 to 290 mL/g VS for yard trimmings and rice straw as substrate, respectively [2].

6.1.2. Continuously fed system

For continuous digesters, unlike the batch bioreactors, the feedstock is constantly fed mechanically or by flow force by the newly entered feedstock, enabling uninterrupted production of biogas [33]. Among the types of continuous digesters, the multiple tank system (or multistage system) stands out, which will now be described.

6.1.2.1. One-stage, two-stage, or multistage continuous fed system

As previously discussed in this chapter (Section 4.1), there are four biochemical reactions in anaerobic digestion: hydrolysis, acidogenesis, acetogenesis, and methanogenesis. When all of these biochemical reactions take place in one reactor, it is called a one-stage continuously fed system (Figure 3B), in contrast, when the biochemical reactions occurs separately in two reactors, it is called a two-stage (or multistage) continuously fed system (Figure 3C) [27].

Organic waste treatment systems that use the two-stage system present advantages over one-stage systems, such as high biogas production rates and yields.

One study demonstrated a 13% increase in methane production from cellulosic material in a process that used a two-stage process compared to a single phase [34]. A similar increase was obtained using olive mill solid residues as the substrate [35]. Another study [36] compared one- and two-stage digestions for the treatment of thin stillage. It obtained approximately 57% total volatile fatty acids to the total chemical oxygen demand ratio, while the digestion obtained from one stage is only 10%. Additionally, the use of two-stage digestion also increased the production of methane, from 0.26 L CH^4/g of the chemical oxygen demand added (one stage) to 0.33 L CH_4/g of the chemical oxygen demand added [36]. This is because the system that performs the separation stages of the biochemical anaerobic digestion benefits the selection and development of different microorganisms for each stage. In addition, the conditions in each respective phase are controlled to generate an optimal environment for the action of each microorganism [37].

Acidogenic bacteria are the prevailing microorganisms in the first stage while the methanogenic group is dominant in the second one. In addition, as previously discussed, the intense production of acids inhibits methane formation in a one-phase system. Hence, the second stage favors bacteria that perform the production of methane gas [28]. The multiple-step system allows a faster, higher performance, and less expensive process than those that use single-stage digester, even though multistage digesters were more expensive to build and maintain [38]. The methane yield from municipal solid waste using a two-stage reactor can be 21% greater than the methane yield obtained from a single-stage process [39].

6.2. Microorganisms retention

In general, the generation time of hydrolytic and acidogenic bacteria ranges from approximately 1–3 days, whereas methanogenic and acetogenic bacteria range from about 1–4 and 5–12 days, respectively [13]. Due to the slow growth of microorganisms during the process of digestion, a reactor operated in a continuous mode can result in washout. Therefore, the rate of loading and unloading cannot exceed the maximum growth rate of microorganisms. In addition, the calculation of this rate is one of goals of process optimization. Additionally, one other way to prevent this type of accident is to use immobilized cells [19]. The use of microbial consortium retention contributes to increased performance of the anaerobic phase [40]. The use of support material such as toasted coconut shells and wood chips produced 720 and 144 L/kg VS of biogas, respectively, while the use of expanded clay showed nearly no production [40].

Anaerobic filters use inert supporting materials such as clay fibers, polyvinyl-chloride sheets, polyurethane foam, polypropylene membranes, carbon fiber textiles, tire rubber, zeolite filters, glass, and polyethylene fibers [40]. It is practical at this point to highlight that not only is the type of support material directly related to the performance of the anaerobic reactor, but so are other factors such as specific surface area, porosity, surface roughness, pore size, and orientation of the packaging material [40].

Microbial immobilization on the surface and in the pores of the inert material allows a reduction in the hydraulic residence time, which can decrease from 30 days to under a week, and it consequently reduces reactor volume and initial cost and increases the yield [32]. Among the used systems are (1) fixed- or packed-bed reactors and (2) fluidized-bed reactors (Figure 4).

6.2.1. Fixed-bed reactors

In its initial application, the fixed-bed system was used as biological filters for sewage treatment, so it is also known as an anaerobic filter (similarly called a biofilm reactor or packed bed). In this system, the particles containing the immobilized cells are fixed or packed into the reactor and the liquid flows through the bed. The fixed-bed reactor (Figure 4A) allows the application of greater organic loads than those applied in the complete mixture of anaerobic digesters. This system uses one kind of reactor that maintains a high biomass density within the reactor through microorganism retention from biofilms that have developed on the support material [32].

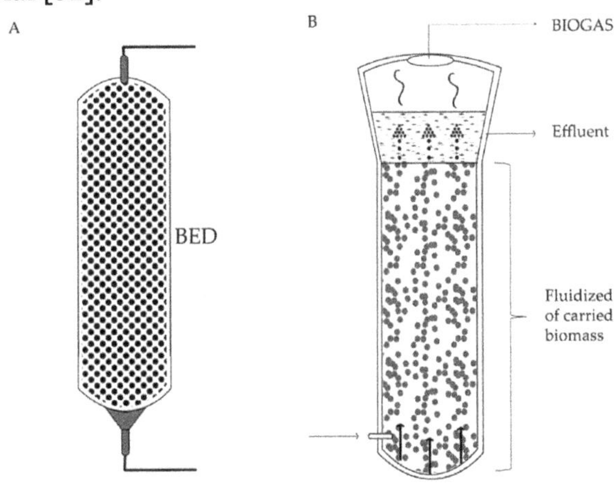

Figure 4. Microorganisms' retention reactors: (A) fixed-bed reactor and (B) fluidized-bed reactor.

6.2.2. Fluidized-bed reactors

In fluidized-bed systems (Figure 4B), the supporting material particles are maintained in suspension within the reactor due to substrate flow. This allows the particles to become unrestricted, and therefore, its entire external surface is available for interaction with the feedstock. This type of system has an advantage over packed-bed because we could substrate particulate packed beds. Furthermore, control of the temperature and the pH is more effective than the packed beds [32].

The performance of both reactors (fixed bed and fluidized bed) was compared with that of a fixed-bed reactor under similar conditions (feed gas to steam ratios of 1.5 and 0.75 at a reactor temperature of 750 °C, GHSV (gas hourly space velocity) of 300 L/min) [41]. This study showed a conversion of 75% CH_4 in a fixed-bed reactor. On the other hand, when using the fluidized-bed reactor, the production was much greater, reaching up to 90% conversion. The authors of this study reported the low yield of the fixed-bed reactor creates points of temperatures below the optimum process temperatures.

7. CONCISE SUMMARY

Presently, a multitude of endeavors are underway to diminish the reliance on oil for energy. This necessity has prompted the advancement of novel technology to harness alternative energy sources, such as the generation of biogas. Biogas production utilizing biomass as raw materials is a significant alternative biofuel that guarantees the provision of clean and cost-effective energy while also aiding in the reduction of trash accumulation. Nevertheless, achieving large yields remains a significant obstacle. An approach to resolve the issue is to enhance the process by fine-tuning certain physical and chemical parameters, such as temperature and pH. This is because the fermentation process entails the involvement of multiple microbial groups, necessitating adjustments to suit the specific environmental conditions of each group. An effective approach is to incorporate fermentation phases, employing multiple reactors to ensure optimal conditions for each microbial group involved in each step. Another obstacle is the hydraulic retention period, which refers to the typical duration that the input substrate remains in the digester before being extracted. One possible option is to use microbe retention, which involves trapping the bacteria within inactive materials, so extending their stay inside the reactor. It is important to do a thorough investigation into the physiology of each microbial gender involved in the process in order to accurately optimize the process parameters. Despite the long history of biogas production, there is still limited knowledge about this process. Hence, additional research on this procedure is important in order to attain higher output and consequently enhance the outcomes of this technique.

REFERENCES

1. Deublein D, Steinhauser A. Biogas from waste and renewable resources: an introduction, 2nd ed. Weinheim, Germany: Wiley-VCH; 2010. doi:10.1002/9783527632794.
2. Mao C, Feng Y, Wang X, Ren G. Review on research achievements of biogas from anaerobic digestion. Renew. Sustain. Energy Rev. 2015; 45:540–555. doi:10.1016/j.rser.2015.02.032.

3. Cuéllar AD, Webber ME. Cow power: the energy and emissions benefits of converting manure to biogas. Environ. Res. Lett. [Internet]. 2008 [cited 2015 Dec 7]; 3(3):034002. doi:10.1088/1748-9326/3/3/034002.

4. Guebitz GM, Bauer A, Bochmann G, Gronauer A, Weiss S, editors. Biogas science and technology [Internet]. 1st ed. New Yourk: Springer International Publishing US; 2015. doi: 10.1007/978-3-319-21993-6.

5. Al Seadi T, Rutz D, Janssen R, Drosg B. Biomass resources for biogas production. In: Wellinger A, Murphy JD, Baxter D, The biogas handbook: Science, production and applications [Internet]; 1st ed. Cambridge: Woodhead Publishing Limited, UK; 2013. p. 19–51. doi: 10.1533/9780857097415.1.19.

6. Mao C, Feng Y, Wang X, Ren G, Morita M, Sasaki K, et al. Inline NH_3 removal from biogas digesters. Appl. Environ. Microbiol. [Internet]. 2014; 50(1):10–18. doi:10.1016/j.biombioe.2012.06.041.

7. Nettmann E, Bergmann I, Pramschüfer S, Mundt K, Plogsties V, Herrmann C, et al. Polyphasic analyses of methanogenic archaeal communities in agricultural biogas plants. Appl. Environ. Microbiol. 2010; 76(8):2540–2548. doi:10.1128/AEM.01423-09.

8. Solli L, Håvelsrud OE, Horn SJ, Rike AG. A metagenomic study of the microbial communities in four parallel biogas reactors. Biotechnol. Biofuels [Internet]. 2014; 7(1):146. doi:10.1186/s13068-014-0146-2.

9. Klocke M, Mähnert P, Mundt K, Souidi K, Linke B. Microbial community analysis of a biogas-producing completely stirred tank reactor fed continuously with fodder beet silage as mono-substrate. Syst. Appl. Microbiol. 2007; 30(2):139–151. doi:10.1016/j.syapm.2006.03.007.

10. Goberna M, Insam H, Franke-Whittle IH. Effect of biowaste sludge maturation on the diversity of thermophilic bacteria and archaea in an anaerobic reactor. Appl. Environ. Microbiol. 2009; 75(8):2566–2572. doi:10.1128/AEM.02260-08.

11. Abendroth C, Vilanova C, Günther T, Luschnig O, Porcar M. Eubacteria and archaea communities in seven mesophile anaerobic digester plants in Germany. Biotechnol. Biofuels [Internet]. 2015; 8:87. doi:10.1186/s13068-015-0271-6.

12. Bochmann G, Montgomery LFR. Storage and pre-treatment of substrates for biogas production. In: Wellinger A, Murphy JD, Baxter D, The biogas handbook: Science, production and applications [Internet]; 1st ed. Cambridge: Woodhead Publishing Limited, UK; 2013. p. 85–103. doi:10.1533/9780857097415.1.85.

13. Gerardi MH. The microbiology of anaerobic digesters. 1st ed. New Jersey: John Wiley & Sons US; 2003. doi:10.1002/0471468967.

14. Cibis KG, Gneipel A, König H. Isolation of acetic, propionic and butyric acid-forming bacteria from biogas plants. J. Biotechnol. [Internet]. 2016; 220:51–63. doi:10.1016/j.jbiotec.2016.01.008.

15. Grabarse W, Mahlert F, Shima S, Thauer RK, Ermler U. Comparison of three methyl-coenzyme M reductases from phylogenetically distant organisms: unusual amino acid modification, conservation and adaptation. J. Mol. Biol. 2000; 303(2):329–344. doi:10.1006/jmbi.2000.4136.

16. Mata-Alvarez J, Macé S, Llabrés P. Anaerobic digestion of organic solid wastes. An overview of research achievements and perspectives. Bioresour. Technol. [Internet]. 2000 [cited 2015 Jan 7]; 74(1):3–16. doi:10.1016/S0960-8524(00)00023-7.

17. Chae KJ, Jang A, Yim SK, Kim IS. The effects of digestion temperature and temperature shock on the biogas yields from the mesophilic anaerobic digestion of swine manure. Bioresour. Technol. 2008; 99(1):1–6. doi:10.1016/j.biortech.2006.11.063.

18. Taconi KA, Zappi ME, Todd French W, Brown LR. Methanogenesis under acidic pH conditions in a semi-continuous reactor system. Bioresour. Technol. 2008; 99(17):8075–8081. doi:10.1016/j.biortech.2008.03.068.

19. Morita M, Sasaki K. Factors influencing the degradation of garbage in methanogenic bioreactors and impacts on biogas formation. Appl. Microbiol. Biotechnol. 2012; 94(3):575–582. doi:10.1007/s00253-012-3953-z.

20. Kadam PC, Boone DR. Influence of pH on ammonia accumulation and toxicity in halophilic, methylotrophic methanogens. Appl. Environ. Microbiol. 1996; 62(12):4486–4492.

21. Stronach SM, Rudd T, Lester JN. Anaerobic Digestion Processes in Industrial Wastewater Treatment. 1st ed. Berlin, Heidelberg: Springer-Verlag; 1986. doi:10.1007/978-3-642-71215-9.

22. Fermoso FG, Collins G, Bartacek J, Lens PNL. Zinc deprivation of methanol fed anaerobic granular sludge bioreactors. J. Ind. Microbiol. Biotechnol. 2008; 35(6):543–557. doi:10.1007/s10295-008-0315-z.

23. Lettinga G, Field J, Van Lier J, Zeeman G, Hulshoff Pol LW. Advanced anaerobic wastewater treatment in the near future. Water Sci. Technol. 1997; 35(10):5–12. doi:10.1016/S0273–1223(97)00222–9.

24. Pullen T. Anaerobic digestion–making biogas—making energy: the earthscan expert guide. London and New York: Routledge; 2014.

25. Fair GM, Moore EW. Time and rate of sludge digestion, and their variation with temperature. Sewage Work. J. 1934; 6(1):3–13.

26. Logan RP. US. Pat. No. 2,605,220 Anaerobic digester. 1952; 1–11.

27. Ward AJ, Hobbs PJ, Holliman PJ, Jones DL. Optimisation of the anaerobic digestion of agricultural resources. Bioresour. Technol. 2008; 99(17):7928–7940. doi:10.1016/j.biortech.2008.02.044.

28. Bachmann N. Design and engineering of biogas plants. In: Wellinger A, Murphy JD, Baxter D, editors. The biogas handbook: science, production and applications. 1st ed. Cambridge: Woodhead Publishing Limited, UK. 2013. p. 85–103. doi:10.1533/9780857097415.1.85.

29. Khalid A, Arshad M, Anjum M, Mahmood T, Dawson L. The anaerobic digestion of solid organic waste. Waste Manag. 2011; 31(8):1737–1744. doi:10.1016/j.wasman.2011.03.021.

30. Nizami AS, Murphy JD. What type of digester configurations should be employed to produce biomethane from grass silage? Renew. Sustain. Energy Rev. [Internet]. 2010; 14(6):1558–1568. doi:10.1016/j.rser.2010.02.006.

31. Banks CJ, Heaven S. Optimisation of biogas yields from anaerobic digestion by feedstock type. In: Wellinger A, Murphy JD, Baxter D, editors. The biogas handbook: science, production and applications. 1st ed. Cambridge: Woodhead Publishing Limited, UK; 2013. p. 131–165. doi: 10.1533/9780857097415.1.85.

32. Duarte JMC, Archer LJ, Bull AT, Holt G, editors. Perspectives in biotechnology. 1st ed. New York: Springer US; 1987.

33. Demirbas A, Ozturk T. Anaerobic digestion of agricultural solid residues. Int. J. Green Energy. 2005; 1(4):483–494. doi:10.1081/GE-200038719.

34. Viñas M, Martinez J, Baselli B. Advantages of an anaerobic reactor for TMP wastewater with separated acidogenic and methanogenic stages. Environ. Technol. 1993; 14(November 2014):995–1000. doi:10.1080/09593339309385375.

35. Rincón B, Borja R, Martín MA, Martín A. Evaluation of the methanogenic step of a two-stage anaerobic digestion process of acidified olive mill solid residue from a previous hydrolytic-acidogenic step. Waste Manag. 2009; 29(9):2566–2573. doi:10.1016/j.wasman.2009.04.009.

36. Nasr N, Elbeshbishy E, Hafez H, Nakhla G, Hesham El Naggar M. Comparative assessment of single-stage and two-stage anaerobic digestion for the treatment of thin stillage. Bioresour. Technol. 2012; 111:122–126. doi:10.1016/j.biortech.2012.02.019.

37. Demirer GN, Chen S. Two-phase anaerobic digestion of unscreened dairy manure. Process Biochem. 2005; 40(11):3542–3549. doi:10.1016/j.procbio.2005.03.062.

38. Nielsen HB, Mladenovska Z, Westermann P, Ahring BK. Comparison of two-stage thermophilic (68°C/55°C) anaerobic digestion with one-stage

thermophilic (55°C) digestion of cattle manure. Biotechnol. Bioeng. 2004; 86(3):291–300. doi:10.1002/bit.20037.

39. Liu D, Liu D, Zeng RJ, Angelidaki I. Hydrogen and methane production from household solid waste in the two-stage fermentation process. Water Res. 2006; 40(11):2230–2236. doi:10.1016/j.watres.2006.03.029.

40. Singh SP, Prerna P. Review of recent advances in anaerobic packed-bed biogas reactors. Renew. Sustain. Energy Rev. 2009; 13(6–7):1569–1575. doi:10.1016/j.rser.2008.08.006.

41. Effendi A, Zhanga Z-G, Hellgardt K, Hondaa K, Yoshidaa T. Steam reforming of a clean model biogas over Ni/Al_2O_3 in fluidized- and fixed-bed reactors. Catal. Today. 2002; 77:181–189. doi:10.1016/j.cattod.2011.07.011.

10 METHODS OF MANUFACTURING MONOCLONAL ANTIBODIES

1. INTRODUCTION: AN OVERVIEW

Monoclonal antibodies (mAbs) have been extensively utilized to effectively target a wide spectrum of extracellular substances with great precision [1]. Monoclonal antibodies (mAbs) are utilized in the diagnosis and treatment of a range of ailments including malignancies, autoimmune disorders, sexually transmitted infections (STIs), and other conditions [2, 3]. Due to notable advancements in design, the utilization of monoclonal antibodies (mAbs) has been extended in recent years. The immunogenicity in humans is being reduced, the bioavailability is being enhanced, the affinity and antigen-binding specificity are being optimized, and other advancements in protein engineering are improving the profiles of therapeutic monoclonal antibodies (Figure 1) [2].

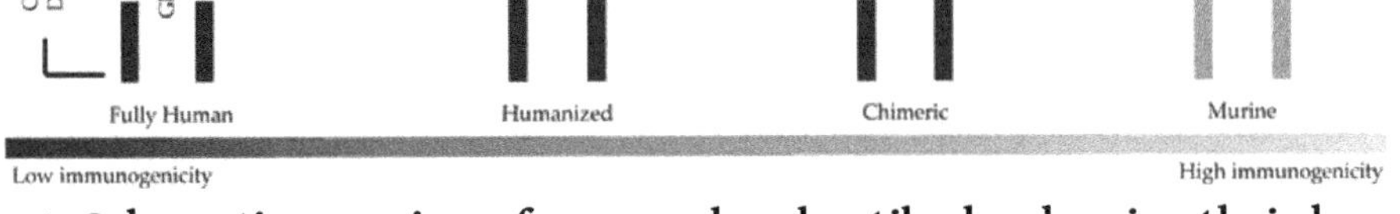

Figure 1. Schematic overview of a monoclonal antibody, showing their heavy and variable chain.

Genetic engineering has enabled the development of novel approaches for obtaining monoclonal antibodies, including the enhancement of humanized antibodies and the creation of production models [4–6]. Progress in molecular and cell biology has facilitated the generation of more effective antibodies, leading to advancements in diagnostic and therapeutic fields. These advancements have caused enhancements in production procedures, resulting in lower production costs and thus leading to a wider adoption of therapies with monoclonal antibodies (mAbs). Implementing process enhancements ensures a reliable and replicable manufacturing of significant volumes of monoclonal antibodies at a reasonable expense [4–6]. The implementation of large-scale production has significantly transformed the monoclonal antibodies market by enhancing its production capacity, rendering it a more viable and efficient production method. Molecular approaches have been the sole reason for a significant advancement in production processes [1, 7]. Typically, the commercial manufacturing of monoclonal antibodies (mAb) starts with either immunizing an animal or using molecular biology techniques to identify and optimize the DNA sequence that codes for the mAb. This is followed by constructing and identifying a stable clone that produces a high amount of the desired mAb. The advancements in cultivation methods used in this context are comparable to those employed in other bioproducts that depend on the growth of microorganisms or cells. These improvements need the creation of a meticulously planned cultivation process that encompasses a comprehensive set of control and related operations. This process is crucial for conducting technical assessments [1, 8]. The production methods of monoclonal antibodies (mAbs) in wave or single-use bioreactors (SUBs) are known for their flexibility and cost-effectiveness, in contrast to the production procedures in traditional stainless steel vats. Utilizing these production platforms in the development of bioprocesses can lead to increased acceptability by the industry [9–11]. Regulatory agencies worldwide have exercised oversight over drugs derived from monoclonal antibodies (mAbs). Hence, it is imperative to develop regulatory regulations that address the growing production and intricacies of this category of medications [10, 11]. The proposed chapter discusses the fundamental principles of monoclonal antibody manufacturing methods, focusing on techniques that involve immobilized cells, wave bioreactor systems, SUBs, and the roller bottles approach. The latest literature has provided descriptions of these approaches, which are used for producing monoclonal antibodies in mice and for producing antibodies from genetically engineered microbes.

2. MABS PRODUCTION TECHNIQUES

2.1. Hybridoma and phage display

Milstein and Köhler described the first technique developed for stable monoclonal antibody production in 1975. This technique consists of creating a hybridoma, a stable hybrid cell capable of producing a single type of antibody against a specific epitope present in an antigen. Hybridoma construction was initially produced from murine models. The technique consists of removing a pool of activated B lymphocytes from an immunized animal spleen and combining them with immortalized myeloma cells unable to produce the enzyme hypoxanthine-guanine-phosphoribosyltransferase (HGPRT), an important enzyme present in the *salvage* pathway, one of the pathways responsible for nucleotide production [1]. To select hybridoma cells, the pool of cells resulting from the fusion (a mix of hybridoma cells and non-fused B lymphocytes and myeloma cells) are cultivated in a selective medium containing aminopterin, which inhibits the nucleotide de novo synthesis. Myeloma cells lack the *salvage* pathway for nucleotide production. When they are exposed to aminopterin present in selective medium, the de novo synthesis is also blocked, and as a result, myeloma cells are no longer viable since all major pathways for nucleotide production are blocked. In contrast, non-fused, activated B lymphocytes can survive as their *salvage* pathway works perfectly and they can continue nucleotide production even if the de novo pathway is blocked by aminopterin. However, these cells are not immortalized and can replicate only a limited number of times after which they eventually die. With this in mind, only cells capable of replicating indefinitely and synthesizing nucleotides through the *salvage* pathway can survive through selection conditions, and these cells are the hybridomas.

In spite of the fact that the primary recombinant mAbs were delivered utilizing this innovation—including the first medication approved by the Food and Drug Administration (FDA) for therapeutic proposes (Table 1)—the great contribution of this technology was mostly to elucidate immune response mechanisms and control in vitro antibody production. Therefore, mAb hybridoma production from murine sources exhibits a genuine downside in human therapeutics (Figure 1).

Table 1. Monoclonal antibody-based therapeutic drugs approved by FDA (Food and Drug Administration) until 2015.

Drug name	Active ingredient	Description	Target	Therapeutic category	approval (FDA)
ACTEMRA®	Tocilizumab	Humanized IgG1κ	IL-6 receptor	Immunological	2010
ADCETRIS®	Brentuximab vedotin	Chimeric IgG1	CD30	Cancer	2011
ARZERRA®	Ofatumumab	Human IgG1κ	CD20	Cancer	2009
AVASTIN®	Bevacizumab	Humanized IgG1	VEGF	Cancer	2004
BENLYSTA®	Belimumab	Human IgG1λ	BLyS	Immunological	2011
BEXXAR	Tositumomab; iodine I 131 tositumomab	IgG2αλ, I131	CD20	Cancer	2003
BLINCYTO	Blinatumomab	BiTE antibody-scFvs	CD19/CD3	Cancer	2014
CAMPATH (LEMTRADA™)	Alemtuzumab	Humanized IgG1κ	CD52	Immunological	2001
CEA-SCAN	Arcitumomab	Murine IgG1 Fab'	CEA	Diagnosys	1996
CIMZIA®	Certolizumab pegol	Humanized Fab', PEG	TNFα	Immunological	2008
COSENTYX®	Secukinumab	Human IgG1κ	IL-17A	Immunological	2015
CYRAMZA	Ramucirumab	Human IgG1	VEGRF-2	Cancer	2014

Drug name	Active ingredient	Description	Target	Therapeutic category	approval (FDA)
DARZALEX	Daratumumab	Human IgG1κ	CD38	Cancer	2015
HERCEPTIN®	Trastuzumab	humanized IgG1κ	HER2	Cancer	1998
EMPLICITI™	Elotuzumab	Humanized IgG1	SLAMF7	Cancer	2015
ENTYVIO	Vedolizumab	Humanized IgG1	α4β7 Integrin	Immunological	2014
ERBITUX®	Cetuximab	Chimeric IgG1	EGFR	Cancer	2004
GAZYVA®	Obinutuzumab	Humanized IgG1	CD20	Cancer	2013
HUMIRA	Adalimumab	Human IgG1	TNF	Immunological	2002
ILARIS	Canakinumab	Human IgG1κ	uman-IL-1β	Immunological/ anti-inflammatory	2009
KADCYLA®	Ado-trastuzumab emtansine	Humanized IgG1; DM1	HER2	Cancer	2013
KEYTRUDA®	Pembrolizumab	Humanized IgG4κ	PD-1	Cancer	2014
LEMTRADA™	Alemtuzumab	Humanized IgG1κ	CD52	Immunological	2001
LUCENTIS	Ranibizumab	Humanized IgG1κ	VEGF-A	Ophthalmic	2006
Muromomab	Orthoclone	Murine IgG2α	CD3	Immunological	1992
Mylotarg®	Gemtuzumab ozogamicin	Humanized IgG4κ, calicheamicin	CD33	Cancer	2000
MYOSCINT®	Imciromab Penlelale	Murine IgG2/4κ Fab';DTPA	Heavy chain of human myosin	Detection of myocardial injury	1996
NUCALA®	Mepolizumab	Humanized IgG1κ	IL-5	Immunological	2015
OPDIVO	Nivolumab	Human IgG4κ	PD-1	Cancer	2014
PERJETA®	Pertuzumab	Humanized IgG	HER2/neu receptor	Cancer	2012
PORTRAZZA	Necitumumab	Human IgGκ	EGFR	Cancer	2015
PRALUENT™	Alirocumab	Human IgG1	PCSK9	Lipid-lowering	2015
PRAXBIND®	Idarucizumab	Humanized IgG1 Fab	Dabigatran (anticoagulant)	Hemostasis	2015
XGEVA®	Denosumab	Human IgG2	RANKL	Bone disorders	2010
ProstaScint®	Capromab pendetide	Murine IgG1κ, GYK-DTPA-HCl	PSMA	Cancer	1996
RAPTIVA®	Efalizumab	Humanized IgG1κ	CD11a	Immunological	2003
RAXIBACUMAB	Raxibacumab	Human IgG1λ	PA of B. Anthracis toxin	Anti-toxin	2012
REMICADE®	Infliximab	Chimeric IgG1κ	TNFα	Immunological	1998
ReoPro®	Abciximab	Chimeric IgG1κ Fab	GPIIb/IIIa	Hemostasis	1993

Drug name	Active ingredient	Description	Target	Therapeutic category	approval (FDA)
REPATHA	Evolocumab	Human IgG2	PCSK9	Lipid-lowering	2015
RITUXAN®	Rituximab	Chimeric IgG1κ	CD20	Cancer	1997
SIMPONI	Golimumab	Human IgG1κ	TNFα	Immunological	2009
SIMULECT®	Basiliximab	Chimeric IgG1κ	IL-2 receptor	Immunological	1998
SOLIRIS®	Eculizumab	Humanized IgG2/4κ	C5	Hemostasis	2007
STELARA®	Ustekinumab	Human IgG1κ	IL-12 and IL 23	Immunological	2009
SYLVANT	Siltuximab	Chimeric IgG	IL-6	Immunological	2014
SYNAGIS®	Palivizumab	Humanized IgG1κ	RSV F	Antiviral	1998
NeutroSpec™	FanolesomaB; technetium Tc 99m	Murine IgM	3-fucosyl-N-acetyllactosamine	Diagnosys	2004
TYSABRI	Natalizumab	Humanized IgG4κ	α4β1/α4β7 integrins	Immunological	2004
UNITUXIN™	Dinutuximab	Chimeric IgG1κ	Glycolipid GD2	Cancer	2015
VECTIBIX®	Panitumumab	Human IgG2κ	EGFR	Cancer	2006
VERLUMA™	Nofetumomab	Murine IgG2b Fab	Glycoprotein antigen expressed in a variety of cancers	Diagnosys	1996
XGEVA	Denosumab	Human IgG2	RANKL	Cancer	2010
XOLAIR®	Omalizumab	Humanized IgG1κ	Human IgE	Immunological	2003
YERVOY®	Ipilimumab	Human IgG1κ	CTLA-4	Cancer	2011
ZENAPAX®	Daclizumab	Humanized IgG1	IL-2 receptor	Immunological	1997
ZEVALIN®	Ibritumomab tiuxetan	murine IgG1κ, Yttrium-90	CD20	Cancer	2002

After a few infusions, murine antibody molecules trigger the human anti-mouse antibody (HAMA) response of the human immune system [1, 12]. To work around this issue, new methodologies have been developed to deliver antibodies similar to human molecules, so the technology evolved to less immunogenic chimeric antibodies (constant regions of human antibodies linked to the variable region of the murine source), creating a new set of therapeutic possibilities (Figure 1). Subsequently, the need for an even less immunogenic alternative boosted the production of humanized antibodies (only the region that interacts with the antigen epitope is from mouse origin) (Figure 1). Even fully human antibodies (Figure 1) can be produced from genetically modified mice [13].

A great improvement in mAb production has come with the development of phage display libraries. This methodology helps to investigate interactions between molecules (protein-protein, protein-peptide, and protein-DNA) and consists, basically, in cloning Fab-region-coding genes amplified from B lymphocytes into bacteriophage plasmid vectors. Then the bacterium can be transformed with these vectors, going on to express the heterologous genes from a viral capsid. This capsid contains viral proteins and proteins encoded by the Fab sequence received by that specific cell. Once the library is complete, the affinity between proteins produced from different Fab regions can be tested against the antigen of interest and the cell transformed with the plasmid that contains those genes can be readily sequenced. The advantages of this methodology are the following: the same library has the potential to generate a great number of new antibodies, it is an in vitro process, so it does not require animal immunizations steps, and because of that, toxic antigens can be tested. Also, a greater variety of antigens can be tested, and antibody molecules can be rapidly obtained [13].

2.2. Culture production factors

2.2.1. Cell lines

Selecting the cell line is a crucial stage in the development of a mAb production system. The cells need to exhibit stability and efficiently produce the target protein in its right shape at elevated levels. The mammalian cell is the preferred expression system for producing monoclonal antibodies (mAbs) based on these criteria. A key benefit of utilizing a mammalian expression system is that the intracellular apparatus is specifically tailored for the synthesis, modification, and release of exceedingly intricate compounds. Most commercial monoclonal antibodies (mAbs) are manufactured using Chinese hamster ovary (CHO) and NS0 cells. These cells are derived from plasmacytoma cells that have been genetically engineered to produce IgG in nonsecreting B cells. Genetic alterations in CHO cells have resulted in the development of cell lines that can produce a large amount of humanized monoclonal antibodies. The cell lines demonstrated the capacity to release a maximum of 100 picograms per cell per day [14]. Further alterations resulted in a significant increase in the synthesis of a chimeric monoclonal antibody, with levels ranging from 80 to 110 picograms per cell per day [15]. Furthermore, there have been alterations in NS0, resulting in increased rates of monoclonal antibody (mAb) synthesis, ranging from 20 to 50 pg/cell/day [16]. Hybridoma cell lines are utilized in industrial monoclonal antibody (mAb) manufacturing, but in smaller amounts. According to a publication, certain hybridoma strains can produce up to 80 picograms per cell each day [16]. However, various mammalian cell lines and unconventional expression systems, including genetically modified plant cells, genetically modified insect cells, and genetically modified microorganism cells, have also been utilized for monoclonal antibody (mAb) production and have become increasingly important in the biopharmaceutical industry [1, 8]. Genetically engineered microorganisms have garnered significant attention in the industry due to their ease of handling and modification compared to animal cells. Additional benefits of manufacturing methods utilizing genetically modified microorganisms include the presence of clearly defined expression systems within these cells, as well as the reproducibility and ease of validation of the production process. Pichia pastoris, a kind of yeast, has significant potential for practical applications due to its ability to secrete large amounts of foreign proteins. Yeast cultivation systems for monoclonal antibody (mAb) manufacturing offer advantages in terms of scalability and cost-effectiveness compared to mammalian cell cultivation systems. They can be grown in conventional stirred tank bioreactors using batch or feed-batch modes of operation. Typically, microorganisms lack the necessary physicochemical and biological traits to properly express and digest monoclonal antibodies (mAbs) [4]. Genetically modified plants have also garnered attention due to their ease of cultivation and propagation. Plant production offers several cultural benefits, including cost-effective medium, little maintenance expenses, and high production yields, making it a more economical option compared to mammalian cell cultures [17]. Nevertheless, the utilization of plant cells can be challenging due to many restrictions, such as distinct glycosylation patterns and post-translational processing [17].

2.2.2. Culture medium

Cultivation media for mammalian cells must have a complex content of ingredients ranging from amino acids to trace elements. To supply the cellular demand of these nutrients, the culture medium uses serum in its composition, however, due to the emergence of diseases caused by defective prions, such as bovine spongiform encephalitis (BSE), there is a great incentive to remove any animal component of culture media composition, especially if the medium is used for industrial production of biopharmaceuticals products. This has led to the emergence of media free from any animal components, including well-defined media for CHO and NS0, the two most utilized cell types in mAb production. The development of a proper medium can be time consuming and very expensive. However, many companies prefer to develop their own production media to maintain the composition between production lots as well as develop an appropriate medium composition for the specific cell type that will be used and to achieve greater control over production. Added to this, the development of downstream processes that meet the requirement for high-purity products and tests to validate the final product quality raises the overall production cost of a drug based on monoclonal antibodies [1].

Despite the complexity of developing a culture medium, much progress has been made in this area, allowing for greater cell growth and increasing cell conservation time in suitable conditions for the growth and production of molecules of interest [8].

2.2.3. Culture conditions

Growing conditions can directly influence the cell growth and production levels of molecules of interest. Usually, mammalian cell culture conditions for mAb production are very well defined: 37 °C, pH 7.15, and dissolved O_2 (OD) levels at 30–60%. CO_2 level is monitored to mimic the physiological standard between 31 and 54 mmHg. However, changes in cellular conditions have shown great potential to change cellular metabolism toward cellular growth or molecule production and this can be used to increase mAb production. Bioprocesses can be designed to occur in two phases. First, cell growth is optimized to reach a certain cell density. Once this density is reached, the second phase begins and the bioreactor conditions are shifted so the cells continue to grow just at a maintenance rate and directing the metabolism toward monoclonal antibody production. Some CHO cell strains and hybridoma cells are sensitive to changes in temperature and pH. When subjected to temperature and pH values lower than those normally used, values between 30 and 35 °C and 6.7–7.0, respectively, cell growth metabolism is reduced and specific production increases. The growth metabolism reduction also contributes to lower production of some metabolic compounds which are toxic for cell cultures, allowing increased cell viability, which spend more time producing molecules of interest. A good way to monitor the growth stage of a cell culture for controlling changes in cultivation is watching the DO and pCO_2 levels, which can also be adjusted to maximize the production of proteins such as mAbs [1].

2.2.4. Production platforms

The cell culture for mAb production can follow three different types of processes. The simplest of them is batch production, which consists of a closed system where a bioreactor is sterilized and prepared with a medium containing all the nutrients needed for cellular growth and product manufacturing and then, cells are inoculated. There is no feeding system with fresh medium or withdrawal of spent medium. As the process runs, nutrient concentration decreases and waste metabolites are produced, lowering cell viability. In spite of being a simple process, batch is not the most suitable type of production platform for mammalian cell cultures, as the environment inside the reactor quickly becomes unfavorable for cell growth and, at the same time, waste product concentration increases. Cultivation factors such as initial nutrient concentration and waste metabolite production directly determine the maximum concentration that cells can reach in a bath culture. Generally, this type of cultivation reaches a maximum density of $1–2 \times 10^6$ cells/mL, and then the cell viability drops rapidly [1]. The production process lasts for 4–7 days, when productivity reaches certain concentration of interest [1]. Supernatant is collected and the product is recovered by downstream processes. The time that each batch takes to finish also depends on the production kinetics. If the production is growth dependent (production occurs concomitantly with cellular growth), batch processes can be stopped as soon as cells reach the stationary phase. But if the product is not associated with growth (production only starts when the growth rate decreases), the culture needs to be carried for a longer period of time since production only starts at stationary phase.

In contrast to batch, a second type of production process utilized is continuous fermentation. There are two types of continuous production: chemostat cultures and perfusion cultures. Concerning chemostat cultures, fresh medium is added to the bioreactor and fermented medium is removed along with cells at a constant flow rate so that the culture volume remains unchanged. The flow rate (dilution rate) controls cellular growth and when these two variables are equal, the bioreactor reaches equilibrium—cell concentration, nutrient concentration, and product concentration are held constant. In this context, the culture can be kept in equilibrium for several months reaching a cell density of 10–30 × 10^6 cells/mL [1]. To avoid viable cell loss along with the constant outflow of the by-products of cell metabolism, many manufacturing plants have developed a cell-recycling system and thus, the perfusion culture method was developed where cells are kept inside the bioreactor. The disadvantages of continuous fermentation are the use of a large amount of expensive culture media and the difficulty in recovering the product, which comes out fairly diluted. These two disadvantages are consequences of the constant medium flow rate. To work around the product dilution problem, some production manufacturing plants have ultrafiltration systems which retain the product inside the bioreactor [18]. Another obstacle of this type of process is that the establishment of culture conditions for a stable industrial production plant can take months. For this to occur, the strain used must be very stable and have its physiological aspects clearly elucidated, such as growth rate, productivity, and response to certain stress conditions. It is not uncommon to hear that numerous attempts are made before the settlement of a stable production plant is achieved, but, once settled, this production process can bring many advantages, since it can be operated in smaller-volume bioreactors, and therefore have greater production flexibility.

The third type of process for producing monoclonal antibodies is by far the most utilized at industrial scale, which is fed-batch process. In this process, the cell density reaches 8–12 × 10^6 cells/mL, and cell viability in the bioreactor is enhanced by controlled nutrient addition at specified intervals [1]. The production process can take 12–20 days [1]. Usually, the same medium used in the initial culture is also used for feeding, but in a more concentrated version. The feeding solution composition can be designed to supply the cells based on their metabolic state at different culture phases by analyzing and identifying the spent medium nutrients that are being more consumed. Furthermore, the medium used in feeding can be modified to promote cell growth or to stimulate molecule production, since different components may modify the behavior of cells, changing the metabolism for different purposes. The feed solution can also be designed to minimize the production of waste metabolites that cause cell stress when in excess. However, their production is not completely avoidable as they eventually reach harmful concentrations. It is relatively easy to scale up and operate this system. More summarized data about the advantages and disadvantages of each process for mAb production can be seen in Table 2.

Table 2. Comparison between different operation modes that can be used for mAbs production.

Production platform	Batch	Feed-batch	Perfusion culture
Advantages	• Simple to scale-up • Control by production lot	• Simple to scale-up • Control by production lot • Production facility is simple • Process is easy to perform and to validate • Higher volumetric production	• Cells are maintained in a relatively optimal biochemical environment • Culture reaches high cell density • Higher volumetric production
Disadvantages	• Difficult to define initial concentration of nutrients • Accumulation of waste metabolites • Degradation of more sensitive products	• Accumulation of waste metabolites • Degradation of more sensitive products	• Lack of homogeneity in the continuous reactor vessel • Challenges regarding long-term operability and maintenance • High cost and long times required for process development experiments • Genetic instability of cells

Adapted from [1, 21] (colocar referencias).

A lot of effort has been made to increase cell longevity in batch and feed-batch modes of operation. It is expected that the longer the cells are maintained viable, the greater the antibodies' production will be. So, in order to maintain cell viability, some culture parameters can be optimized, such as culture media, feed solution, and mAb secretion rates and by-product production. To improve mAb titers in the batch platform, the start medium can be supplemented with glucose and amino acids, increasing mAb production up to eightfold when compared with regular media [9, 26]. Improvements for the fed-batch platform can be achieved by adjustments in feed solution, as mentioned before. Feed solutions containing glucose and aminoacids/glutamine have been reported to increase mAb titers from two to fourfold, reaching production of up to 2 g/L, when compared with the batch production platform [19].

The optimization of the antibody secretion rate can be achieved by high-density cell cultivation. On a fed-batch platform, a high cell cultivation culture can reach an mAb productivity rate of 0.94 g/L/day and a final titration of 17 g/L, while a continuous culture performed at high density conditions can reach final titration and productivity rates of 0.8 and 1.6 g/L/day, respectively [20]. Optimizing mAb secretion highly depends on the cell line chosen for production. Each cell strain can be influenced by the manufacturing conditions and respond differently to increasing or decreasing mAb production and secretion [19]. The accumulation of toxic by-products is a great bottleneck in manufacturing processes since they can inhibit cell growth and then directly affect mAb production. Although a few strategies to minimize this by-product accumulation have shown to be promising, some are not applicable for a large-scale production. Optimizing medium composition and feed solutions with substrates that reduce toxic compound production is the most common strategy used at industrial scales of production [19].

Although most mAbs are produced by fed-batch process, there are tendencies indicating that in the future many bioprocesses will be operated in continuous platforms, especially for the production of biopharmaceuticals. On these platforms, the production system will be coupled to upstream and downstream processes [21]. However, for this to actually happen, a great improvement in technological development still needs to be achieved.

2.3. Production systems

Large-scale synthesis of monoclonal antibodies for therapeutic purposes surpasses the production capacity of laboratory-scale production (Figure 2). Multiple production methods have been created and improved over time, with new options appearing. Commercial-scale production of monoclonal antibodies (mAbs) can be carried out using adherent cells or suspension cells. However, suspension cells are the preferred choice due to their widespread use and more advanced production methods for cell cultivation, resulting in higher efficiency. Therefore, the process of increasing the scale using suspension cells is more straightforward. One additional benefit of the suspension production system is that it eliminates the need for a bioreactor with a high surface area for cell adhesion. This is because the productivity of adherent cells is directly proportional to the size of the bioreactor [22].

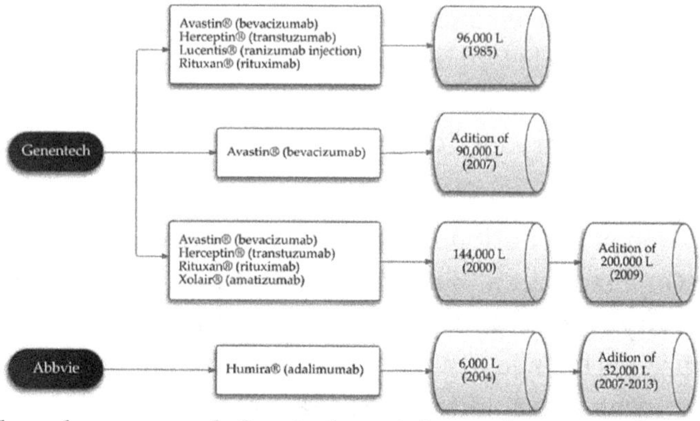

Figure 2. Work volumes used for industrial production of some commercial monoclonal antibodies [27, 28].

Some cultivation issues and worries have arisen regarding the production scale increase, maintenance of product quality, contamination control, demand for oxygen supply, and control over DO and CO_2 removal, among others. Regarding suspension cell cultures, aeration is in part dependent on the agitation of the culture inside the bioreactor, which can lead to cell shear stress. To work around cultivation problems, major advances have been made in the process itself by developing better culture control and conditions, as well as the improvement and development of new bioreactors [7, 23].

2.3.1. Production systems for cells in suspension cultures

The various bioreactors frequently employed for monoclonal antibody (mAb) synthesis in submerged mammalian cells include stainless steel stirred tank bioreactors (STR), air-lift reactors, and disposable bioreactors. Further elaboration on each of these bioreactors will be provided as under.

2.3.2. Stainless steel stirred tank bioreactors

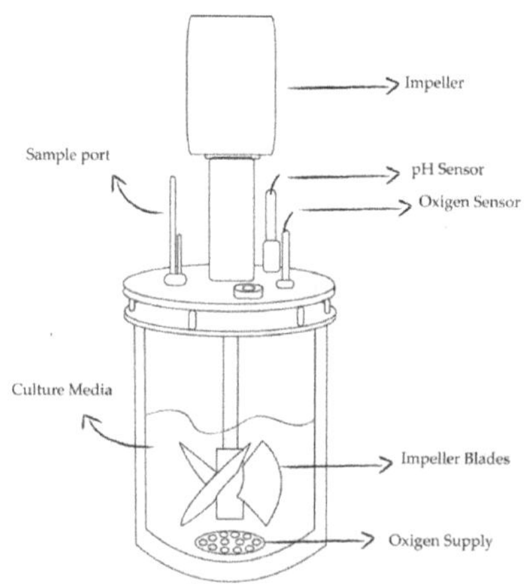

Figure 3. Schematic representation of a stainless steel stirred tank bioreactor. Showing the main components in a cell cultivation.

Stainless steel stirred tank bioreactors are the most consolidated type of bioreactor used for industrial mAb production and consist of baffle-stirred tanks linked to rotor systems (Figure 3). It is a consolidated system, and there is a lot of knowledge and experience surrounding this technology, acquired by its vast industrial use beyond production using mammalian cells.

The cultivation in this bioreactor allows for wide flexibility of working volumes, ranging from 1.0 to 25.0 L [1], since this system is easily scalable to larger volumes due to its high control over production conditions and extensive handling knowledge. The mechanisms and cleaning and sterilization protocols are well defined. Additionally, cultivation parameters for this system, such as gas transfer coefficient, agitation, aeration, temperature maintenance, pH, and others are well controlled and regulated when compared to other production systems. Another advantage of the STR is that it can be used for cultivation of various cell types and in addition, the products obtained from the cultivation in this type of bioreactor are easily approved for therapeutic use, as regulatory terms are well defined for this type of production [11].

However, the biggest disadvantage for the use of STR is the stress caused by shear. It can cause cell lysis and lead to loss in mAb productivity.

2.3.3. Air-lift reactors

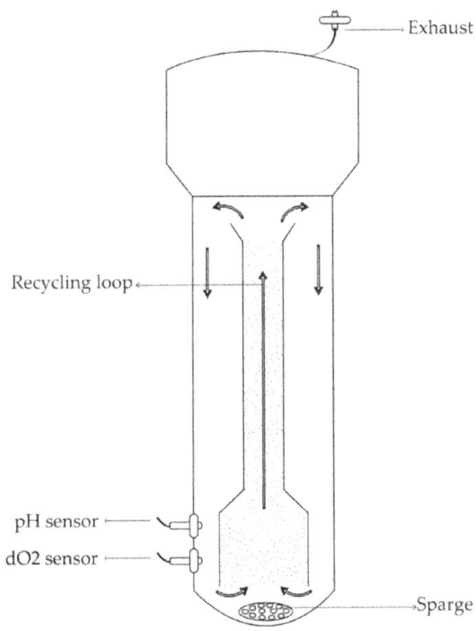

Figure 4. Schematic representation of an air-lift bioreactor. Showing the main components in a cultivation process.

Air-lift reactors are also broadly used for the industrial production of mAbs. The reactor consists of tanks with a bubble column inside, and air is injected into the column base (Figure 4). The air flows through the column's length to the top of the bioreactor as degassed culture medium flows in the opposite direction to the reactor bottom. This creates a constant gentle mixing of the medium as well as proper culture aeration, annulling part of the shear stress caused by other stirring systems. Other advantages of this operation system are that it is easier to scale-up, contamination problems are more unlikely to occur, and the equipment is simpler. In spite of these advantages, this system is less utilized than STR reactors because the working volume ranges only from 2.0 to 5.0 L [1] and the air-lift reactor handling is not so well elucidated [11].

2.3.4. Disposable bioreactors

The first single-use bioreactors emerged in the late 1990s with the launch of a wave reactor system. After that, disposable stirred tank bioreactors were developed [11].

This method brought many advantages for mAb manufacturing. At the end of the process, the bioreactor is discarded and replaced by a new clean and sterile one. This eradicates cross contamination between batches and decreases the time consumed with the equipment preparation between batches. When all the advantages of this process are taken in account, the savings made regarding production and investment capital are highly significant when compared with other process methods. The great disadvantage of this production system is the small work volume supported, ranging from 50 to 2000 L [1].

The wave system consists of a sterile plastic bag (CellBag™) lying on a rocking platform (Figure 5). The bag is half filled with cultivation medium and half filled with a gas mix of interest. The platform motion creates an undulation movement in the culture, ensuring efficient aeration and culture mixing without causing shear damage [10, 11, 13]. The other available systems combine the convenience of a disposable system with the well-known stirred tank system and they are HyClone S.U.B®, Millipore® (CellReady™), or Xcellerex® (XDR™).

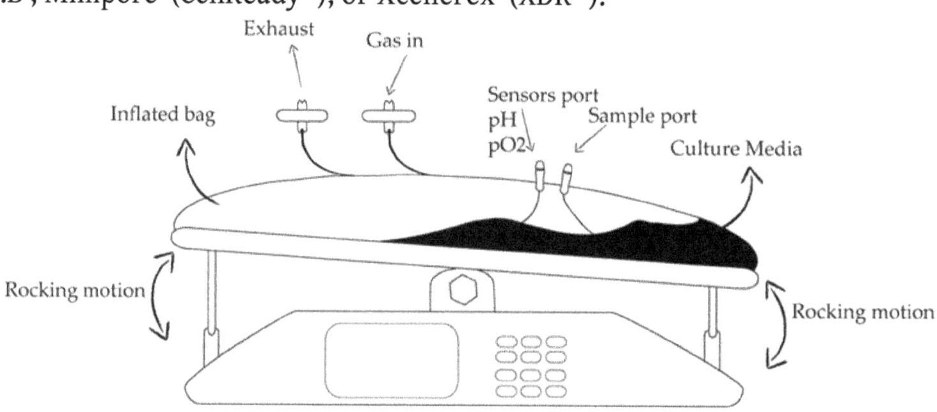

Figure 5. Schematic representation of a disposable wave bioreactor. Showing the main components in a cell cultivation process.

The main features of SUBs are related to their technical characteristics similar to those of stainless steel bioreactors, that is, aeration rate, agitation, reactor geometry, and ease of monitoring internal conditions, a process similar to stainless steel bioreactors [9].

SUBs are being widely used to replace many processes for the production of bioproducts. SUBs may be a cheaper and more efficient alternative from an industrial point of view, and its principle can easily replace any bioprocess to adapt the method to the platform of interest to be replaced, such as large tanks and stainless steel or the motion rocking platforms [9, 24].

SUBs have been used in bioprocesses for monoclonal antibody production involving several expression systems, including mammalian cells, microorganisms, plants, mammary glands, etc. Animal cell culture technology is one of the oldest techniques for the production of mAbs.

There is also the production of bottles known as roller bottles, consisting of mammalian cells growing in nutritional and physical conditions controlled in bottles which remain in rotational movement.

2.3.5. Roller bottles

Roller bottles are a rotary motion system for growing cells and for the production of some bioproducts. It has been an alternative to other monoclonal antibody production systems (Figure 6). Roller bottles provide conditions that favor the transfer of oxygen and temperature control without aeration, agitation propellers, or circulation pumps. The bottle is mounted on a turntable which gives homogeneity of growth and aeration of the culture medium [11, 25, 28].

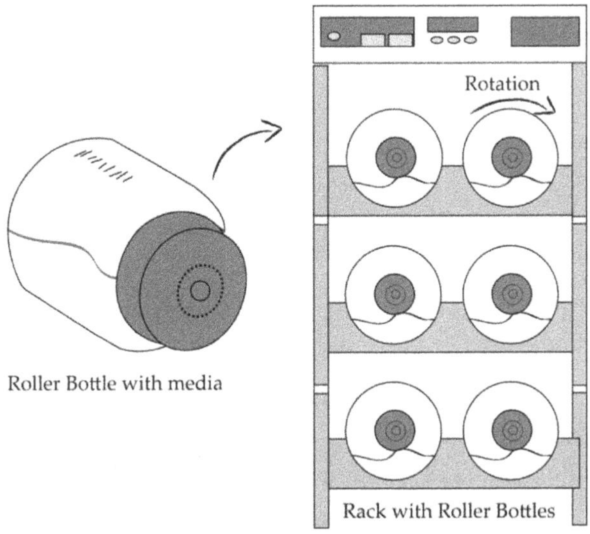

Figure 6. Schematic representation of roller bottles bioreactor and a rack with the rotational motion system in a cultivation for mAb production.

For the production of monoclonal antibodies at commercial scale, the roller bottle technique can be adapted to racks containing tens of bottle in a production line. The advantages of this technique is the high growth potential linked to ease of handling and monitoring of certain conditions such as temperature and rotation. However, the scale of view requires a large physical footprint, which can make the process less economical [11, 25].

3. CONCISE SUMMARY

Monoclonal antibodies constitute 50% of the biopharmaceuticals that are currently being sold, with a market value of $75 billion. In recent years, there have been significant advancements in the development methodology of antibodies, particularly in the production processes. These advancements have enabled the scaling up of production and facilitated the introduction of antibodies into the market. It is anticipated that the need for high-quality biologics will continue to rise in the future decades. In general, the processes used for scheduling bioproducts/biosimilars that depend on culturing microorganisms or cells are similar to those used for other bioproducts/biosimilars. These processes involve developing a carefully planned culturing process that includes a comprehensive range of control and associated operations to facilitate technical evaluations. We are currently confronted with a significant problem due to the growing demands of regulatory authorities for improved quality and reduced process costs within the healthcare systems. Modifications to the plant design are needed in order to provide highly adaptable and versatile facilities capable of accommodating small-scale production volumes.

REFERENCES

1. Chartrain M, Chu L. Development and production of commercial therapeutic monoclonal antibodies in Mammalian cell expression systems: an overview of the current upstream technologies. Curr Pharm Biotechnol 2008;9:447–67. doi:10.2174/138920108786786367.
2. Ecker DM, Jones SD, Levine HL. The therapeutic monoclonal antibody market. MAbs 2015;7:9–14. doi:10.4161/19420862.2015.989042.
3. Reichert JM. Metrics for antibody therapeutics development. MAbs 2010;2:695–700. doi:10.4161/mabs.2.6.13603.
4. Potgieter TI, Cukan M, Drummond JE, Houston-Cummings NR, Jiang Y, Li F, et al. Production of monoclonal antibodies by glycoengineered Pichia pastoris. J Biotechnol 2009;139:318–25. doi:10.1016/j.jbiotec.2008.12.015.
5. Li F, Vijayasankaran N, Shen A, Kiss R, Amanullah A. Cell culture processes for monoclonal antibody production. MAbs vol. vol. 2(5) 2010;2:466–79. doi:10.4161/mabs.2.5.12720.

6. Jones ML, Seldon T, Smede M, Linville A, Chin DY, Barnard R, et al. A method for rapid, ligation-independent reformatting of recombinant monoclonal antibodies. vol. 354. 2010. doi:10.1016/j.jim.2010.02.001.
7. Raven N, Rasche S, Kuehn C, Anderlei T, Klöckner W, Schuster F, et al. Scaled-up manufacturing of recombinant antibodies produced by plant cells in a 200-L orbitally-shaken disposable bioreactor. Biotechnol Bioeng 2015;112:308–21. doi:10.1002/bit.25352.
8. Chon JH, Zarbis-Papastoitsis G. Advances in the production and downstream processing of antibodies. N Biotechnol 2011;28:458–63. doi:10.1016/j.nbt.2011.03.015.
9. Diekmann S, Dürr C, Herrmann A, Lindner I, Jozic D. Single use bioreactors for the clinical production of monoclonal antibodies – a study to analyze the performance of a CHO cell line and the quality of the produced monoclonal antibody. BMC Proc 2011;5:P103. doi:10.1186/1753-6561-5-S8-P103.
10. Tang YJ, Ohashi R, Hamel JFP. Perfusion culture of hybridoma cells for hyperproduction of IgG 2a monoclonal antibody in a wave bioreactor-perfusion culture system. Biotechnol Prog 2007;23:255–64. doi:10.1021/bp060299a.
11. Singh V. Disposable bioreactor for cell culture using wave-induced agitation. Cytotechnology 1999;30:149–58. doi:10.1023/A:1008025016272.
12. Liu JKH. The history of monoclonal antibody development – progress, remaining challenges and future innovations. Ann Med Surg 2014;3:113–6.
13. Chames P, Van Regenmortel M, Weiss E, Baty D. Therapeutic antibodies: successes, limitations and hopes for the future. Br J Pharmacol 2009;157:220–33.
14. Page MJJ, Sydenham MA. High level expression of the humanized monoclonal antibody Campath-1H in Chinese hamster ovary cells. Nat Biotechnol 1991;9:64–8.
15. Fouser LA, Swanberg SL, Lin BY, Benedict M, Kelleher K, Cumming DA, et al. High level expression on a chimeric anti-ganglioside GD2 antibody: genomic kappa sequences improve expression in COS and CHO cells. Biotechnology (NY) 1992;10:1121–7.
16. Savinell JM, Palsson BO. Network analysis of intermediary metabolism using linear optimization. J Theor Biol 1992;154:455–73. doi:10.1016/S0022-5193(05)80162-6.
17. Moussavou G, Ko K, Lee J-H, Choo Y-K. Production of monoclonal antibodies in plants for cancer immunotherapy. Biomed Res Int 2015;2015:306164. doi:10.1155/2015/306164.

18. Rodrigues ME, Costa AR, Henriques M, Azeredo J, Oliveira R. Technological progresses in monoclonal antibody production systems. Biotechnol Prog 2009;26:332–51. doi:10.1002/btpr.348.
19. Bibila TA, Robinson DK. In pursuit of the optimal fed-batch process for monoclonal antibody production. Biotechnol Prog 1995;11:1–13. doi:10.1021/bp00031a001.
20. Chang HN, Jung K, Choi J-D-R, Lee JC, Woo H-C. Multi-stage continuous high cell density culture systems: a review. Biotechnol Adv 2014;32:514–25.
21. Croughan MS, Konstantinov KB, Cooney C. The future of industrial bioprocessing: batch or continuous? Biotechnol Bioeng 2015;112:648–51. doi:10.1002/bit.25529.
22. Shukla AA, Thömmes J. Recent advances in large-scale production of monoclonal antibodies and related proteins. Trends Biotechnol 2010;28:253–61. doi:10.1016/j.tibtech.2010.02.001.
23. Munro TP, Mahler SM, Huang EP, Chin DY, Gray PP. Bridging the gap: facilities and technologies for development of early stage therapeutic mAb candidates. MAbs 2011;3:440–52. doi:10.4161/mabs.3.5.16968.
24. Ferrara N. Pathways mediating VEGF-independent tumor angiogenesis. Cytokine Growth Factor Rev 2010;21:21–6. doi:10.1016/j.cytogfr.2009.11.003.
25. Chu L, Robinson DK. Industrial choices for protein production by large-scale cell culture. Curr Opin Biotechnol 2001;12:180–7. doi:10.1016/S0958-1669(00)00197-X.
26. Jain E, Kumar A. Upstream processes in antibody production: evaluation of critical parameters. Biotechnol Adv 2008;26:46–72. doi:10.1016/j.biotechadv.2007.09.004.
27. Genentech Manufacturing 2009. http://www.gene.com/media/company-information/manufacturing (accessed March 10, 2016).
28. Tebbey PW, Varga A, Naill M, Clewell J, Venema J. Consistency of quality attributes for the glycosylated monoclonal antibody Humira® (adalimumab). MAbs 2015;7:805–11. doi:10.1080/19420862.2015.1073429.

11 LIPOPEPTIDES ARE SYNTHESIZED USING FERMENTATION PROCESSES: ENDOPHYTIC BACTERIA, FERMENTATION STRATEGIES AND EASY METHODS FOR BACTERIAL SELECTION

1. INTRODUCTION: AN OVERVIEW

The biotechnological application of microbial lipopeptides (LPs) has received significant attention in several fields such as the pharmaceutical industry, food preservation, and agriculture. Lipopeptides possess a wide range of highly organized variations and have the capacity to reduce both surface and interfacial tension. Structurally, they comprise of a peptide that is hydrophilic and a fatty acid acyl chain that is hydrophobic. The range of amino acids typically falls between 7 and 25, whereas the length of fatty acid chains vary from 13 to 17 carbons. A single strain has the capability to generate many isoforms of a certain polypeptide. The lipopeptides produced by Bacillus and Paenibacillus (firmicutes) as well as Pseudomonas (Proteobacteria) have been extensively researched [1, 2]. In addition, LPs can also be synthesized by Streptomyces [3, 4] and fungal strains [5]. The LPs exhibit significant variability, with their structural equivalents arising from numerous amino acid changes. Some of the most well-documented lipopeptides (LPs) produced by Bacillus strains include surfactin, iturin, fengycin, and lichenysin. However, Pseudomonas strains produce viscosin, tensin, arthrofactin, massetolid, pseudodesmin, xantholysin, pseudofactin, and syringomicin. These lipopeptides, like many others, are promising candidates for replacing antibiotics and antifungals that have no impact on controlling harmful microbes.

Lipopeptide surfactants are naturally occurring compounds that consist of several macromolecules from the same family or class. The nutritional parameters have the ability to affect the characteristics of the generated LPs [6]. Nevertheless, the primary constraints on their manufacturing process are the expenses associated with production and the quantity of output. Various carbon sources and culture conditions have been documented to enhance the production of iturins, surfactins, and fengycins. Currently, a diverse range of inexpensive raw materials such as rice bran, soybean, potato peels, and molasses are being utilized in their manufacturing process. Furthermore, research has shown that divalent cations, specifically Mn2+ and Fe2+, play a significant role in LP synthesis. The inclusion of manganese in the medium culture resulted in a significant increase in yield rates, rising from 0.33 to 2.6 g/L [7]. In addition, the inclusion of ZnSO4, FeCl3, and MnSO4 resulted in an augmentation in surfactin production in Bacillus subtilis [8]. In this context, we will focus on the essential factors for maximizing the production of LPs and developing future techniques to optimize both liquid and solid-state fermentations (SSFs). Both types of fermentation play a crucial role in large-scale industrial manufacturing processes.

Another crucial aspect to consider here pertains to the strains that produce microorganisms. The genetic burden of the microorganisms plays a crucial role in influencing the production yields of LPs, as the ability to produce a metabolite is regulated by genes. Hyperproducer strains are necessary. However, how can we identify these bacteria that are creating an excessive amount? The formation of lipopeptides can be detected using many approaches, including: (i) culture-dependent techniques, (ii) surface analysis and emulsifying activity-based methods, (iii) assessment of cell surface hydrophobicity, and (iv) chemical identification [9]. An ideal and universally acknowledged study should include the genetic identification and the structural and genetic characterization of the generated LPs by the specific isolate. This methodology ensures the morphological and genotypic characteristics of the microbial isolates for the generation of LPs. At present, PCR and gene sequencing are efficient methods for rapidly screening and identifying microbial producers, as well as determining the genes responsible for LP production. Moreover, liquid chromatography coupled to mass spectrometry (LC-MS) and the more recent MALDI-TOF mass spectrometry techniques have been regarded as the swiftest and most effective methods for isolating and identifying LPs in mixtures and sequencing peptides, respectively. These approaches are also valuable for identifying the most innovative LPs and even the minor alterations in the sequence of amino acids that will define its qualities. Undoubtedly, these technologies will enhance and motivate the advancement of fermentation processes in this field.

Our laboratory has a specific focus on bacterial endophytes. An endophyte is a microorganism, specifically a bacteria or fungus, that resides within the interior tissues of plants without causing any signs of disease. Endophytes are present in almost all terrestrial plants. As the understanding of the beneficial relationships between plants and microbes grows, the importance of endophytes is also expanding. These microorganisms are now being targeted for biotechnological advancements in order to combat plant diseases such as fungi and bacteria. We possess evidence indicating that a substantial assemblage of endophytic bacteria possesses the capacity to eliminate their rivals (pathogens) from their habitat through the utilization of LPs. This chapter focuses on the exploration of lipopeptide-producing endophytic bacteria that extend outside the Bacillus species. This information is valuable for alternative research in agricultural microbiology and for the manufacture of antibacterial compounds. Due to the tight association between certain endophytic bacteria and human infections, the use of these bacteria as biopesticides on commercial crops is subject to strict regulations. This final aspect has a detrimental effect on the utilization of endophytic microorganisms as instruments for disease management and the creation of novel bioinoculants for agriculture. Consequently, it is imperative that such LPs are generated through the process of fermentation.

It is important to note that the generation of biosurfactants is influenced by the physiological state of the bacteria, with quorum sensing likely playing a role. Quorum sensing is a potentially underestimated factor in fermentation processes. This chapter aims to elucidate the ways in which this phenomena and other factors can modify the efficiency of LP production.

2. LIPOPEPTIDES: CLASSIFICATION, MICROORGANISMS THAT CREATE THEM, FERMENTATION METHODS, AND SUBSEQUENT PROCESSING TECHNIQUES

2.1. Classes of lipopeptides and their applications

The production of LPs in their active form requires transcriptional induction, translation and post-translational modifications. The main machinery for their synthesis is multi-modular and consists of non-ribosomal peptide synthetases (NRPSs) [10]. Synthetases are organized on modules; each module permits the incorporation of a specific amino acid, subsequent condensation, termination and cyclization of the peptide chain. The synthesized peptides contain d-amino acids, β-amino acids and hydroxyl- or *N*-methylated amino acids. The integrated system introduces heterogeneity among LPs. The peptide moiety is inactive until it is coupled to a fatty acyl chain. The lipid aliphatic chain, of variable length, fuses with the N-terminal residue of the peptide chain, and then the bioactive LP is generated. After the biosynthesis of the LPs is finished, the molecule is modified by glycosylation or halogenation by specific enzymes associated to the synthetases [11].

Table 1. Important and the most studied lipopeptides from *Bacillus*, *Pseudomonas* and *Streptomyces* and their activity roles.

	Lipopeptides	*Microbial-producers*	Activity roles
Surfactin family	Surfactin linchenysin pumilacidin WH1 fungin	*B. subtilis* *B. polyfermenticus* *B. megaterium* *B. licheniformis* *B. pumilus* *B. amyloliquefaciens*	-Enhanced oil recovery Antibacterial Antiviral Antimycoplasma Antitumoral Anticoagulant Enzyme inhibition
Iturin family	Iturin Bacillomycins Mycosubtilin Subtulene (contains a unique Iso C15-long chain β -amino acid)	*B. subtilis* *B. megaterium*	-Antifungal -Biopesticides
Fengycin family	Fengycin Plipastatin Agrastatin1	*B. subtilis* *B. thuringiensis* *B. circulans* *B. megaterium*	-Strong fungitoxic agent against filamentous fungi -Immunomodulating activities
Pseudomonas sp. Lipopeptides	Viscosin Massetolide Entolysin	*Pseudomonas sp*	-Antibacterial and Antifungal activity against; *Mycobacterium tuberculosis*, Gram positive bacteria, *G. candidum* and *R. pilimanae*
Streptomyces sp. Lipopeptides	Daptomycin	*Streptomyces roseosporus*	-Broad spectrum activity against *staphylococci* (MRSA), beta-hemolytic *Streptococcus spp., Pneumococci, Clostridium spp., and Enterococci sp.* –MRSA *Staphylococcus aureus* -Antiparasitic -Immunosuppressor

Many bacteria and some fungi produce LPs, which have several roles including activity against bacteria, fungi, virus and more recently, it has been discovered their antitumor activity. Lipopeptides are also are involved in bacterial motility, in the swarming behavior and in the attachment to surfaces [1]. On the other hand, the extensive use of chemicals to control pathogens (bacteria and fungi) has modified the behavior of these microorganisms in humans and plants. The growing drug resistance, in these pathogens, urges for alternative antimicrobial molecules for clinical and crop protection, as well as for food preservation. As we mentioned above, LPs can be cyclic or linear based on the topology of the peptide chain. Here, we present an updated overview on the bioactive LPs and their uses, being the cyclic lipopeptides the most biologically relevant with proved activity and more market applications in several fields. The characteristics of the LPs are discussed below, and their properties, structures and uses have been summarized in Table 1 and Figure 1.

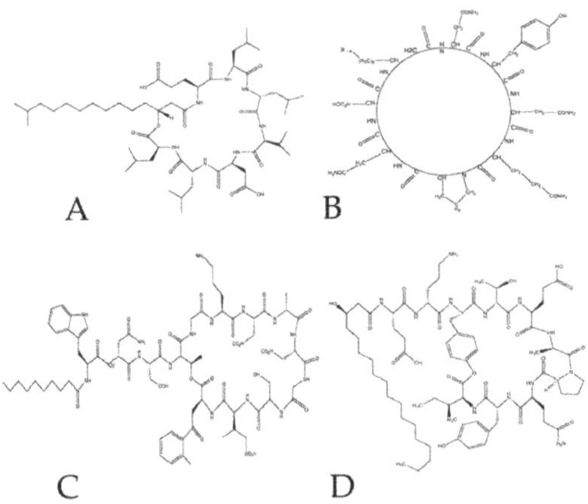

Figure 1. Pharmaceutically and economically important lipopeptides. Structures of representative member of lipopeptide synthesized by Bacillus and Streptomyces. (A) Daptomycin, (B) surfactin, (C) iturin and (D) fengycin.

Daptomycin. This structure is a cyclic decanoyl lipid chain attached to 13 amino acids (a 10-member macrolactone and three exocyclic residues) peptide. It is produced by *Streptomyces roseosporus*, a Gram-positive bacterium. It has potent antimicrobial properties and it has been clinically approved for its use as antibiotic since 2003. It is marketed under the tradename Cubicin. The mode of action of daptomycin is still unclear, but two hypotheses have been proposed, the first one states the inhibition of lipoteichoic acid synthesis (proteoglycan component of the cell wall of Gram-positive bacteria); the second states the disruption of bacterial membrane potential (depolarization) via pore formation and its calcium ion dependence. Concomitantly, the bacterial cell loses the ability to accumulate amino acid substrates while leaving glucose uptake intact [12, 13]. It has been successfully used to control skin infections, endocarditis, osteomyelitis and soft-tissue infections [14]. The cost of daptomycin in the market is approximately €125/day at dose of 6 mg/kg/day.

Surfactin. Surfactins constitute a major class of antibiotic LPs produced by *Bacillus* spp. They are highly active biosurfactants able to reduce the surface tension of water to 27 mN m^{-1} at 20 μmol. This group consist of a heptapeptide bonded to a C_{13}–C_{15} fatty acyl chain [15]. Surfactins are able to permeate the lipid membranes as dimer and form ion channels in planar lipid bilayer membranes. These compounds are effective against Gram-positive and Gram-negative bacteria and also have antimycoplasma, antiviral and antitumor activity and suppress inflammatory responses through inhibition of phospholipase A2 [16, 17]. Surfactin also inhibits phytopathogens such as *Pseudomonas syringae*, *Xanthomonas axonopodis*, *Sclerotinia sclerotium*, *Botrytis cinerea*, *Colletotrichum gloeosporioides* and stimulates plant defense [18, 19]. Since surfactins have hemolytic activity their medical applications are limited.

Iturins. Iturin is an antifungal cyclic lipopeptide produced by *Bacillus* spp. These amphiphilic compounds are characterized by a peptide ring of seven amino acid residues including an invariable d-Tyr2, with the constant chiral sequence LDDLLDL closed by a C14–C17 aliphatic β-amino acid. Iturins have a high polymorphism due to amino acid variations. These variants including iturin A, iturin C, iturin D, iturin E, Bacillomycin D, Bacillomycin F, Bacillomycin Lc, Mojavensin A and mycosubtilin [20]. Iturin A has been shown to form potassium ion-conducting channels in lipid bilayers. Iturins can act as biocontrol agents of plant pathogens [14]. They exert their fungicidal action by interacting with sterol components in the fungal membrane. Mojavensin A, a new member of the family, is cytotoxic [21]. From a clinical perspective, a disadvantage associated with iturins is their haemolytic activity.

Fengycins. This class of cyclic lipopeptides includes fengycins and plispastatins produced by some *Bacillus* and *Paenibacillus* strains. Fengicins are decapetides acylated with a β-hydroxylipid tail (C14–C18) and cyclized between the phenol side chain of Tyr 3 and the C-terminus. They act on plasma membrane of fungal cells and have been suggested for agriculture. They have antitumor activity because of the production of reactive oxygen species and mitochondria-dependent apoptosis [22]. They are good candidates for medical applications due their milder haemolytic activity.

Pseudofactin I and II. These compounds are cyclic octapeptides bonded to palmitic acid produced by *Pseudomonas fluorescens* BD5. The C-terminal of the carboxylic group forms a lactone with the –OH of threonine. Their emulsification activity and stability are greater in comparison to other synthetic surfactants, thus have a great potential for bioremediation or biomedicine. For example, Pseudofactin II exerts cytotoxicity in human melanoma [23].

Viscosin is obtained from *Pseudomonas fluorescens*. It has antibiotic activity, is highly surface active and is able to inhibit the migration of cancer cells. In *Pseudomonas*, this LP protects it from protozoan predation. Viscosin increases the efficiency of surface spreading over plant roots and protects germinating seedlings in soil infected with plant pathogen [24].

Linear cationic lipopeptides. Limited research has been performed on this small group. This is surprising as they have the potential to be more accessible than cyclic lipopeptides. This group includes saltavalin (which was named because contains serine, alanine, leucine, threonine, valine and 2,4-diaminobutyric acid, isolated from *P. polymyxa*), jolipeptin (*B. polymyxa*), cerexins (*B. cereus*, *B. mycoides*), tridecaptins (*P. terrae*). All exhibit antibacterial activity against Gram-negative and Gram-positive bacteria.

2.2. Endophytic microorganisms that make lipopeptides

Endophytic microorganisms form a durable and stable interaction with plants because of their endophytic lifestyle. Within this mutually beneficial relationship, the plant supplies nutrients and protection to the microbes. In return, the endophyte enhances the plant's ability to withstand both living and non-living stress factors, while also stimulating its growth. Certain endophytes are recognized for their ability to synthesize insecticidal chemicals. These bioactive secondary metabolites can have a direct role in antibiosis and/or in initiating induced systemic resistance (ISR). Several data have been reported regarding the production of LPs by endophytic bacteria, which may elucidate the antifungal or antibacterial activity against plant diseases. In a recent study, Gond et al. discovered a type of bacteria called Bacillus that lives inside plants and generates substances called antifungal LPs. This bacteria also activates the expression of genes that help the plant defend itself against pathogens in maize [25]. Gaining insight into the process of biological control enables the manipulation of the environment to establish optimal circumstances for enhanced biocontrol. In Mexico, certain groups are currently conducting research that emphasizes the practical use of LPs instead of bacteria. This is because many of the bacteria utilized in research are associated with human infections. Our research group has discovered strains of bacteria capable of producing LPs, which have been extracted from several plants such as agave, banana, and maize (Beltran-Gracia, unpublished publication). Table 2 is a compilation of endophytic strains that have been reported to produce lipopeptides.

Table 2. Endophytic bacteria reported as lipopeptide-producers.

Endophytic lipopeptide-producers	Secreted lipopeptides		
B. subtilis	Surfactin	Iturin	Linchenysin
	Pumilacidin	Bacillomycins	Mycosubtilin
	Fengycin	Plipastatin	
B. amyloliquefaciens	Surfactin	Linchenysin	Pumilacidin
	Fengycin		
B. megaterium	Surfactin	Linchenysin	Pumilacidin
	Iturin	Bacillomycins	Mycosubtilin
	Subtulene	Fengycin	Plipastatin
B. circulans	Fengycin	Plipastatin	Iturin
B. tequilensis	Surfactin	Iturin	Fengycin
B. polymyxa	Polymyxin		
B. macerans	Surfactin	Iturin	Fengycin
Streptomyces sp.	Daptomycin		
Pseudomonas sp.	Viscosin	Massetolide	Entolysin

2.3. Lipopeptide manufacturing through the fermentation process involves optimizing both culture conditions and operational variables

In order to incorporate LPs into industrial processes and for medical, pharmaceutical and agricultural uses, it is required their production by fermentation and their posterior downstream. It is clear that one of the main limitations for commercial applications of LPs are the high production costs and the low yield. To overcome these barriers, many efforts have been focused in improving the fermentation process, which represents a fundamental stage in the global production. Lipopeptides have been reported as growth-associated metabolites. In contrast to other bacterial secondary metabolites, production of LPs is induced when the cells have exhausted one or more essential nutrients, in example, surfactin production is induced in actively growing cells during the transition from exponential to stationary phase (SP); fengycin synthesis is related to the early SP, and iturins only accumulate in the later SP [1].

The production of LPs can be achieved by liquid fermentation (LF) or solid-state fermentation (SSF) and now, both methods have been proposed for scale up their industrial production. The LF is an advantageous and typical process used for LPs production in controlled bioreactors, while SSF is still in evolution but has gained attention owing its priority to LF, including lower investment for production, less time and higher secondary metabolite yields.

A critical factor into industrial LP production is media optimization. In fact, the nature of the carbon substrate, N, P, Na, Mg, Fe, Zn and Mn ions concentration in the medium, have been shown to influence enormously the nature and quantity of the LP produced by several bacterial strains. An orderly and planned statistical procedure to screen the effect of each component of the media is very useful. For example, a Plackett-Burman procedure was applied to find that glucose, K_2HPO_4, and urea concentrations had the most influence into LPs production by *Bacillus subtilis* of 11 tested variables (glucose, urea, ammonium sulfate, NaCl, $MgSO_4$, KH_2PO_4, K_2HPO_4, $MnSO_4$, $FeSO_4$, $ZnSO_4$). After, a Central Composite Design was conducted to optimize the three selected factors, finding a maximum biosurfactant concentration of 3.1 g/L when using 15 g/L glucose, 6 g/L urea and 1 g/L K_2HPO_4, keeping the other parameters at their minimum values [26]. A similar statistical procedure was also applied to determine the effect of sucrose, ammonium nitrate, NaH_2PO_4, K_2HPO_4, $MgSO_4$, $MnCl_2$, extract yeast as components of culture media to growth *Bacillus amyloliquefaciens*, using the Plackett-Burman design in the production of C15-surfactin, indicated a significant effect of sucrose, ammonium nitrate and NaH_2PO_4. The optimum values of the tested variables were 21.17 g/L sucrose, 2.50 g/L ammonium nitrate and 11.56 g/L NaH_2PO_4 with a production of 134.2 mg/L LP [27]. In a third case, a five-level four-factor Central Composite Design was employed to determine the maximum LP yield by *Bacillus subtilis* testing sucrose, ammonium chloride, ferrous sulphate and zinc sulphate. Optimum fermentation components were 22.431 g/L of sucrose, 2.781 g/L of ammonium chloride, 6.7879 mM of $FeSO_4$ and 0.0377 mM of $ZnSO_4$ to produce 1.712 g/L of LP. Only the ammonium chloride had no significant effect [28].

2.3.1. Carbon sources to optimize lipopeptide production

It is clear the importance of carbon source in any fermentation process due to its impact in the bacterial metabolism as well as in production costs. The nature and quantity of the carbon source are the most important factors that would affect LPs production. Structural and compositional diversity of LPs is substrate dependent. For example, *Bacillus amyloliquefaciens* was grown in a minimal salt medium with different carbon sources (sucrose, dextrose, maltose, lactose, glycerol and sorbitol) where a C:N ratio remained constant at 15.55. The surfactin, iturin and fengycin were detected when dextrose, sucrose and glycerol were used as carbon source. However, in the presence of maltose, lactose and sorbitol only iturin was produced. Also, these carbon sources significantly influenced the antifungal activity of the molecules. Those bacteria grown in media supplemented with dextrose or sucrose produced LPs with the higher antifungal activity. The maximum biosurfactant activity was observed when the growing minimal salt medium was supplemented with sucrose [29]. In a similar study, it was reported that among several carbon sources: glucose, sucrose, galactose, maltose, sucrose, glycerol, mannitol, soluble starch and dextrin, evaluated for C15-surfactin production, sucrose was the best carbon source [27].

2.3.2. Nitrogen sources

Several inorganic nitrogen compounds have been tested in LP production trials, i.e. ammonium nitrate, ammonium sulphate, sodium nitrate, urea and glutamic sodium. And looking for cheaper raw materials, complex compounds such as soybean flour, peptone and casein acid hydrolysate have been assayed too [27]. When using organic nitrogen sources it was observed that tryptone enhances the lipopeptide production because it contains several homologous L-amino acids to those found in LPs [30]. A similar behavior was reported in the modified Landy medium, where l-glutamic acid was replaced with various L-α-amino acids at the same concentration (5 g/L). Cottonseed-derived medium (Pharmamedia) proved to be a suitable substrate for the production of 220 mg/L of surfactin from *Bacillus subtilis*; this medium is suitable to achieve high production yields at low cost, which in turn makes it profitable for large scale usage. Moreover, supplementing Pharmamedia with $Fe^{2}+$ (4.0 mM) and sucrose (2 g/L) leads to a maximum production of about (300 mg/L) [31]. Other interesting nitrogen source is rapeseed meal, a low-cost material that was used to synthetize iturin A by *Bacillus subtilis*. The maximum iturin A concentration was 0.60 g/L after 70 h of incubation, which was 20% and 8.0 times higher than that achieved with peptone and ammonium nitrate media, respectively [32].

2.3.3. Divalent ions

To optimize the trace element composition of culture media the use of statistical experimental designs is preferred. Such is the case, where was doubled surfactin concentration when was applied the statistical method Taguchi to determine cation effect of culture medium [33]. The role of $Fe^{2}+$ in the synthesis of LPs is crucial; there are some reports where the supplementation of this cation enhanced the yield of biosurfactants [31, 33]. The addition of $Fe^{2}+$ into fermentation medium was utilized to optimize surfactin production from *Bacillus subtilis*, reaching yields up to 3 g/L into minimal salt medium; the optimal $Fe^{2}+$ dosage (4.0 mM) leading to 8-fold and 10-fold increments in cell concentration and surfactin yield, respectively, as compared to those media without $Fe^{2}+$ [34]. *Bacillus amyloliquefaciens* in a culture medium supplemented with 0.2 mM of iron was able to produce 92.78 mg/L of iturin after 5 days with no pH control in the culture. Moreover, if the starting pH at 6.64 and 0.2 mM of ferrous sulphate, an iturin A production of 121.28 mg/L was obtained [35].

2.3.4. Operational conditions

The pH, temperature, dissolved oxygen concentration and degree of aeration affect cellular growth, and consequently, biosurfactant production. Optimal operational conditions vary from strain to strain, and the better growing condition for each particular strain must be determined experimentally. For example, surfactin synthesis by *Bacillus subtilis* can be achieved at temperatures ranging from 25 to 37°C; the optimal temperature for the surfactin production by *Bacillus subtilis* DSM 3256 was 37.4°C. In contrast, for thermophilic *Bacillus* spp. surfactins were produced at temperatures above 40°C without detriment on their activity [36]. Regarding pH, a greater LP activity was observed when the pH was adjusted to between 3.0 and 8.0.

Statistical tools had been used to optimize LP biosynthesis considering operational conditions. The response surface methodology has been used to determine the maximum LP production varying the temperature, initial pH and culture cycle. Another important condition to be controlled is the dissolved oxygen. The oxygen acted at different levels, suggesting a complex system for regulating the synthesis of LP in B. *subtilis* ATCC6633. So, the oxygen transfer is one of the critical parameters for process optimization and scaling-up of production of surfactin [36]. Varying the oxygen transfer conditions, the synthesis could be oriented to mixed production or to surfactin monoproduction. The fraction of surfactin towards total LPs produced and the maximal surfactin production both increased with $k_L a$ increase (surfactin concentration about 2 g/L at $k_L a = 0.04–0.08$ s^{-1}), while the maximal fengycin production (fengycin concentration about 0.3 g/L) was obtained at moderate oxygen supply ($k_L a = 0.01$ s^{-1}). The production of LP represents a challenge due to its surface properties. Foam causes a severe decrease of oxygen transfer. A significant decrease of k_La (up to 27%) was measured during fermentation process reduces LP biosynthesis [37].

Classical bioreactors aerated by gas bubbling are not suitable for production of LP biosurfactants due to excessive foaming. The use of antifoam agents is not appropriate because it can affect the bacterial physiology and downstream processing. The alternatives are bioreactor with foam collector, rotating discs biofilm reactor [38], bubbleless membrane aerated bioreactor [39] and three-phase inverse fluidized bed bioreactor [40]. Another alternative to foam control is the use of a strictly anaerobic bioreactor cultivation to produce surfactin. Most interesting, the product yields exceeded classical aerobic fermentations, in which foam fractionation was applied. Additionally, values for specific production rate surfactin (0.005 g/(g h)) and product yield per consumed substrate ($Y_{P/S}$ = 0.033 g/g) surpass results of comparable foam-free processes [41]. The bioreactor design is still a challenge to get better productivities in industrial processes; the LPs synthesis is not an exception.

2.4. Solid-state fermentation (SSF)

Solid-state fermentation (SSF) is an alternative technology for the production of high-value molecules. The SSF could be an alternative method for the LPs production. The SSF uses agro-industrial wastes as substrates, which contributes to reduce the production costs. The productivity of LPs synthetized by SSF depends on initial moisture content, incubation temperature, fermentation time, substrates used and supplementary nutrients such as mineral salts. The temperature is an important parameter for both bacterial growth and LP production. Within a range from 25 to 40°C, the optimal temperature for growth was found to be 30°C, but the biosynthesis of LPs is favored at 37°C. At the beginning, in the firsts 24 h, temperature should be maintained at 30°C, then shifted up to 37°C to enhance LP production [42].

The selection of raw materials to formulate a culture medium, as the previously described, requires experimental tests. Savings in time and resources can be achieved using statistical methods, which help in optimizing components and concentrations in the formulation of a culture medium. The optimal composition of culture medium in solid-state fermentation to get the highest LP production had been determined using a 'Central Composite Design'. First, a screen to select the major solid substrates was performed, where rapeseed meal, corn flour, soybean flour, bean cake, wheat bran, rice hull and rice straw were considered as candidates to support bacterial growth and biosurfactant production. For the election, a quantity of each solid substrate was supplemented with 1.0 ml of mineral solution with initial pH 7.5 and moisture content 55%. To increase the porosity of the substrates and improve its availability, each of them was mixed with an inert substrate (perlite, vermiculite, beads). After that, both easily digestible carbon sources (glucose, sucrose, starch, l-glutamic, maltose, glycerol, d-galactose) and nitrogen sources (tryptone, peptone, yeast extract, urea, NH_4NO_3, NH_4Cl, $(NH_4)_2SO_4$ at 2% (w/w) were added. The substrates selected to the optimization of the medium composition for LP production by 'Response Surface Methodology' were soybean flour, rice straw, starch and yeast extract. The optimal conditions were 1.79% starch and 1.91% yeast extract by employing 5.58 g soybean flour and 3.67 g rice straw as the solid substrate with initial pH 7.5, moisture content 55% and a 10% inoculum level at 30°C for 2 days. Under these conditions, the experimental yield of LPs reached 50.01 mg/gram of dry substrates [43]. SSF many times is compared with submerged fermentation. By this reason, a comparative study was performed to determine the compositions and properties of LP products purified and the transcription values of LP genes under submerged fermentation and SSF. Results revealed no significant differences in the polarity and structure of the two LP products. But, LP obtained by submerged fermentation possessed higher amino acid proportions, better emulsification activity and antagonistic activity than that from solid-state fermentation. For solid-state fermentation, the transcription accumulation levels of the LP synthetic genes srfA and sfp were higher than for submerged fermentation at the same stage. Transcripts for ituD and lpa-14 remained elevated for a longer period of time under solid-state fermentation conditions, accounting for differences in the production and fermentation periods between both fermentation techniques [44].

2.5. Downstream processing involves the extraction and refinement of lipopeptides from the fermentation process.

Due to different applications of LPs, we need different levels of purity. Crude LPs can straightaway be used in bioremediation related applications, where the overall economy of the process is the most important concern. On the other hand, partially purified fractions (about 60–80% pure) can suit applications in microemulsion based nanoparticle synthesis, laundry and food industry. However, the requirement for ultrahigh pure product is indispensable, if the LPs are to be considered for pharmaceuticals and human healthcare [45]. As we mentioned before, the intense foaming produced during aerobic liquid fermentation is a big obstacle for the commercialization making their recovery and purification difficult. A great deal of monetary input will be required for purification, this account 60% of the total production cost. Different techniques have been developed to extract and purify LPs. Among the most used techniques to extract are acid precipitation (HCl 6N), solvent extraction (chloroform, ethyl acetate, dichloromethane or mixtures of chloroform-methanol), ammonium sulphate precipitation with dialysis to remove small molecules and salts, and foam fractionation (utilized for continuous retention process and high purity). For purification normally have been used membrane ultrafiltration techniques, ionic exchange chromatography and adsorption-desorption on resins (XAD-4, XAD-7 HP, HP-2MG, HP-20) or activated carbon. The High Performance Liquid Chromatography(HPLC) (an excellent method for separation of this class of molecules) uses a reverse phase with C18 columns. The LPs separated can be detected using ultraviolet absorbance or diode arrays detectors and each peak separated is collected using a fraction collector for further analysis of their structure.

2.5.1. Example of techniques used for lipopeptide recovery, purification and identification

Membrane ultrafiltration. This technique serves essentially as an intermediate process for the recovery and purification of LPs. The separation of LPs by membrane filtration depends on their molecular aggregation behavior and on their ability to form micelles, since a process become economic feasible when a high MWCO (molecular weight cut-off) membranes were used. In general, the use of low MWCO membranes requires high maintenance due to low permeate fluxes through smaller pores that get easily plugged by monomers and progressive reduction in flux caused by the mechanism of concentration polarization [45].

A two-step ultrafiltration process using large pore size membranes (up to MWCO = 300 kDa) was investigated to separation of LPs aggregated in single and mixed solutions from fermentation culture. In single solutions of LP both surfactin and mycosubtilin formed micelles of different size depending on their concentration. However when the LPs were in the same solution, they formed mixed micelles of different size and probably conformation to that formed by the individual LPs, this prevents their separation according to size. An effective rejection in the first ultrafiltration step was achieved by membranes with MCWO= 10–100 kDa but poor rejection by the 300 KDa membrane [46]. The rejection is a measure of retention capacity of a membrane. However, some properties of LP micelles such as poor stability and non-uniform size distribution limit the use of readily scalable high MWCO membranes for the purification of LP, as smaller sized micelles and monomers can easily pass through the pores of these membranes. An addition of $Ca^{2}+$ ions causes the structural transformation of surfactin monomers to larger micellar aggregates, showing excellent features such as compact structure, narrow size distribution, and improved stability [45].

Chromatographic technique for daptomycin. Daptomycin was purified from clarified fermentation broth using anion exchange chromatography and reverse phase chromatography. The anion resin was a highly cross-linked agarose with dextran surface extender. Daptomycin was eluted from the anion exchange column with a NaCl gradient from 0.2 to 1.0 M in water. The semi-purified daptomycin was then added to a reversed phase column and was washed with water containing

15% of alcohol. The reverse phase resin was a mono-sized, porous resin made of polystyrene and divinyl benzene (source: RPC 30). After, daptomycin was eluted with 40–70% of ethanol. Two reverse phase columns were involved to improve the purity of the final product at different pH. The first column was run at pH 7.5–8.0, while the second one was eluted at pH 3.0–3.1. The purified daptomycin is then filtered and lyophilized under standard conditions with at least 95% of purity [47].

Resins. Macroporous adsorption resin (MAR) chromatography has been successfully used for separation of bioactive molecules on the basis of hydrophobic/hydrophilic interactions between solute and resin surface. MAR has been resolved problems related with low efficiency to separate LP mixtures into individual families employing a simple stepwise solvent gradient elution under optimal conditions. The adsorption and desorption of solutes on MARs depend upon the properties of the resins such as particle size, pore diameter, surface area and polarity. An example is the performance of a non-polar resin (HP-20) that combine features such as higher surface area, pore size and appropriate polarity, allowing it to have a superior adsorption capacity over other resins. Dual gradient MAR was applied to a cell free broth diluted until its total crude LP concentration was 3 g/L, then it was pumped at a flow rate of 1 ml/min into the column pre-packed with HP-20 resin (15 g) until breakthrough point. After one run of adsorption and desorption, the three LP families were successfully enriched in separate fractions and the recovery yields were 79.5% for iturin, 94.4% for fengycin and 89.4% for surfactin. Their purities in the enriched fractions were found to be 68.3, 77.6 and 91.6%, respectively. This process represents a basis for *in situ* recovery of LPs from the culture broth in continuous mode [48].

2.5.2. HLPC-MALDI-TOF

HPLC is an excellent method for the separation of individual LP separation. The most employed technique is reverse phase chromatography, due to this method can separate this metabolite based on its polarity. The separated products are detected by UV absorbance detection and each individual peak can be collected for further analysis of their structure. Also, use of a diode array detector is recommended. This detector can measure simultaneous wavelengths in a range of 200–600 nm, this means that we can detect LPs as they are eluted from the column in several wavelengths. For mobile phases reported, the methanol:water (80:20) is the most commonly employed, due this phase can elute several LPs as fengicins and iturins. Also, another mobile phase used is acetonitrile:water, as the methanol:water phase, this mobile phase can elute LPs as surfactins. The proportions for a acetonitrile:water can change depending on the LP that you want to separate. The typical column for LP's separation is a C-18 column, the length can vary from 150 to 250 mm, and this depends on the resolution and separation desired. In terms of particle size, 5 μm is the most adequate for the stationary phase. Table 3A shows some conditions of HPLC most widely used for separation as well as quantification of LPs derived from a fermentative process [49, 50].

MALDI-TOF (matrix-assisted laser desorption/ionization-time of flight) is a mass spectrometry technique that allows the identification of intact compounds. Samples to be analyzed are mixed with a matrix and dried on a stainless steel plate, onto which a laser with various degrees of energy is fired to forming gaseous ions, which can be separated in a time of flight (TOF) analyzer and detected. Now, MALDI-TOF-MS has come to be regarded as a very fast and reliable tool for identification of LPs when compared to the conventional methods like culturing and purifying the LPs. The MALDI-TOF-MS for the identification of LPs has been previously reported [51, 52]. For LP identification, we can use CHCA (α-cyano-4-hydroxycinnamic acid), SA (sinapic acid) and DHB (2,5-dihydroxybenzoic acid). To the best of our knowledge, DHB matrix is better than CHCA, because we obtain a good quality spectra with intensities above 2E4 and LP isomers can be observed. In Table 3B are shown relevant information of the mass range for specific LPs identification derived from MALDI-TOF analysis.

Table 3. HPLC strategies and conditions (A) used for lipopeptides separation from extracts of culture filtrate and (B) ranges of m/z of typical peaks obtained by MALDI-TOF analysis. In bold, the reference peaks for lipopeptide identification.

TECHNIQUE	*CONDITIONS*		*LIPOPEPTIDE SEPARATED*
(A) HPLC	COLUMN	C-18, 150mm–250mm length, size particle 5μm	ITURIN, FENGYCIN SURFACTIN AMPHISIN LOKISIN HODERSIN, TENSIN VICOSINAMIDE DAPTOMYCIN VISCONSIN MASSETOLIDE ENTOLYSIN
	MOBILE PHASE	Methanol: water (80:20), Acetonitile:water (80:20 and 40:60), acetonitrile: acetic acid (68:32)	
	RETENTION TIMES	3-10 min iturin 10-16 min fengycin 16-25 min surfactin 28.5 min amphisin 29 min lokisin 29.2 min hodersin 29.9 min tensin 31.6 min vicosinamide	
	DETECTORS	UV-VIS (205, 235, 278, 285nm), diode array.	

TECHNIQUE	*CONDITIONS*		*LIPOPEPTIDE SEPARATED*
(B) MASS SPECTROMETRY	PEAKS (m/z)	IDENTIFICATED LIPOPEPTIDE	
	1001.42, 1029.42; 1016.56, 1030.58, 1044.59, 1058.61, 1072.62; 1030.64, 1044.65, 1058.66, 1072.67	ITURIN A	
	1052.63,1094.45, 1122.47, 1136.55; 1052.64, 1066.57; 1052.62, 1066.60, 1080.59	BACILLOMYCIN	
	1435.58, 1449.63, 1463.68 , 1477.66; 1421.61, 1435.66, 1459.69, 1463.71, 1477.72; 1421.75, 1435.76, 1449.77, 1463.79, 1477.60, 1491.81	FENGYCIN	
	1488.70, 1502.71, 1516.62; 1488.71, 1502.72; 1488.79, 1502.79	FENGYCIN (POTASSIUM ADDUCT)	
	1074.58, 1088.57, 1102.55	BACILLOMYCIN (SODIUM ADDUCT)	
	1485.70 , 1499.71, 1513.71; 1471.75, 1485.76, 1499.77, 1513.79	FENGYCIN (SODIUM ADDUCT)	
	878.47, 892.41, 906.50	KURSTAKINS	
	1202.75, 1188.73, 1143.70, 1129.68, 1168.77, 1154.75	POLYMYXINS	
	1421.75, 1407.73, 1393.72, 1379.70	BACITRACINS	

2.6. Identification of lipopeptide-producing endophytes using the integration of molecular biology and chromatographic-mass spectrometry techniques

To search for endophytic bacteria in nature that can produce LPs, researchers must optimize their screening and identification times. Actually, the strategies for the selection of endophytic microorganisms as candidates for biocontrol or lipopeptide-producers combines antifungal-antibacterial screenings, a molecular analysis of the genes involved in the LP synthesis and then an analysis of the extracts obtained from the culture medium by HPLC and mass spectrometry.

In practice, after endophyte isolation, we identify the purified strain by MALDI-TOF and confirm its identity by 16s rDNA sequencing. Once identified, we select the strain as a function of its antagonistic ability against plant pathogenic strains. For screening antifungal activity of the isolates, we use a dual culture test against fungal strains collected from plants and previously identified. Normally, the plates are incubated for 3–5 days to observe inhibition of the fungal mycelium (Figure 2). For screening the presence of LP product biosynthetic gene clusters present in the antifungal isolate, PCR-based screening methods are used for different genes involved. In Table 4, we show the summarized sequences and methods for PCR analysis most used for some lipopeptide-genes, then each amplification product is analyzed by agarose gel electrophoresis to compare the predicted base-pair number (Figure 3). The sequence data obtained must be analyzed by BLAST and detailed *in silico* phylogenetic analysis to confirm the PCR product.

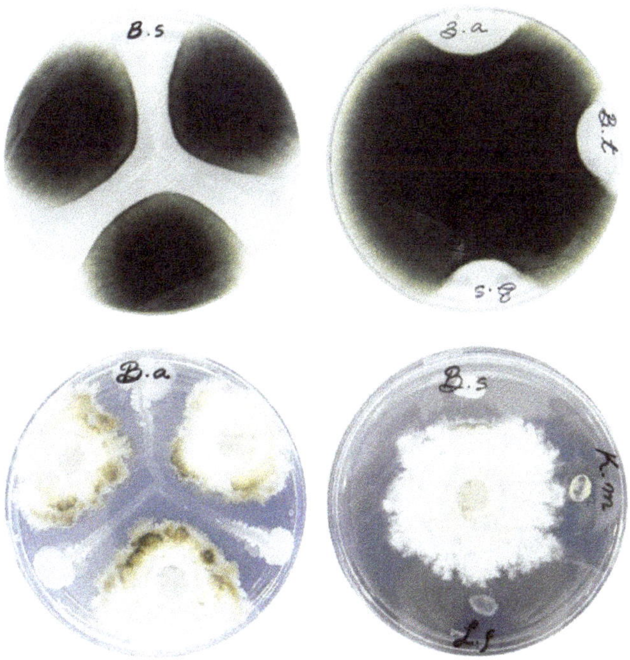

Figure 2. Antagonism of endophytic Bacillus subtilis (Bs), B. amyloliquefaciens (Ba) and B. tequilensis (Bt), Kocuria marina (Km) and Lysinibacillus fusiformis (Lf) isolated from agave, banana and maize against Rhizotocnia sp. and Colletotrichum sp. incubated by 5 days. It is noted that Bacillus strains used are the only ones that have antifungal activity, so the other strains should be discarded for identification analysis.

Table 4. PCR primers commonly used for amplification of lipopeptide genes.

LIPOPEPTIDES	GENES	PRIMERS	PRIMERS SECUENCES (5′–3′)	AT*	PCR++
Iturin	ituD	ITUD-F1	TTGAAYGTCAGYGCSCCTTT	55	482
	ituC	ITUD-R1	TGCGMAAATAATGGSGTCGT	55	594
	ituA	ITUC-F1	CCCCCTCGGTCAAGTGAATA	55	647
		ITUC-R1	TTGGTTAAGCCCTGATGCTC		
		ITUD1F	GATGCGATCTCCTTGGATGT		
		ITUD1R	ATCGTCATGTGCTGCTTGAG		
Surfactin	srfA	SRFA-F1	AGAGCACATTGAGCGTTACAAA	55	626
	srfP	SRFA-R1	CAGCATCTCGTTCAACTTTCAC	52	675
	srf/Ich	SFP-F1	ATGAAGATTTACGGAATTTA	43	428
		SFP-R1	TTATAAAAGCTCTTCGTACG		
		As1-F	CGCGGMTACCGVATYGAGC		
		Ts2-R	ATBCCTTTBTWDGAATGTCCGCC		
Mycosubtilin	Myc/itu	Am1-F	CAKCARGTSAAAATYCGMG	45	419
		Tm1-R	CCDASATCAAARAADTTATC		
Fengycin	fen	Af2-F	GAATAYMTCGGMCGTMTKGA	45	452
		Tf1-R	GCTTTWADKGAATSBCCGCC		
Piplastin	pps	Ap1-F	AGMCAGCKSGCMASATCMCC	58	1029
		Tp1-R	GCKATWWTGAARRCCGGCGG		

*AT: ANNEALING TEMPERATURE (°C).
+PCR: PCR PRODUCT SIZE EXPECTED (bp).

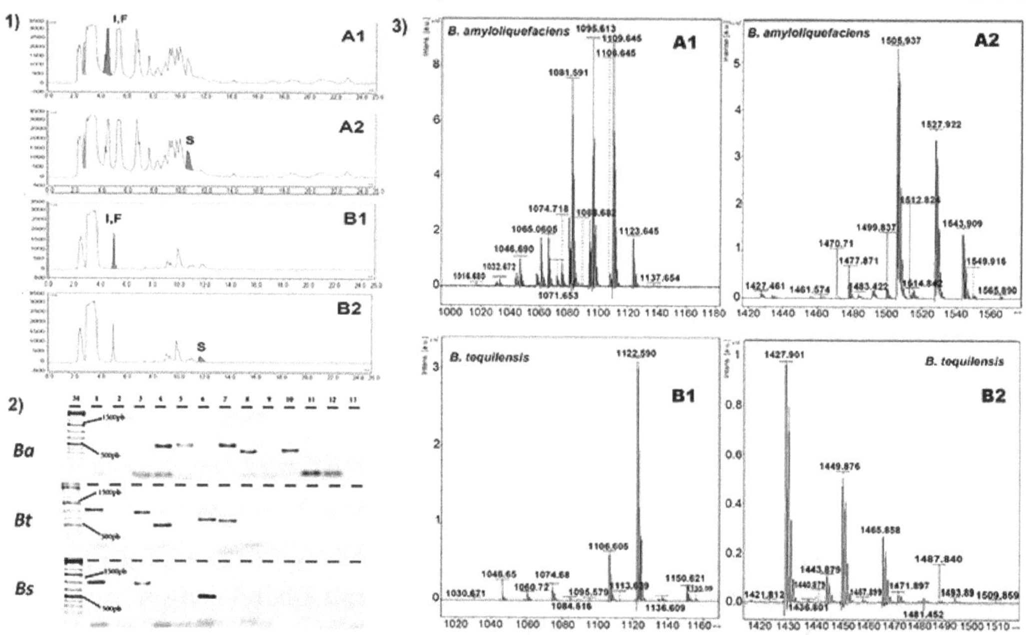

Figure 3. Agarose gel electrophoresis of PCR products of Bacillus endophytes (Below left) and separation and identification of the antifungal lipopeptides from acidic methanol extract by using reversed-phase HPLC(Above left) and MALDI-TOF MS analysis (Right)

The crude extracts obtained from culture media must be subject to HPLC separation and their collection for mass identification. The chromatographic profile of acidic methanol extracts and MALDI-TOF spectra of LPs from endophytic *Bacillus amyloliquefaciens* and *B. tequilensis shown* the presence of LPs groups iturins, fengycins and surfactin (Figure 3). Mass ranges (*m/z*) are 1001.42–1072 for iturin, 1471–75–1513.79, 1488.70–1516.62 for K and Na adducts of fengycin respectively and 1000–1100 for surfactin.

Section 2: PCR detection of lipopeptide and biosynthesis genes from Bacillus amyloliquefaciens (A), B. tequilensis (B), B. subtilis (C). M DNA ladder, Lane 1 Subtilosin, Lane 2 Sublancin, Lane 3 Plipastatin, Lane 4 Iturin D, Lane 5 Iturin C, Lane 6 Iturin A, lane 7 Surfactin A, Lane 8 Surfactin F, Lane 9 Mycosubtilin, Lane 10 Fengycin, lane 11 Subtilin A Lane 12 Ericin and Lane 13 Surfactin P.

Chromatogram profile and MALDI-TOF mass spectrum of B. amyloliquefaciens (A) and B. tequilensis (B). The chromatograms were obtained under the following conditions: 0-3min: 45%-50% acetonitrile; 3-8min: 50%-80% acetonitrile; 8-25min: 80-100% acetonitrile, temperature of 38°C and a C-18 column (5µm particle size, 250 mm). Mass spectra were obtained with a RP 700-3500 Da, DHB as matrix. In section 1 (A1 and B1), we can observe the mass spectrum of the HPLC fraction collected (within 4 min for A1 and 5 min for B1) that represent iturin (Section 3: with mass of 1016.68m/z and surfactin (1065.06 m/z) indicated that both strains produce this lipopeptides. In A2 and B2 we observe fengycin, in both cases the mass spectrum contains the representative mass of fengycin (1477 m/z) in section 3.

2.7. Quorum sensing in fermentative processes and lipopeptide production

Quorum sensing (QS) is a form of cell-cell communication by which bacteria communicate by secreting signaling molecules called autoinducers that help regulate gene expression. The QS molecule as N-acylhomoserine lactones in Gram-negatives or AIP in Gram-positives regulates different bacterial functions such as antibiotic biosynthesis, production of virulence factor, bacterial swarming, sporulation, competence and transition to the stationary growth phase.

How are LP production and quorum sensing associated? Surfactin a LP widely mentioned in this chapter book was proposed as a quorum-sensing molecule that activates the process of sporulation and production of biofilm [53]. For regulation of surfactin production by a cell density-responsive mechanism *B. subtilis* utilized a peptide pheromone Com X. Com X accumulates in the growth medium. So, the QS control the *srf* operon expression via Com X. However, few studies have been reported relating QS with fermentative processes. The production of putisolvins, also cyclic lipopeptides, in *Pseudomonas putida* occur at the end of the exponential growth phase, which indicates that the production of putisolvins is mediated through a quorum sensing-mechanism [54, 55]. Another example that links QS with LP production is rhamnolipid a glycolipid biosurfactant produced also by *Pseudomonas spp*. The *rhl* quorum sensing system in *P. aeruginosa* regulates the production of rhamnolipid type biosurfactants. RhlA, a rhamnosyltransferase, catalyses the synthesis of fatty acid dimers that subsequently serve as the precursor for RhlB to form monorhamnolipids and dirhamnolipids catalyzed by RhlC. The genes (*rhlA*, *rhlB* and *rhlC*) for these catalyses are under the control of QS. Other studies link rhamnolipids synthesis to nutritional conditions, such as nitrogen exhaustion and the alternative sigma factor σ^{54} for nitrogen limitation. Schmidberger et al. [56], reports an interesting study of *P. aeruginosa* and rhamnolipid synthesis. Using PCRq, gene expression was monitored over entire course of fermentation. They observed until late deceleration phase (or ending log phase), an increase in relative gene expression of the *las*, *rhl* and *pqs* quorum-sensing regulon under nitrogen limitation.

Batch fermentation procedures present a biological enigma when it comes to the transcriptional activity of the genes involved in LP synthesis circuits in both Gram-positive and Gram-negative bacteria. Further investigations into the molecular biology involved in the process of LP fermentation are required. Tracking the gene expression of LPs during the whole duration of the fermentation process yields insights into regulatory events that may or may not be associated with Quorum Sensing (QS). QS, a variable, has been largely neglected in investigations on fermentative processes. Information on quorum sensing (QS) in fermentation is expected to assist in optimizing bioreactor conditions, addressing nutritional restrictions, and perhaps utilizing QS signal molecules to enhance production yields of LPs and other microbial products.

3. CONCISE SUMMARY

The examples shown briefly in this chapter are not comprehensive. These examples aim to demonstrate the possible utilization of bacterial LPs and emphasize their potential applications in the sectors of medical and agriculture. We also highlight the capacity of endophytic bacteria to create lipopeptides, which presents research prospects for comprehending the mechanisms underlying the biological regulation that takes place in the specific environments they occupy as endophytes. Acquiring knowledge about this and related subjects will facilitate the development of new compounds that are non-toxic to people, especially in cases when we are unable to directly utilize bacteria in agricultural settings. It is evident that further research is necessary to enable the widespread utilization of microbial LPs. This research should specifically concentrate on enhancing production efficiency to get better yields at a reduced cost. Solid-state fermentation is a crucial field of investigation within the realm of fermentation processes. This field is still in its nascent phase, and advancements will be achieved using a combination of ecological, physiological, structural, genetic, and fermentative methodologies.

REFERENCES

1. Ongena M, Jacques P. Bacillus lipopeptides: versatile weapons for plant disease biocontrol. Trends Microbiol. 2007;16(3):115–125. DOI: 10.1016/j.tim.2007.12.009.
2. Van Der Voort M, Meijer HJ, Schmidt Y, Watrous J, Dekkers E, Mendes R, Dorrestein PC, Gross H, Raaijmakers JM. Genome mining and metabolic profiling of the rhizosphere bacterium Pseudomonas sp. SH-C52 for antimicrobial compounds. Front Microbiol. 2015;6:693. DOI: 10.3389/fmicb.2015.00693.
3. Singh LS, Sharma H, Talukdar NC. Production of potent antimicrobial agent by actinomycete Streptomyces sannaensis strain SU118 isolate from phoomdi in Loktak Lake of Manipur, India. BMC Microbiol. 2014;14:278. DOI: 10.1186/s12866-014-0278-3
4. Li L, Ma T, Liu Q, Huang Y, Hu C, Liao G. Improvement of daptomycin production in Streptomyces roseosporus through the acquisition of pleuromutilin resistance. BioMed Res Int. 2013;2013:479742. DOI: 10.1155/2013/479742
5. Meca G, Sospedra I, Valero MA, Mañes J, Font G, Ruiz MJ. Antibacterial activity of the enniatin B, produced by Fusarium tricictum in liquid culture, and cytotoxic effects on Caco-2 cells. Toxicol Mech Methods. 2011;21(7):503-512. DOI: 10.3109/15376516.2011.556202
6. Li Yi-M, Haddad NIA, Yang S-Z, Mu B-Z. Variants of lipopeptides produced by Bacillus licheniformis HSN221 in different medium components evaluated by a rapid method ESI-MS. Int J Pept Res Ther. 2008;14(3):229–235. DOI: 10.1007/s10989-008-9137-0.
7. Wei Y-H, Chu I-M. Mn^{2}+ improves surfactin production by Bacillus subtilis. Biotechnol Lett. 2002;24(6):479–482. DOI: 10.1023/A:1014534021276
8. Abdel-Mawgoud AM, Aboulwafa MM, Abdel-Haleem HN. Optimization of surfactin production by Bacillus subtilis isolate BS5. Appl Biochem Biotechnol. 2008;150(3):305–325. DOI: 10.1007/s12010-008-8155-x.
9. Plaza G, Chojniak J, Rudnicka K, Paraszkiewicz, BP. Detection of biosurfactants in Bacillus species: genes and products identification. J Appl Microbiol. 2015;119(4):1023–1034. DOI: 10.1111/jam.12893.
10. Roonsawang N, Washio K, Morikawa M. Diversity of nonribosomal peptide synthetases involved in the biosynthesis of lipopeptide biosurfactants. Int J Mol Sci. 2010;12(1):141–172. DOI: 10.3390/ijms12010141.

11. Duitman EH, Hamoen LW, Rembold M, Venema G, Seitz H, Saenger W, Bernhard F, Reinhardt R, Schmidt M, Ullrich C, Stein T, Leenders F, Vater J. The mycosubtilin synthetase of Bacillus subtilis ATCC6633: a multifunctional hybrid between a peptide synthetase, an amino transferase, and a fatty acid synthase. Proc Natl Acad Sci U S A. 1999;96(23):13294–13299. DOI: 10.1073/pnas.96.23.13294.

12. Bayer AS, Schneider T, Sahl, HS. Mechanisms of daptomycin resistance in Staphylococcus aureus: role of the cell membrane and cell wall. Ann N Y Acad Sci. 2013;1277:139–158. DOI.10.1111/j.1749-6632.2012.06819.x.

13. Zhang, T, Muraih JK, MacCormick B, Silverman J, Palmer M. Daptomycin forms catión-and size-selective pores in model membranes. Biochim Biophys Acta. 2014;1838(10):2425–2430. DOI: 10.1016/j.bbamem.2014.05.014

14. Patel S, Ahmed S, Eswari SJ. Therapeutic cyclic lipopeptides mining from microbes: latest strides and hurdles. World J Microbiol Biotechnol. 2015;31(8):1177-1193. DOI: 10.1007/s11274-015-1880-8.

15. Shao C, Liu L, Gang H, Yang S, Mu B. Structural diversity of the microbial surfactin derivatives from selective esterification approach. Int J Mol Sci. 2015;16(1):1855–1872. DOI: 10.3390/ijms16011855.S.

16. Kim K, Jung SY, Lee DK, Jung JK, Park JK, Kim DK, Lee CH. Suppression of inflammatory responses by surfactin, a selective inhibitor of platelet cytosolic phospholipase A2. Biochem Pharmacol. 1998;55(7):975–985. DOI:10.1016/S0006-2952(97)00613-8.

17. Hamley IW. Lipopeptides: from self-assembly to bioactivity. Chem Commun. 2015;51(41):8574–8583. DOI: 10.1039/C5CC01535A.

18. Meena KR, Kanwar S. Lipopeptides as the antifungal and antibacterial agents: applications in food safety and therapeutics. BioMed Res Int. 2015;2015:473050. DOI: 10.1155/2015/473050.

19. Cawoy H, Mariutto M, Henry G, Fisher C, Vasilyeva N, Thonart P, Dommes J, Ongena M. Plant defense stimulation by natural isolates of Bacillus depends on efficient surfactin production. Mol Plant-Microbe Interact. 2014;27(2):87–100. DOI: 10.1094/MPMI-09-13-0262-R.

20. Ali S, Hameed S, Imran A, Iqbal M, Lazarovits G. Genetic, physiological and biochemical characterization of Bacillus sp. strain RMB7 exhibiting plant growth promoting and broad spectrum antifungal activities. Microb Cell Fact. 2014;13:144. DOI: 10.1186/s12934-014-0144-x.

21. Ma Z, Wang N, Hu J, Wang S. Isolation and characterization of a new iturinic lipopeptide, mojavensin a produced by a marine-derived bacterium Bacillus mojavensis. J Antibiot. 2012;65:317–322. DOI: 10.1038/ja.2012.19.

22. Yin H, Guo C, Wang Y, Liu D, Lv Y, Lv F, Lu Z. Fengycins inhibits the growth of the human lung cáncer cell line 95D through reactive oxygen species production and mitochhondria-dependent apoptosis. Anti-Cancer Drugs. 2013;24(6):587–598. DOI: 10.1097/CAD.0b013e3283611395.
23. Janek T, Krasowska A, Radwanska A, Lukaszewicz M. Lipopetide biosurfactant pseudofactin II induced apoptosis of melanoma A 375 cells by specific interaction with the plasma membrane. PLoS ONE. 2013;8(3):e57991. DOI: 10.1371/journal.pone.0057991.
24. Alsohim AS, Taylor TB, Barrett GA, Gallie J, Zhang XX, Altamirano-Junqueira AE, Johnson LJ, Rainey PB, Jackson RW. The biosurfactant viscosin produced by Pseudomonas fluorescens SBW25 aids spreading motility and plant growth promotion. Environ Microbiol. 2014;16(7):2267–2281. DOI: 10.1111/1462-2920.12469.
25. Gond SK, Bergen MS, Torres MS, White JF. Endophytic Bacillus spp. produce antifungal lipopeptides and induce host defence gene expression in maize. Microbiol Res. 2015;172:79–87 DOI: 10.1016/j.micres.2014.11.004
26. Mnif I, Chaabouni-Ellouze S, Ghribi D. (2012) Optimization of the nutritional parameters for enhanced production of B. subtilis SPB1 biosurfactant in submerged culture using response surface methodology. Biotechnol Res Int. 2012;2012:795430. DOI: 10.1155/2012/795430.
27. Liu X, Ren B, Gao H, Liu M, Dai H, Song F, Yu Z, Wang S, Hu J, Kokare CR, Zhang L. Optimization for the production of surfactin with a new synergistic antifungal activity. PLoS One. 2012;7(5):e34430. DOI: 10.1371/journal.pone.0034430.
28. Gu X-B, Zheng Z-M, Yu H-Q, Wang J, Liang F-L, Liu R-L. Optimization of medium constituents for a novel lipopeptide production by Bacillus subtilis MO-01 by a response surface method. Process Biochem. 2005;40:3196–3201. DOI: 10.1016/j.procbio.2005.02.011.
29. Singh AK, Rautela R, Cameotra SS. Substrate dependent in vitro antifungal activity of Bacillus sp strain AR2. Microb Cell Fact. 2014;13:67. DOI: 10.1186/1475-2859-13-67.
30. Zhao P, Quan C, Jin L, Wang L, Wang J, Fan S. Effects of critical medium components on the production of antifungal lipopeptides from Bacillus amyloliquefaciens Q-426 exhibiting excellent biosurfactant properties. World J Microbiol Biotechnol. 2013;29:401–409. DOI: 10.1007/s11274-012-1180-5.
31. Al-Ajlani MM, Sheikh MA, Ahmad Z, Hasnain S. Production of surfactin from Bacillus subtilis MZ-7 grown on pharmamedia commercial medium. Microb Cell Fact. 2007;6:17. DOI: 10.1186/1475-2859-6-17.

32. Jin H, Zhang X, Li K, Niu Y, Guo M, Hu C, Wan X, Gong Y, Huang F. Direct bio-utilization of untreated rapeseed meal for effective iturin A production by Bacillus subtilis in submerged fermentation. PLoS One. 2014;9(10):e111171. DOI: 10.1371/journal.pone.0111171.

33. Wei Y-H, Lai C-C, Chang J-S. Using Taguchi experimental design methods to optimize trace element composition for enhanced surfactin production by Bacillus subtilis ATCC 21332. Process Biochem. 2007;42:40–45. DOI: 10.1016/j.procbio.2006.07.025.

34. Wei Y-H, Wang L-F, Chang J-S. Optimizing iron supplement strategies for enhanced surfactin production with Bacillus subtilis. Biotechnol Progr. 2004;20:979–983. DOI: 10.1021/bp030051a.

35. Lin H-Y, Rao YK, Wu W-S, Tzeng Y-M. Ferrous ion enhanced lipopeptide antibiotic iturin A production from Bacillus amyloliquefaciens B128. Int J Appl Sci Eng. 2007;5(2):123–132.

36. Chen W-C, Juang R-S, Wei Y-H. Applications of a lipopeptide biosurfactant, surfactin, produced by microorganisms. Biochem Eng J. 2015;103:158–169. DOI: 10.1016/j.bej.2015.07.009.

37. Fahim S, Dimitrov K, Gancel F, Vauchel P, Jacques P, Iordan Nikov I. Impact of energy supply and oxygen transfer on selective lipopeptide production by Bacillus subtilis BBG21. Bioresour Technol. 2012;126:1–6. DOI: 10.1016/j.biortech.2012.09.019.

38. Chtioui O, Dimitrov K, Gancel F, Dhulster P, Nikov I. Rotating discs bioreactor, a new tool for lipopeptides production. Process Biochem. 2012;47(12):2020–2024. DOI: 10.1016/j.procbio.2012.07.013.

39. Coutte F, Lecouturier D, Leclère V, Béchet M, Jacques P, Dhulster P. New integrated bioprocess for the continuous production, extraction and purification of lipopeptides produced by Bacillus subtilis in membrane bioreactor. Process Biochem. 2013;48(1):25–32. DOI:10.1016/j.procbio.2012.10.005.

40. Fahim S, Dimitrov K, Vauchel P, Gancel F, Delaplace G, Jacques P, Nikova I. Oxygen transfer in three phase inverse fluidized bed bioreactor during biosurfactant production by Bacillus subtilis. Biochem Eng J. 2013;76:70–76. DOI:10.1016/j.bej.2013.04.004.

41. Willenbacher J, Rau J-T, Rogalla J, Syldatk C, Hausmann R. Foam-free production of surfactin via anaerobic fermentation of Bacillus subtilis DSM 10(T). AMB Express. 2015;5:21. DOI: 10.1186/s13568-015-0107-6.

42. Zhu Z, Li R, Yu G, Ran W, Shen Q. Enhancement of lipopeptides production in a two-temperature-stage process under SSF conditions and its bioprocess

in the fermenter. Bioresour Technol. 2013;127:209–215. DOI: 10.1016/j.biortech.2012.09.119.

43. Zhu Z, Zhang G, Luo Y, Ran W, Shen Q. Production of lipopeptides by Bacillus amyloliquefaciens XZ-173 in solid state fermentation using soybean flour and rice straw as the substrate. Bioresour Technol. 2012;112:254–260. DOI: 10.1016/j.biortech.2012.02.057.

44. Zhu Z, Zhang J, Wu Y, Ran W, Shen Q. Comparative study on the properties of lipopeptide products and expression of biosynthetic genes from Bacillus amyloliquefaciens XZ-173 in liquid fermentation and solid-state fermentation. World J Microbiol Biotechnol. 2013;29(11):2105–2114. DOI: 10.1007/s11274-013-1375-4.

45. Rangarajan V, Dhanarajan G, Sen R. Improved performance of cross-flow ultrafiltration for the recovery and purification of $Ca^{2}+$ conditioned lipopeptides in diafiltration mode of operation. J Membr Sci. 2014;454:436–443. DOI: 10.1016/j.memsci.2013.12.047.

46. Jauregi P, Coutte F, Catiau L, Lecouturier D, Jacques P. Micelle size characterization of lipopeptides produced by B. subtilis and their recovery by the two-step ultrafiltration process. Sep Pur Technol. 2013;104:175–182. DOI: 10.1016/j.seppur.2012.11.017.

47. Mansson M, Dale EK, Hauge S, Overballe-Petersen C, Hirth KA, Hansen DB. Process for purifying lipopeptides. 2015. US Patent 9150894 B2.

48. Dhanarajan G, Rangarajan V, Sen R. Dual gradient macroporous resin column chromatography for concurrent separation and purification of three families of marine bacterial lipopeptides from cell free broth. Sep Pur Technol. 2015;143:72–79. DOI: 10.1016/j.seppur.2015.01.025.

49. Yang H, Li X, Li X, Yu H, Shen Z. Identification of lipopeptide isoforms by MALDI-TOF-MS/MS based on the simultaneous purification of iturin, fengycin, and surfactin by RP-HPLC. Anal Bioanal Chem. 2015;407:2529–2542. DOI: 10.1007/s00216-015-8486-8.

50. Torres MJ, Brandan CP, Petroselli G, Erra-Balsells R, Audisio MC. Antagonistic effects of Bacillus subtilis subsp. subtilis and B. amyloliquefaciens against Macrophomina phaseolina: SEM study of fungal changes and UV-MALDI-TOF MS analysis of their bioactive compounds. Microbiol Res. 2016;182:31–39. DOI: 10.1016/j.micres.2015.09.005.

51. Urajová P, Hájek J, Wahlsten M, Jokela J, Galica T, Fewer D, Kust A, Zapomelová-Kozlíková E, Delawská K, Sivonen K, Kopecký J, Hrouzek PA. Liquid chromatography-mass spectrometric method for the detection of

cyclic β-amino fatty acid lipopeptides. J Chromatogr A. 2016;1438:76–83. DOI: 10.1016/j.chroma.2016.02.013.

52. Sajitha KL, Dev SA, Florence EJM. Identification and characterization of lipopeptides from Bacillus subtilis B1 against Sapstain fungus of rubberwood through MALDI-TOF-MS and RT-PCR. Curr Microbiol. 2016. DOI: 10.1007/s00284-016-1025-9.

53. Lopez D, Vlamankish H, Losick R, Kolter R. Cannibalism enhances biofilm development in Bacillus subtilis. Mol Microbiol. 2009;74:609–618. DOI: 10.1111/j.1365-2958.2009.06882.x.

54. Dubern JF, Lugtenberg BJJ, Bloemberg GV. The ppuI-rsaL-ppuR Quorum sensing system regulates biofilm formation of Pseudomonas putida PCL1445 by controlling biosynthesis of the cyclic lipopetides putisolvin I and II. J Bacteriol. 2006;188:2898–2906. DOI: 10.1128/JB.188.8.2898-2906.2006.

55. Das P, Mukherjee S, Sen R. Genetic regulations of the biosynthesis of microbial surfactants: an overview. Biotech Genet Eng Rev. 2008;25:165–186 DOI: 10.5661/bger-25-165.

56. Schmidberger A, Henkel M, Hausmann R, Schwartz T. Expression of genes involved in rhamnolipid synthesis in Pseudomonas aeruginosa PAO1 in a bioreactor cultivation. Appl Microbiol Biotechnol. 2013;97:5779–5791. DOI: 10.1007/s00253-013-4891-0

12 THE ROLE OF LACTIC ACID BACTERIA IN THE FERMENTATION PROCESS OF CEREALS AND PSEUDOCEREALS

1. INTRODUCTION: AN OVERVIEW

Throughout the ages, human society has employed many methods to safeguard and maintain the quality of diverse food items. Throughout history, various types of foods have been prepared by different cultures. For instance, the ancient Egyptians were known for making alcoholic beverages, while nomadic people from central Asia prepared yoghurt and kefir. Germanic tribes fermented meat, Eskimos fermented fish, and ancient Persians made boza. Native tribes in pre-Columbian America fermented maize. [1] The earliest documented evidence of the fermentation process dates back to 6000 BC, making it one of the oldest means of preserving food [2, 3]. It is likely that the ancient people lacked any grasp of microbiology, but in the mid-19th century, Louis Pasteur made substantial contributions to the comprehension of the fermentation process. He pioneered the recognition of microbes' significance and demonstrated the existence of diverse forms of fermentation [3]. The initial and main objective of fermentation was to achieve a preservation effect. As a result, numerous fermented foods were produced using various preservation procedures due to their distinct flavors, smells, and textures that were highly valued by consumers [4, 5]. The fermentation process has resulted in the production of various traditional food products, including dairy goods (such as cheese, butter, and yogurt), fermented meats, plant and fruit products (such as sausages, silage, sauerkraut, olives, and grapes), and fermented cereal products like bread and beer [6]. Fermented food and beverages are food and drinks that have undergone a biochemical transformation caused by microbial enzymes, specifically amylases, proteases, and lipases. This transformation converts polysaccharides, proteins, and lipids into non-toxic products that have desirable tastes, aromas, and textures, which are appealing to consumers [4, 7].

During food fermentations, specific organisms thrive due to the treatment and storage conditions, resulting in a beneficial impact on the food instead of causing spoilage. Fermented foods are primarily created through the action of lactic acid bacteria (LAB) and fungi, particularly yeasts and to a lesser degree, molds. Both kinds of organisms have a shared biological niche and can thrive in environments with low pH and reduced water activity. However, only lactic acid bacteria and facultative yeasts can flourish under anaerobic circumstances. These microorganisms are commonly found together in fermented products, both dairy and non-dairy. However, in certain instances, they can act as agents that cause spoiling. The fermentation process can be initiated by microorganisms that are naturally present in the substrate, or by intentionally introducing starter and adjunct cultures [9].

2. LACTIC ACID BACTERIA

Lactic acid bacteria (LAB) represent an ubiquitous and heterogeneous species with common feature of lactic acid production as a result of sugar metabolism which leads to an acidification of the environment down to a pH of 3.5 [10]. The monograph by Orla-Jensen (1919) formed the basis of the present classification of LAB that take into account the cellular morphology, mode of glucose fermentation, growth temperature and sugar utilization possibilities [11]. Taxonomically, LAB are divided into two distinct phyla: *Firmicutes* and *Actinobacteria*. Within the *Firmicutes* phylum genera such as *Lactobacillus, Lactococcus, Leuconostoc, Oenococcus, Pediococcus, Streptococcus, Enterococcus, Tetragenococcus, Aerococcus, Carnobacterium, Weissella, Alloiococcus, Symbiobacterium* and *Vagococcus* belong. Within the *Actinobacteria* phylum, lactic acid bacteria belong to the *Atopobium* and *Bifidobacterium* genera [12].

Lactic acid bacteria are Gram-positive, non-sporulating, non-pigmented and non-motile rods and cocci, most of which are non-respiring but aerotolerant anaerobes. They lack cytochromes and porphyrins and are therefore catalase- and oxidase-negative. LAB tend to be nutritionally fastidious, often requiring specific amino acids, B-vitamins and other growth factors. Some do take up oxygen through the mediation of flavoprotein oxidases, thus producing hydrogen peroxide and/or re-oxidizing NADH during dehydrogenation of sugars. The cellular energy is derived from the fermentation of carbohydrates to produce major lactic acid. They use one of two different pathways and this provides a useful diagnostic feature in their classification. Since many species of lactic acid bacteria (LAB) and other food-associated bacteria had a long historical association with human foods, they are recognized as generally regarded as safe (GRAS) bacteria. Infections by LAB are characterized as opportunistic that rely on host factors rather than on intrinsic pathogenicity. Only rare cases of clinical infections have been reported in humans, for example, in patients with endocarditis or with immune deficiency [8, 12–15]. Homofermentative organisms produce only lactic acid from the glucose fermentation during the Embden-Meyerhof-Parnas glycolytic pathway. Heterofermenters produce roughly equimolar concentration of lactate, ethanol/acctatc and carbon dioxide from glucose (Table 1).

Table 1. Principal genera of the lactic acid bacteria [8].

Genera	Morphology	Fermentation	Lactate isomer	DNA (mole % G-C)
Lactococcus	Cocci	Homo	l	33–37
Lactobacillus	Rods	Homo/hetero	d, l, dl	32–53
Leuconostoc	Cocci	Hetero	d	38–41
Streptococcus	Cocci	Homo	l	40
Pediococcus	Cocci	Homo	dl	34–42

2.1. Microbial inoculants employed in the process of lactic acid fermentation

Categories Lactobacillus is known for its significant evolutionary heterogeneity, which is evident from its wide range of % G-C content. Typically, they are described as bacteria that are Gram-positive, require low levels of oxygen, do not produce spores, and do not have flagella. They have a rod-like or coccobacilli shape. These microorganisms are frequently present in a wide range of habitats, including dairy and meat products that have undergone fermentation, as well as fermented and pickled vegetables. They can also be found on the surfaces of human mucous membranes, such as the gastrointestinal and vaginal tracts, as well as in soil and plants [16, 17]. Intestinal lactobacilli, including Lb. rhamnosus, Lb. acidophilus, Lb. reuteri, Lb. plantarum, and Lb. paracasei, engage with the host and have been associated with various health advantages [18–20]. Lactobacillus reuteri is a highly beneficial type of bacteria that is commonly included in infant dried milk formula. It is specifically used to help newborns who have lactose intolerance or who need to recover from diarrhea [21]. Lactobacilli are naturally found in breast milk, particularly species of Lb. fermentum, Lb. rhamnosus, Lb. gasseri, and Lb. salivarius [18, 22–25]. Liptáková and her colleagues [26] regularly detected Lb. plantarum in the breast milk of healthy women.

Lactobacillus species are categorized into three groups according to the end products of fermentation: obligate homofermenters, facultative heterofermenters, and obligate heterofermenters [17, 27]. Obligate homofermenters, such as Lb. helveticus, Lb. acidophilus, Lb. delbrueckii, and others, primarily ferment hexoses into lactate, but they are incapable of fermenting pentoses or gluconate. Lactobacillus acidophilus strains are widely recognized as the most effective lactobacilli for supporting health. They naturally occur in the human gut microbiome. Probiotic strains are incorporated into dairy meals due to their physiological advantages. The facultative heterofermenters utilize the Embden-Meyerhof-Parnas (EMP) pathway to ferment hexoses and employ phosphoketolase activity to ferment pentoses, resulting in the production of lactate, acetate, formic acid, and ethanol. This metabolic process is observed in Lactobacillus plantarum and Lactobacillus casei. Obligate heterofermenters, such as Lb. brevis, Lb. reuteri, Lb. fermentum, or Lb. kefir, utilize the phosphoketolase pathway to ferment hexoses and pentoses. The primary byproducts of this fermentation process are lactic acid, acetic acid (or ethanol), and carbon dioxide [13, 15, 28].

Categories Lactococcus is the primary group of mesophilic bacteria that are commonly employed in the manufacture of lactic acid, particularly in dairy fermentations such as sour milk and cream, lactic butter, and many types of artisanal and commercial cheeses. Some of them are appropriate for fermentations involving cereal and pseudocereal grains [29, 30]. In 1873, Joseph Lister was the first person to isolate and identify the bacterium responsible for milk fermentation. The culture was initially termed Bacterium lactis and was later changed to S. lactis. In 1919, Orla-Jensen classified mesophilic lactic streptococci into two distinct species: S. lactis and S. cremoris. These species were then categorized as part of the Group N Streptococci [29]. Currently, the genus Lactococcus consists of nine species: L. lactis (including the subspecies lactis, cremoris, and hordniae), L. garvieae, L. piscium, L. plantarum, L. raffinolactis, L. chungangensis, L. fujiensis, L. formosensis, and L. taiwanensis [31, 32]. Compared to L. lactis, L. cremoris lacks the ability to ferment maltose and ribose, grow in the presence of 4% salt, and hydrolyze arginine. Lactococcus lactis subsp. lactis var. diacetylactis metabolizes citrate into diacetyl, carbon dioxide, and acetone, which contribute to the creamy and buttery aroma found in fermented milks, cream, butter, and certain types of cheeses such as Camembert, Emmental, and Cheddar [13, 15, 33].

Several varieties of L. lactis produce bacteriocins, which exhibit antibacterial properties, particularly against a limited range of Lactococci. However, nisin and lacticin 3147 have a far wider range of activity, affecting a broader spectrum of Gram-positive bacteria. Nisin has been approved as a food additive for the purpose of regulating harmful microorganisms [11, 29]. Sadiq et al. [34] identified three bacteriocin-producing strains of L. lactis, namely TI-4, CE-2, and PI-2. These strains were found to be effective against B. subtilis and S. aureus. The highest levels of bacteriocins (specifically nisin A and nisin Z) were produced at temperatures of 25°C and 30°C, and at pH levels of 5 and 8, respectively.

Leuconostocs, specifically Ln. mesenteroides subsp. cremoris, are the primary heterofermentative lactic acid bacteria utilized in the production of various fermented dairy products and cheeses. They are known for their ability to enhance the flavor of these products. The process of citrate fermentation plays a crucial role in the production of diacetyl and carbon dioxide in some varieties of cheese. The genus Leuconostoc comprises 12 species that have been found in various sources such as plants, fermented foods (including meats, vegetables, and dairy products), vacuum-packaged cold-stored meat, honcy, Ethiopian coffee fermentation, kimchi, palm wine, cane juice, and human clinical samples. These species are Ln. mesenteroides, Ln. pseudomesenteroides, Ln. carnosum, Ln. citreum, Ln. fallax, Ln. gasicomitatum, Ln. gelidum, Ln. holzapfelii, Ln. inhae, Ln. kimchii, Ln. lactis, and Ln. palmae [35-38].

There are currently twelve species of the Pediococcus genus that have been identified. These species include P. acidilactici, P. pentosaceus, P. argentinicus, P. cellicola, P. claussenii, P. damnosus, P. ethanolidurans, P. inopinatus, P. lolii, P. parvulus, P. siamensis, and P. stilesii. Pediococci, unlike other cocci in the LAB, typically do not create chains of cells [35, 39]. Pediococci are commonly found in dairy products and the dairy environment, and they can have a significant effect on texture because they produce exopolysaccharides. Garai-Ibabe et al. identified and separated two strains of P. parvulus (CUPV1 and CUPV22) that have the ability to create a large amount of 2-substituted (1,3)-β-d-glucan, which significantly increases the viscosity of the growth media [40]. Pediococci are commonly present in numerous fermented meat and fish items, fermented beans, cereals, olives, and sauerkraut. Certain strains are suggested to possess probiotic action because of their capacity to survive and attach to the gastrointestinal tract, as well as their claimed potential to modulate the immune system [13, 39, 41].

Streptococcus is a phrase that originated from the Greek words 'streptos', meaning easily twisted like a chain, and 'kokkos', meaning grain or seed. It was first used in 1874 by Billroth to describe bacteria that form chains and have a coccoid structure. In 1884, Rosenbach coined the term Streptococcus to describe S. pyogenes, a type of coccus that forms chains and was isolated from a suppurative abscess in humans. In 1906, Andrewes and Horder conducted an examination of 1200 streptococci that were obtained from human, air, and milk sources. Through the analysis of sugar metabolism, decrease of neutral red, and growth characteristics in milk, they were able to differentiate eight distinct types. In 1937, Sherman created the initial all-encompassing and organized categorization of streptococcal samples obtained from various sources, including the environment, commensal organisms, and hospitals. In order to account for their high sensitivity to bile, he removed all absolutely anaerobic cocci and pneumococci from the Streptococcus genus. Instead, he introduced four main divisions: pyogenic, enterococcus, lactic, and viridans group [42]. The findings from molecular taxonomy research have led to significant revisions in the classification of Streptococcus spp. Specifically, the group of streptococci previously referred to as 'lactic' streptococci have now been reclassified as the genus Lactococcus. Additionally, several members of Sherman's 'enterococcus' division have been designated as foundational members of the genus Enterococcus [43]. The classification of Streptococci into seven groups is determined by analyzing the sequence data of the 16S rRNA gene. This classification method has been found to be consistent with the results of DNA-DNA re-association experiments and numerical taxonomic investigations [44-46].

The sole species of Streptococcus that is beneficial in the fermentation of milk for the creation of yoghurt and certain types of cooked cheeses such as Grana Padano, Gorgonzola, Mozzarella, and Fontina is S. thermophilus var. salivarius. It is classified as GRAS (Generally Recognized as Safe) in the USA and has a Qualified Presumption of Safety in the European Union, thanks to its extensive track record of being safely used in food production. The final outcomes of lactose fermentation consist of lactate, acetaldehyde, and diacetyl. Certain strains have the capability to make thermophilins, which are protein compounds that can suppress the growth of listeria and clostridia bacteria. In particular, thermophilin 13 and 1277 exhibit a wide range of inhibitory effects [15, 27, 28, 47–50].

Bacteria having a distinct Y shape, belonging to the genus Bifidobacterium, were first discovered in 1899 by Tissier in feces samples of breast-fed children. These bacteria were designated B. bifidus. In 1924, Orla-Jensen identified the genus Bifidobacterium as a distinct taxonomic group. However, due to the similarities between bifidobacteria and the genus Lactobacillus, they were classified inside the latter genus. In 1957, Dehnart discovered that there are numerous varieties of Bifidobacterium and suggested a method to distinguish these bacteria by examining their hexose fermentation pathway [51]. The predominant strains commonly discovered in the human gastrointestinal system consist of B. adolescentis, B. bifidum, B. breve, B. catenulatum, B. pseudocatenulatum, B. longum subsp. infantis, B. longum subsp. longum, B. dentium, and B. angulatum [52]. Bifidobacteria make up to 25% of the fecal microbiota in adults and 80% in babies [53]. Matsukiand and co-workers [54] found that the most commonly found bifidobacteria in the adult digestive tract are B. catenulatum, B. longum, and B. adolescentis. In contrast, B. breve, B. infantis, and B. longum are the predominant species in the infant gut.

The key Bifidobacterium species for probiotic use include B. longum, B. bifidum, and B. animalis. Children who consumed milk-based formula enriched with Bifidobacterium (specifically the B. lactis Bb-12 strain) were shielded from experiencing symptomatic rotavirus infection. The daily intake of three cups per day of B. longum yoghurts resulted in a reduction of gastrointestinal symptoms related with erythromycin. B. bifidum NCFB 1454 exhibited antimicrobial activity against certain strains of Listeria, Bacillus, Enterococcus, Pediococcus, and Leuconostoc, attributed to the synthesis of bifidocin B [15, 16, 28, 51, 53, 55, 56].

3. ANTIMICROBIAL COMPOUNDS PRODUCED BY LACTIC ACID BACTERIA

Lactic acid bacteria may produce substances and thus create conditions harmful for undesired bacteria, yeasts and moulds which lead to the increase of food shelf life [57]. Temperature and incubation period are the main factors modulating production of antimicrobial substances. Sathe et al. [58] in their study evaluated the impact of the growth phase on antimicrobial activity of *Lb. plantarum* at 30°C. The evaluated strain showed maximal antimicrobial activity at the end of exponential phase of growth, and in stationary phase after 48 h of cultivation, decline in antimicrobial activity was observed. These results are consistent with the study of Batish et al. [59] who observed maximal antimicrobial activity of the same strain after 48 h of incubation at 30°C. The main product of fermentation by lactic acid bacteria is mostly lactic acid. However, under aerobic conditions, carbon dioxide and acetic acid are created as a result of oxidative dissimilation, while hydrogen peroxide as an intermediate product is formed [27]. Most of the isolated and identified antimicrobial substances produced by lactic acid bacteria are with low-molecular weight composed of organic acids, reuterin, hydrogen peroxide, hydroxyl fatty acids, phenolic and proteinaceous compounds [60].

When lactic acid is produced, the pH decreases and consequently the organic acids or small fatty acids (SFAs) become undissociated and represent the main antimicrobial activity of the LAB [61]. It has been shown that organic acids penetrate bacterial membrane of the target microorganism and inhibit transport mechanism in the cell by reducing pH values [62]. The effect of acids depends not only in combination with lowering pH and reduction of redox potential but also on the type and concentration of acid presented in the environment [63]. Acetic acid in comparison to lactic acid was described as being more effective, and is able to inhibit growth of moulds, yeasts and bacteria [5]. Propionic acid inhibits moulds and selected Gram-positive microorganism [62]. Phenyllactic acid and pyroglutamic acid are able to inhibit growth of *Aspergillus niger*, *A. flavus* and *Penicillium expansum*, and both were isolated from cell-free extract of *L. plantarum* and *Lb. rhamnosus* GG (LGG) [60, 64]. Liptáková et al. [64] mathematically predicted the inhibitory effect of *Lb. rhamnosus* GG (LGG) on the growth dynamics of *Candida maltosa* YP1 and *Geotrichum candidum* yeasts. At 18°C, growth rates of the yeasts in mixed cultures decreased about 50% compared with rates of its pure cultures. The effectiveness of growth inhibition of *C. maltosa* was dependent on initial LGG concentration; the most antagonistic activity of lactobacilli was determined at log 4 and log 6 initial concentration (Figure 1). Greifová et al. [65] described the inhibitory effect of D,

L-phenyllactic acid on moulds such as *Alternaria alternata*, *A. flavus*, *Cladosporium herbarum*, *Fusarium nivale*, *Mucor racemosus* and *P. funiculosum*.

Liptáková and co-workers [66] focused on the growth of yoghurt contaminant *C. maltosa* YP1 in milk as influenced with initial different numbers of *Lb. rhamnosus* VT1 (ranged from 1 to 15% v/v) and temperature. The growth parameters of yeast in dependence on the lactobacilli counts at 17°C are summarized in Table 2. The antagonistic relationship between *C. maltosa* YP1 and *Lb. rhamnosus* VT1 was based not only on the lactic acid but it was consequence of the other antimicrobial, non-proteinaceous and non-saccharidic substances, identified by Plocková et al. [67] and also pyroglutamic acid, later identified by Liptáková et al. [64].

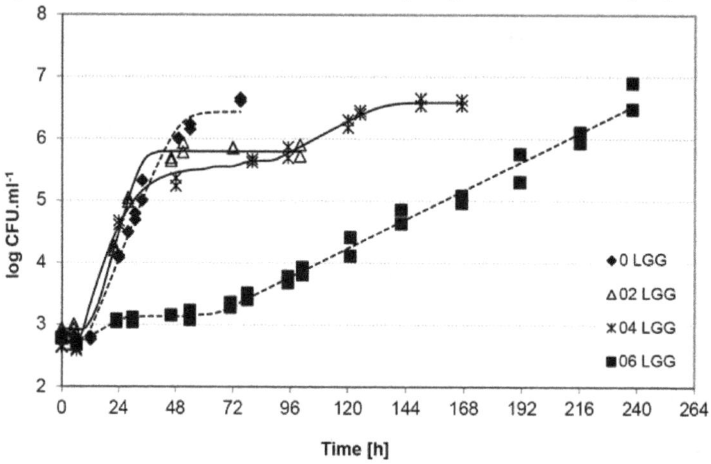

Figure 1. Growth dynamics of C. maltosa YP1 in co-culture with Lb. rhamnosus GG at 18°C in dependence on various initial lactobacilli concentration (♦ without LGG addition, △ 2 log LGG initial counts, ✻ 4 log LGG initial counts, ■ 6 LGG initial counts).

Table 2. Values of growth rate (Gr) and lag time (λ) of *C. maltosa* YP1 in milk in dependence of initial numbers of *Lb. rhamnosus* VT1 at 17 ± 0.5°C.

Initial inoculation of *Lb. rhamnosus* VT1 (% v/v)	Growth rate (log CFU ml^{-1} h^{-1})	Lag-phase duration (h)
1.0	0.062	0.1
	0.064	0.1
5.0	0.055	5.5
	0.052	8.1
10.0	0.046	72.9
	0.046	74.2
15.0	0.041	76.4
	0.043	73.1

3.1. Hydrogen peroxide

Most lactic acid bacteria produce hydrogen peroxide in the presence of oxygen. After its accumulation, inhibitory effect is mediated through oxidizing effect on membrane lipids and cell proteins of targeted microorganism. The antimicrobial activity of the compound in lower concentrations mostly in food is enhanced by treatment with the formation of hypothiocyanite catalysed by lactoperoxidase system [68]. Fitzsimmons and Berry [69] reported in their study the inhibitory effect of hydrogen peroxide on the growth of *C. albicans*. The minimum inhibitory concentration is less than 0.025% [60].

3.2. Carbon dioxide

Carbon dioxide at low concentrations may stimulate the growth of selected bacteria. Creating an anaerobic environment may be toxic to some aerobic food microorganisms through its action on cell membranes and its ability to reduce internal and external pH values [5].

3.3. Bacteriocins

Bacteriocins are a diverse collection of antimicrobial peptides that are generated by ribosomes and have the ability to kill or hinder the growth of other bacterial strains. Usually, LAB bacteriocins exhibit a limited range of antibacterial activity, while certain strains may also produce bacteriocins with a wide range of antibacterial activity. Specific strains of lactic acid bacteria have the ability to hinder the proliferation of harmful Gram-positive bacteria and spoilage bacteria, as well as yeasts. According to reports, bacteriocins are known to hinder the growth of certain Gram-negative species. Lozo et al. [70] shown that the strain Lb. paracasei subsp. paracasei BGBUK2-16, isolated from typical home cheese, produces bacteriocin 217 (Bac 217), which has an inhibiting impact on Staphylococcus aureus. Strains of Lactobacillus fermentum, Lactobacillus pentosus, Lactobacillus paracasei, and Lactobacillus rhamnosus, which were obtained from a traditional wheat-based corn drink, have been found to produce active bacteriocins that can inhibit the growth of Escherichia coli, Pseudomonas aeruginosa, and Enterococcus faecalis [71]. Valdés-Stauber and Scherer [72] identified and described Linocin M18, a bacteriocin generated by B. linens M18 during the stationary growth phase. This bacteriocin shown the ability to effectively hinder the growth of Listeria spp., particularly L. monocytogenes, L. innocua, L. ivanovii, and several coryneforms. However, it did not show any inhibitory effects on Gram-negative bacteria. In their work, Corsetti et al. [73] characterized the antimicrobial compounds present in sourdough and classified them as bacteriocins-like inhibitory chemicals. Leuconostocs, specifically Ln. mesenteroides subsp. mesenteroides Y105 and UL5, have the capability to generate bacteriocins that have antilisterial activity [37, 74].

Certain strains of Pediococcus spp. can exhibit antimicrobial properties through the synthesis of pediocins, which can target and inhibit the growth of undesired and pathogenic microbes such as Listeria spp. and Clostridium perfringens [75]. Gurira and Buys [76] identified P. acidilactici and P. pentosaceus as non-starter lactic acid bacteria in Bouquet and Gouda cheeses. These bacteria were found to have the ability to suppress the growth of L. monocytogenes ATCC 7644 and B. cereus ATCC 1178 through the production of pediocins. In their work, Altuntas et al. [77] verified the antilisterial efficacy of the pediocin-producing strain P. acidilactici 13.

3.3.1. Reuterin

Reuterin is a product of glycerol fermentation produced during stationary phase by *Lb. reuteri*, *Lb. brevis*, *Lb. buchneri*, *Lb. collinoides* and *Lb. coryniformis* under anaerobic conditions, which enables to suppress ribonuclease activity [60]. Reuterin has a wide inhibitory spectrum against Gram-negative and Gram-positive bacteria, yeasts, fungi and protozoa: *Salmonella*, *Shigella*, *Clostridium*, *Staphylococcus*, *Listeria*, *Candida* and *Trypanosoma* [78]. An inhibitory effect on the growth of genus *Aspergillus* and *Fusarium* has been reported. The addition of glycerol to the media containing lactic acid bacteria producing reuterin increased its antifungal activity [60].

4. PROBIOTICS AND FUNCTIONAL FOODS

4.1. Probiotics

The term "probiotic" has its origins in Greek, where it means "for life". In 1954, Vergin provided the initial definition of probiotics as the antithesis of antibiotics. One year later, Kolb suggested that probiotics may be used to correct the microbial imbalance caused by antibiotic treatment in the human body. In 1974, Parker provided a definition for probiotics as organisms and chemicals that help maintain a healthy balance of microorganisms in the gut. The definition most commonly referenced is that of Fuller's (1992), which states that probiotics are "live microbial feed supplements that have a beneficial impact on the host animal by improving its intestinal microbial balance." The definition proposed by a working committee from the Food and Agriculture Organization/World Health Organization (FAO/WHO) states that probiotics are living bacteria that, when given in sufficient quantities, provide a health advantage to the host (2002). Health benefits must be substantiated by rigorous clinical trials conducted on human subjects and then published in peer-reviewed scientific journals [79]. Several types of bacteria (Lactobacillus, Bifidobacterium, Lactococcus, Leuconostoc, Pediococcus, S. salivarius subsp. thermophilus, E. faecium, E. faecalis, E. coli, B. cereus, B. subtilis, B. clausii, B. coagulans, B. licheniformis, and B. polyfermenticus) and yeast Saccharomyces boulardii is primarily utilized as a probiotic in dairy goods, including milk, yogurt, and probiotic cheeses. It is also found in non-dairy food and beverages such as dry sausages, soy milk drinks, juices, and fermented cereal products like Boza, Bushera, Mahewu, and Pozol [57, 80–85]. Lee and colleagues [86] examined the probiotic capabilities of B. polyfermenticus KU3, which was obtained from kimchi. The spore suspension exhibited resistance to fake gastric juice and maintained viability for a duration of 24 hours in artificial bile acid. Furthermore, it demonstrated high adhesion to the HT-29 cell line and shown anti-carcinogenic activity, indicating the presence of B. polyfermenticus KU3. Cell B. polyfermenticus exhibited potent anti-proliferative effects on multiple cancer cell lines including HeLa, LoVo, HT-29, and MCF-7 (inhibition percentage ranging from 90.5% to 96.9%). Liptáková et al. [87] found that the probiotic strains Lb. acidophilus 145 and Lb. rhamnosus GG had a similar inhibitory effect on the growth of HeLa and Caco-2 cells. This effect was attributed to the adhesion and metabolism of the probiotics, with Lb. acidophilus 145 inhibiting cell proliferation by 95-96% and Lb. rhamnosus GG inhibiting it by 68%.

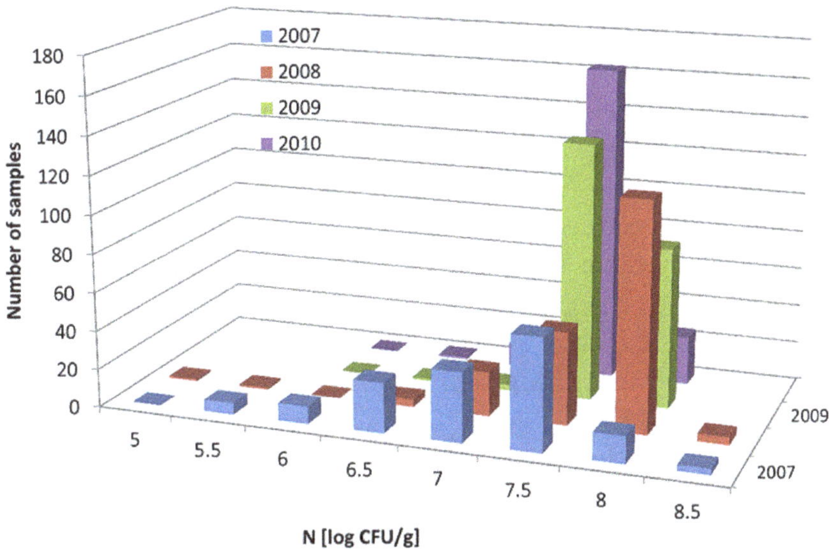

Figure 2. Lactobacillus acidophilus contents in the fermented milk at the end of shelf life.

The selection of a probiotic microbe is influenced by several variables. Probiotics must be species that are safe, non-pathogenic, and non-toxic. They should also be able to survive the journey through the intestinal tract and stick to the intestinal mucosa. Additionally, they should be capable of producing organic acids, specifically lactic and acetic acids [57, 79]. Tripathi and Giri [85] state that the survival of probiotics in food is influenced by various parameters including pH, water activity, redox potential of foods, presence of salt, sugar, hydrogen peroxide, bacteriocins, aroma and colouring compounds, processing, packing, and storage conditions. It is advisable to keep probiotic foods at a temperature ranging from 4 to 5°C. The maximum survival rate of Lb. acidophilus LA-5 in yoghurt was found to be 20 days at a temperature of 2°C. However, for B. lactis BB 12, the most favorable storage temperature was determined to be 8°C [88, 89]. at order for probiotic bacteria to have a positive impact on the host's health, they need to be alive and present at a high concentration of approximately 106 to 107 CFU/ml or g at the end of the product's shelf life. Therefore, the minimum recommended daily dose for therapeutic purposes is typically considered to be 108 to 109 CFU/ml or g [16]. Liptáková et al. [90] measured the amount of Lb. acidophilus 145 in acidophilus milk at the end of its shelf life when it was stored at temperatures of 6, 8, and 10°C. The count of probiotic Lb. acidophilus 145 varied from 6 to 7 logarithmic units. Valík et al. [91] observed the levels of Lb. acidophilus in fermented milk at the end of its shelf life over a span of 5 years (2007-2011). The mean values in logarithmic colony-forming units per gram (log CFU/g) varied from 6.85 to 7.47, respectively. During the years 2007 and 2008, 9.87% and 1.01% of the samples had less than 106 CFU/g of Lb. acidophilus at the end of consumption. In the remaining years, no samples were found to have a count lower than 106 CFU/g (Figure 2).

The precise processes underlying the health-enhancing benefits of probiotics are not yet fully comprehended. However, it is known that probiotics exhibit anti-carcinogenic and anti-mutagenic action, suppress allergies, lower serum cholesterol levels, and reduce blood pressure [12, 80, 92, 93]. Lactobacillus bulgaricus and Streptococcus thermophilus possess the ability to undergo lactose fermentation, making them advantageous for those with lactose intolerance [94]. Strains of Lactobacillus rhamnosus GG, Bifidobacterium lactis Bb-12, Lactobacillus acidophilus, Lactobacillus casei Shirota, and Lactobacillus reuteri have been found to have positive effects in treating, decreasing the duration of, or preventing acute diarrhea caused by rotavirus [95-97]. S. boulardii, when administered as a non-pathogenic biotherapeutic yeast, is crucial in treating or preventing antibiotic-associated diarrhea caused by C. difficile [83, 98-101]. Probiotics, specifically Lb. acidophilus, Lb. plantarum, Lb. rhamnosus, and Bifidobacterium, have the ability to decrease the activity of certain enzymes in the feces that convert procarcinogens into carcinogens (such as β-glucuronidase, azoreductase, urease, nitroreductase, and glycocholic acid reductase). This is achieved through the production of short-chain fatty acids and may contribute to a reduced risk of colorectal carcinoma [79, 102]. Additional mechanisms for the anti-carcinogenic effects of probiotics are discussed in the studies conducted by Commune et al. [92] and Faghfoori et al. [103], respectively.

4.2. Fermented cereals and pseudocereals functional products

Recently, there is an explosion of consumer's interest in functional foods; therefore, a key priority for food industry is the development of such products with a high quality and safety [104]. The aim of these products is to have beneficial effect on host health affecting gut microbial composition subsequently with reducing the risk of chronic diseases [105]. Cereals have been investigated in recent years regarding their potential use in the production of functional foods [106].

Possible application of cereals in functional food can be summarized as follows:

- As fermentable substrates for the growth of probiotic bacteria (lactobacilli, bifidobacteria);
- As prebiotics due to their content of non-digestible oligosaccharides (galacto-oligosaccharides and fructo-oligosaccharides);
- As dietary fibre promoting beneficial effects on human host;
- As encapsulation matter for probiotics to enhance their stability [104, 107].

Cereals have been and still are one of the most important sources of human diet [108] and are grown over 73% of total harvest area [109]. A number of cereals are grown in different countries, including wheat, barley, oat, corn, rye, rice and millet, particularly important from an economical point of view. According to FAO's latest forecast, cereal production in 2015 stands at close to 2525 million tonnes but is still 1.4% below than the record in 2014 [110]. Cereal grains and their derivatives represent an important nutritive component both in developed and in developing countries [111]. They are considered as one of the most important sources of dietary proteins, carbohydrates (starch and fibre), vitamins (B group) and minerals for people all over the world [112].

4.2.1. Nutritional value of cereals

Cereal grains are primarily a source of carbohydrates, and thus a good source of energy [113]. They form about two-thirds up to three-quarters of dry matter [114]. Monosaccharides are the basic components of oligo- and polysaccharides and are most represented in the forms of hexoses (fructose, glucose and galactose) and pentoses, arabinose and xylose [115]. Starch, the major component of cereal grains, occurs in starch granules of different sizes in endosperm.

Within common varieties, 25–27% of starch is presented as amylase and 72–75% represents amylopectin. However, in cereals a portion of the presented starch is not digested and absorbed in the small intestine. This is referred to as resistant starch and it appears to act in a similar way to a dietary fibre [113]. A wide variety of biochemical processes occur in cereals during fermentation as a result of lactic acid bacteria. Fermentation process itself may lead to an increase in the content of reducing sugars, which was confirmed also in a study by Marko et al. [116]. Simple carbohydrates are metabolized directly to organic acids and the glucose as a final product of starch metabolism is utilized immediately [116]. Lambo et al.

[117] described the decrease in starch content during fermentation of barley with lactobacilli.

Cereals are in general good sources of proteins. The proportions of essential amino acids and their digestibility mainly determine protein nutritional quality. Because of different production systems, environmental factors, as well as genotype, it is difficult to obtain comparative values of protein contents of different cereals. Thus, ranges of 5.8–7.7% of protein on a dry weight have been measured for rice, 8.0–15.0% for barley and 9.0–11.0% for maize. The amount of lysine, which is the limiting amino acid for all cereals, varies between species with the highest values in oat and rice and lowest in wheat and maize [118]. The most represented is glutamic acid in the form of glutamine [119]. Degradation and depolymerization of proteins during fermentation process depend not only on the metabolic activity of presented bacteria but also on enzymes that naturally occurred in cereals. Peptides are converted to amino acids by the activity of lactic acid bacteria by the specific intracellular peptidases that are subsequently converted to the specific products influencing the aroma and taste of final products [120]. Antony and co-workers [121] in their study pointed out that the fermentation process does not generally significantly change the total protein content of cereals. However, in the case of yeast corn fermentation, Cui et al. [122] found a significant increase ($P < 0.05$) in the total protein content.

Lipids are only a minor component of cereal grains with the amount varying from 1.7 to 7.0% on a dry mass basis, dependent on the type of cereal grain. The germ is the richest source of lipids. In particular, cereals are rich in essential fatty acids and contain only trace amounts of saturated fatty acids [123]. Oxidation of lipids during fermentation process creates volatiles that contribute to the flavour of final products. Linoleic, oleic and linolenic acids are oxidized by lipoxygenases by forming hydroperoxides that are formed to aldehydes [124]. Aldehydes are converted to alcohols by alcohol dehydrogenases during fermentation process [125]. Antony et al. [121] in their study did not record any changes in the total lipid content during the millet fermentation with the endogenous microorganisms.

Cereals may contribute to vitamin intake due to the presence of most B-vitamins and appreciable amounts of vitamin E. Wholegrain cereals also contain considerable amount of calcium, magnesium, iron, zinc, as well as lower levels of many trace elements, for example, selenium. The content of minerals ranges from 1.0 to 2.5% [113, 126]. Cereals contain relatively high levels of phytate (0.2–1.4%), concentrated mostly in the aleurone layer, which can bind minerals and there is an evidence of its decreased absorption in the presence of phytate, so minerals are not available to microorganisms. However, at a pH values less than 5.5, phytates are hydrolysed by endogenous phytases, thus minerals are released from the complex [9]. In our investigation, changes in chemical composition of maize flours before and after expiry date were determined (Table 3). The percentage of starch and reducing sugars is one of the most important aspects showing the suitability of the tested substrate in fermentation technologies. A decline in the content of reducing sugars (60.1%) and starch (7.9%) was observed. Matejčeková and co-workers [30] recorded a decline of reducing sugars in amaranth flours before and after expiry date of about 31% in their study.

In comparison to milk and dairy products, the nutritional quality of cereals and their products is sometimes inferior, or poor. The reason is the lower protein content in comparison to milk, limitations in the amounts of certain amino acids, notably lysine, and the presence of antinutritive compounds (phytic acid, tannins and polyphenols) and a coarse nature of grains [7, 127]. Cereals typically undergo a range of processes that change the nutritional content. Milling is the main process associated with cereals; also, extrusion is used to produce a variety of different types of products [128].

Table 3. Chemical composition of maize flours before and after expiry date (%).

	Proteins	Lipids	Starch	Reducing sugars
Corn flour 1	3.21 ± 0.00	1.59 ± 0.03	68.71 ± 0.12	4.24 ± 0.01
Corn flour 2	4.46 ± 0.07	2.49 ± 0.00	63.30 ± 0.24	1.69 ± 0.01

Corn flour 1 (before expiry date), corn flour 2 (after expiry date), the results are means ± standard deviation of two determinations.

Helland et al. [106] studied the growth and metabolism of four selected probiotic strains in rice- or maize-based puddings with milk or water. All four tested strains showed good growth and survival in cereal-based puddings.

4.2.2. Food and beverages made from fermented cereals and pseudocereals

Fermented food and beverages are defined as those products that have been subjected to the effect of microbial enzymes, particularly amylases, proteases and lipases that causes biochemical transformation of polysaccharides, proteins and lipids to non-toxic variety of desirable products with tastes, aromas and textures attractive to a consumer [4, 7]. Microorganisms responsible for the fermentation process may be presented naturally in the substrate, or may be added as a starter culture [9].

Traditional cereal- and pseudocereal-fermented products are made of various kinds of substrates all over the world, mainly widespread in Asia and Africa. Fermentation may have multiple effects on the nutritional value of food [129].

- The development of non-dairy-fermented products is a challenge to the food industry by producing high-quality functional products. The main aims of cereal fermentation can be summarized as follows:
- Preservation, which relies mainly on acidification (production of lactic, acetic and propionic acid) and/or alcoholic production often in combination with reduction of water activity [130];
- Enhances the safety of final products by the inhibition of pathogens [131];
- Affecting sensory properties (taste, aroma, colour and texture);
- Improves the nutritional value by removing antinutritive compounds (phytic acid, enzyme inhibitors, tannins and polyphenols) and enhances the bioavailability of components;
- Reduces the level of carbohydrates as well as non-digestible poly- and oligosaccharides [9].

Cereal fermentations affected by characteristic variables include the following:

- The type of cereal determining the content of fermentable substrates, growth factors, nutrients, minerals, nitrogen sources and buffering capacity;
- Duration and temperature of fermentation process;
- Water content;
- Additional components (sugars, salt and exposure to oxygen);
- Sources of amylolytic activity to gain fermentable sugars from starch or other polysaccharides [9, 132].

Fermented cereal-based products are prepared in different parts of the world, mainly in developing countries—Asia and Africa, in combination with legumes to improve overall protein quality of the final fermented products [7]. Petruláková and Valík [133] evaluated the growth and metabolic activity of *Lb. rhamnosus* GG during fermentation of leguminous porridges. Cell density during 21-day cold storage was stable except whole soybean, yellow pea and red bean. Metabolic activity of observed strain caused decrease in pH values to the final 5.6–6.0 and subsequently during cold storage decreased. Fermented products are usually prepared in the form of beverages, gruels or breakfast meals. Most of the fermented products are made in Asia (soy sauce), India (idli and dosa) and in the Middle East (kishk). In America, as a basic raw material for the production of cereal-fermented foods, corn is used, in products such as tesgüino (alcoholic beverage of Mexico) or jamin-bang-bread made in Brazil [134]. An overview of traditional fermented food and beverages is summarized in Table 4.

Table 4. Overview of traditional fermented products and beverages [7, 135].

Fermented food/country	Raw material/ substrate	Microorganism
Idli—South India, Sri Lanka	Rice	*Leuconostoc mesenteroides*, *Enterococcus*, *Torulopsis*
Dosa—India	Rice	*Leuconostoc mesenteroides*, *Streptococcus faecalis*, *Torulopsis candida*
Kishk—Egypt, Syria	Wheat, milk	*Lactobacillus plantarum*, *Lb. casei*, *Lb. Brevis*, yeasts.
Tarhana—Turkey	Wheat, yoghurt	*Lactobacillus bulgaricus*, *Streptococcus thermophilus*, *Saccharomyces cerevisiae*
Ogi—West Africa	Maize, millet, sorghum	*Lactobacillus plantarum*, yeasts, moulds
Pozol—Mexico	Maize	Moulds, yeasts, bacteria
Injera—Ethiopia	Sorghum, maize	*Candida guilliermondii*
Sake—Japan	Rice	*Saccharomyces sake*
Bouza—Egypt	Wheat	LAB
Boza—Albania, Turkey	Wheat, millet	LAB, *Saccharomyces cerevisiae*, *Leuconostoc*
Mahewu—South Africa	Maize	*Lactococcus lactis*
Chicha—Peru	Maize	*Aspergillus*, *Penicillium*, yeasts, bacteria
Uji—Kenya, Uganda	Maize, millet, sorghum	*Lactobacillus plantarum*, *Leuconostoc mesenteroides*

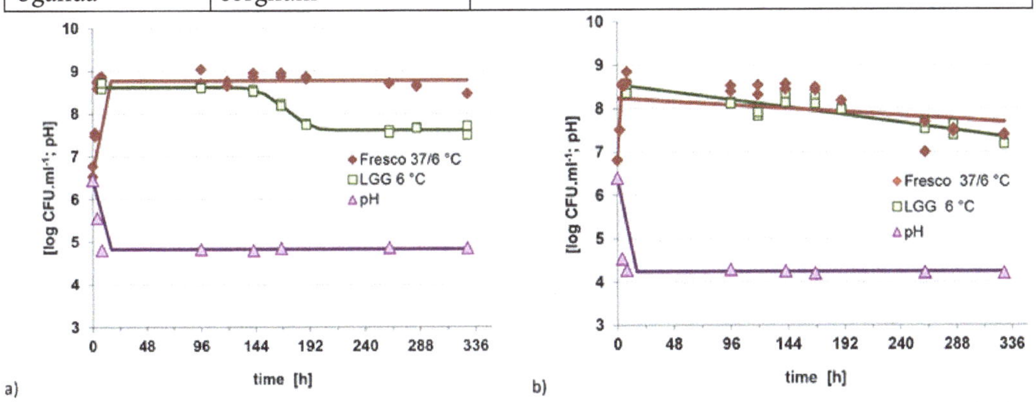

Figure 3. Presumptive counts of the cocci from Fresco DVS 1010 culture and Lb. rhamnosus GG (LGG) content in milk-based (a) and water-based (b) maize mashes.

As an example, the growth of Fresco DVS 1010 culture at 37 °C and the survival of probiotic strain *Lb. rhamnosus* GG (6 °C) in milk- and water-based maize mashes with sucrose are demonstrated in Figure 3 as well as the growth parameters in

Table 5. In general, the obtained maximal counts of monitored Fresco DVS 1010 culture after 8 h of fermentation process were $N = 10^8–10^9$ CFU ml-1 from initial $N_0 = 10^6–10^7$ CFU ml-1, which shows the suitability of tested sweet corn mashes for the growth and survival of lactic acid bacteria. During the refrigerated storage at 6°C (Table 6), a decline in the number of probiotic strain *Lb. rhamnosus* GG was observed, but not under the levels of 10^6 CFU ml-1 necessary from the legislation point of view.

Table 5. Growth parameters of Fresco culture, 8-h fermentation at 37°C in maize mashes.

Microorganism	Substrate corn flour	Gr_f (log CFU ml-1 h-1)	λ (h)	k_{pH} (h-1)
Fresco DVS 1010	Milk	0.522	–	-0.231
	Milk + caramel	0.446	–	-0.345
	Milk + chocolate	0.563	–	-0.172
	Water	0.445	–	-0.481
	Water + caramel	0.508	0.59	-0.298
	Water + chocolate	0.540	–	-0.462

Gr_f, growth rate; λ, lag-phase duration; k_{pH}, rate constant for the decrease of pH. The growth data were fitted using DMFit tool, kindly provided by Dr. J. Baranyi.

In botanical terms, amaranth, quinoa and buckwheat are not true cereals. They are dicotyledonous plants, and thus not cereals (monocotyledonous). Their seeds are in function and composition similar to true cereals, so they are referred as pseudocereals [136, 137]. Gluten-free pseudocereals increased attention worldwide, because they represent alternative to conventional gluten-containing cereals and industrially are used for the production of gluten-free products, especially for celiac patients. They enrich the nutrition of health people and contribute to their balanced diet. In comparison to cereals, pseudocereals are characterized by the increased availability of proteins, as well as its higher content. Moreover, pseudocereals are the major source of minerals and vitamins, and in comparison to cereals, the content of essential amino acid lysine is higher [138–141].

Due to its chemical composition, amaranth is considered as one of the most nutritious plants that is easy to grow and over 60 species of amaranth are known worldwide [142]. Grains are characterized with balanced composition of essential amino acids, especially lysine and methionine, higher content of proteins (15–17%) and starch (60–65%) [143, 144]. Compared to other cereals, the fat content is higher, ranging from 7 to 8%. Overall, amaranth is a good source of vitamins (riboflavin, niacin and vitamin E) and minerals such as calcium and magnesium [138]. A growing number of studies have investigated the usage of amaranth in cereal technology not only in the production of nutrient-rich gluten-free products but also to enrich diet of health people [145]. Several studies have also reported the possibility to enrich wheat-based products with amaranth to improve the quality and overall nutritional value of final products [140]. Matejčeková et al. [146] confirmed in their study the growth of probiotic and potentially probiotic strains (*Lb. acidophilus* 145, *Lb. rhamnosus* GG, *Lb. rhamnosus* VT1 and *Lb. paracasei* subsp. *paracasei*) in water- and milk-based amaranth mashes. The same authors [30] studied the growth and survival of probiotic strain *Lb. rhamnosus* GG in flavoured amaranth mashes, which were acceptable not only from the microbiological point of view but also from the sensory evaluation. Kocková and Valík [147] evaluated the suitability of selected cereals and pseudocereals for the development of new probiotic foods fermented by *Lb. rhamnosus* GG. The highest growth rate was calculated in the case of amaranth flour (0.589 log CFU $g^{-1} h^{-1}$) and the longest lag phase was observed.

Table 6. Parameters evaluating the behaviour of *Lb. rhamnosus* GG in fermented maize mashes during storage at 6°C when added after fermentation.

Substrate corn flour	k_d (log CFU $ml^{-1} h^{-1}$)	λ (h)	N_0 (log CFU ml^{-1})	N_{end} (log CFU ml^{-1})
Milk	-0.0212	139.14	8.61	7.61
Milk + caramel	-0.0031	–	8.57	7.91
Milk + chocolate	-0.0200	–	7.91	7.04
Water	-0.0036	–	8.47	7.19
Water + caramel	-0.0033	141.78	8.27	7.65
Water + chocolate	-0.0093	–	7.47	7.03

k_d, rate constant for decrease of the *Lb. rhamnosus* counts; N_0, initial counts; N_{end}, final counts after 14 days of storage period. The growth data were fitted using DMFit tool kindly provided by Dr. J. Baranyi.

Table 7. Growth parameters of *Lb. rhamnosus* GG in fermented buckwheat-flavoured mashes during fermentation at 37°C.

Substrate buckwheat flour	Gr_f (log CFU $ml^{-1} h^{-1}$)>	λ (h)	N_0 (log CFU ml^{-1})	N_{max} (log CFU ml^{-1})
Milk + vanilla	0.251	2.7	6.80	8.02
Milk + caramel	0.641	1.1	6.25	8.49
Milk + chocolate	0.332	3.0	6.74	8.32
Water + vanilla	0.275	2.4	6.93	8.48
Water + caramel	0.580	–	6.12	8.40
Water + chocolate	0.258	1.3	6.76	8.49

Gr_f, growth rate; λ, lag-phase duration; N_0, initial counts; N_{max}, counts after storage period. The growth data were fitted using DMFit tool kindly provided by Dr. J. Baranyi.

Together with amaranth, buckwheat and its products are studied in connection with celiac disease. Buckwheat was initially grown mainly in Asia and later has spread to Europe, Australia as well as to USA and Canada. The total carbohydrate content is 67–70%, of which 55% represents starch stored in the endosperm, as in common cereals. Buckwheat has a good content of thiamine, riboflavin and pyridoxine, and also represents a good source of minerals—magnesium, copper and potassium. It is characterized by a unique concentration of phytochemicals, in particular rutin, which has a positive effect on health especially in the prevention of cardiovascular diseases [148, 149]. Pelikánová et al. [109] evaluated the growth dynamics of *Lactobacillus* spp. in sweet buckwheat gruels. The population density of tested lactobacilli reached counts 10^8–10^9 CFU ml^{-1} after 8 (10) h of fermentation, and after a 3-week-refrigerated storage period, the number of lactobacilli slightly increased except *Lb. acidophilus* 145. Liptáková et al. [87] in their study examined the pressed buckwheat products. The most suitable strain for fermentation was *Lb. rhamnosus* GG. Pressed fermented buckwheat water product with vanilla flavour was after 24 h of fermentation and after 5 days of storage evaluated with higher points according to the final evaluation of overall sensory acceptance.

Table 8. Parameters of *Lb. rhamnosus* GG in fermented buckwheat-flavoured mashes during storage at 6°C.

Substrate buckwheat flour	k_d (log CFU ml^{-1} h^{-1})	λ (h)	N_{end} (log CFU ml^{-1})
Milk + vanilla	0.0006	–	8.54
Milk + caramel	-0.0002	–	8.42
Milk + chocolate	0.0009	–	8.89
Water + vanilla	0.0000	–	8.38
Water + caramel	-0.0002	–	8.41
Water + chocolate	0.0000	–	8.49

k_d, rate constant for decrease of the *Lb. rhamnosus* counts; λ, lag-phase duration; N_{end}, final counts after 21 days of storage period. The growth data were fitted using DMFit tool kindly provided by Dr. J. Baranyi.

As for the example, growth and fermentative metabolism of probiotic strain *Lb. rhamnosus* GG in buckwheat mashes with caramel/vanilla/chocolate flavour is summarized in Tables 7 and 8. Investigated probiotic strain showed sufficient growth and survival in prepared flavoured mashes with the growth rates ranging from 0.251 to 0.641 log CFU ml^{-1} h^{-1}. At the end of cold storage, densities of *Lb. rhamnosus* GG maintained above the minimum limit of 10^6 CFU ml^{-1}.

The interest of consumers in fermented cereal- or pseudocereal-based products is growing. The development of non-dairy-fermented products including probiotics may lead to enrichment of the diet in patients suffering from celiac disease, people with allergies, or intolerances, but it may contribute to the balanced diet of healthy subjects [149]. If the cereal or pseudocereal products are presented with an attractive sensory taste, it may represent a suitable option for the development of new probiotic foods. Thus, in our study we evaluate the overall sensory acceptability of maize-flavoured (chocolate/caramel) mashes (Figures 4 and 5). The overall acceptability was evaluated from 2.80 to 3.30 (four-point scale) that indicated pleasant acceptance except caramel water mash (2.56). Kocková and Valík [135] noted negative effect of a 21-day storage period on overall acceptability buckwheat product with salt fermented by probiotic strain *Lb. rhamnosus* GG from values 3.31 to 2.44. In our study, no decline of overall acceptance during storage period was observed.

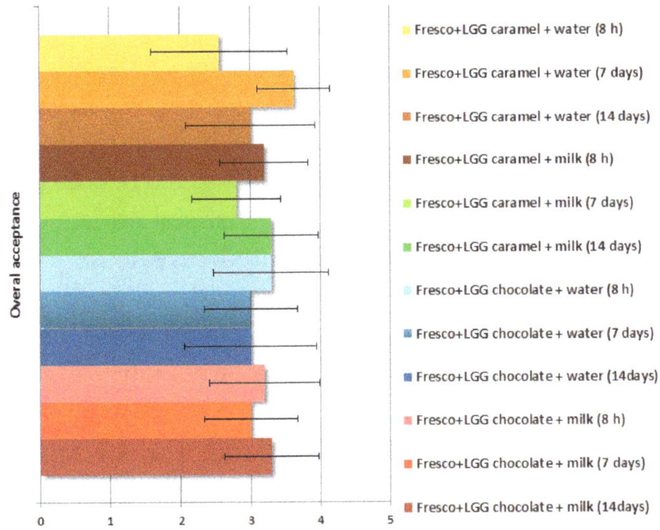

Figure 4. Evaluation of overall acceptability maize caramel/chocolate mashes (LGG—Lactobacillus rhamnosus GG).

Figure 5.Photo-documentation of flavoured final maize products.

5. CONCISE SUMMARY

The creation of novel functional foods in the world is primarily driven by sustainable diets and the growing interest of consumers in personal health. Across the globe, a diverse array of chemicals and microorganisms are utilized in the production of numerous fermented meals. To guarantee the superior quality and safety of such a product, a profound comprehension of the fermentation process, the various types and functions of microorganisms employed, and the specific characteristics of the end result is essential. Lactic acid bacteria are used as a substitute for food preservation because they produce weak organic acids and other compounds that impede the growth of microorganisms. This is achieved by lowering the pH and reducing the redox potential. The LAB and their metabolites have the ability to decelerate or hinder the growth of unwanted bacteria, yeasts, and toxigenic fungi in food. There is data indicating that LAB (lactic acid bacteria) have the ability to decrease the gluten content in cereals, which is a growing concern for approximately 0.5-1% of the global population. Several lactic acid bacteria and other microbial strains, including E. coli Nissle, B. cereus, B. subtilis, and S. boulardii, are classified as probiotics due to their well-documented beneficial effects on human health. The incorporation of probiotics into fermented cereal- or pseudocereal-based products offers a viable option for individuals with celiac disease, milk protein allergies, lactose intolerance, or other metabolic limitations, as an alternative to milk products. Additionally, it can contribute to a well-rounded diet for individuals who are generally healthy.

REFERENCES

1. Todorov SD, Holzapfel WH. Traditional cereal fermented foods as sources of functional microorganism. In: Holzapfel WH, editor. Advances in Fermented Foods and Beverages. 1st ed. Cambridge: Woodhead Publishing Ltd; 2014. p. 123–153. DOI: 10.1016/B978-1-78242-015-6.00006-2
2. Rivera-Espinoza Y, Gallardo-Navaro Y. Non-dairy probiotic products. Food Microbiology. 2010;27:1–11. DOI:10.1016/j.fm.2008.06.008
3. Prajapati JB, Nair BM. The history of fermented foods. In: Farnworth ER, editor. Handbook of Fermented Functional Foods. 2nd ed. London: CRC Press; 2008. p. 1–24.
4. Steinkraus HK. Classification of fermented foods: worldwide review of household fermentation techniques. Food Control. 1997;8:311–317. DOI: 10.1016/S0956-7135(97)00050-9
5. Caplice E, Fitzgerald GF. Food fermentations: role of microorganisms in food production and preservation. International Journal of Food Microbiology. 1999;50:131–149. DOI: 10.1016/S0168-1605(99)00082-3

6. Hugenholtz J. The lactic acid bacterium as a cell factory for food ingredient production. International Dairy Journal. 2008;18:466–475. DOI: 10.1016/j.idairyj.2007.11.015

7. Blandino A, Al-Aseeri ME, Pandiella SS, Cantero D, Webb C. Cereal-based fermented foods and beverages. Food Research International. 2003;36:527–543. DOI: 10.1016/S0963-9969(03)00009-7

8. Adams MR, Moss MO. Food Microbiology. 1st ed. Cambridge: The Royal Society of Chemistry; 1995. 398 p.

9. Hammes WP, Brandt MJ, Francis KL, Rosenheim J, Seitter MFH, Vogelmann SA. Microbial ecology of cereal fermentations. Trends in Food Science and Technology. 2005;16:4–11. DOI: 10.1016/j.tifs.2004.02.010

10. Charlier C, Cretenet M, Even S, Le Loir, Y. Interactions between Staphylococcus aureus and lactic acid bacteria: an old story with new perspectives. International Journal of Food Microbiology. 2009;131:30–39. DOI: 10.1016/j.ijfoodmicro.2008.06032

11. Von Wright A, Axelsson L. Lactic acid bacteria: an introduction. In: Lahtinen SJ, Ouwehand AC, Salminen S, Von Wright A, editors. Lactic Acid Bacteria. Microbiological and Functional Aspects. 4th ed. Boca Raton: CRC Press; 2012. p. 1–16.

12. Wedajo B. Lactic acid bacteria: benefits, selection criteria and probiotic potential in fermented food. Journal of Probiotics and Health. 2015;3:129–138. DOI: 10.4172/2329-8901.1000129

13. Limsowtin GKY, Broome MC, Powell IB. Lactic acid bacteria, taxonomy. In: Roginski H, Fuquay JW, Fox PF, editors. Encyclopedia of Dairy Sciences. 1st ed. Oxford: Academic Press; 2003. p. 1470–1479.

14. Pfeiler EA, Klaenhammer TR. The genomics of lactic acid bacteria. Trends in Microbiology. 2007;15:546–553. DOI: 10.1016/j.tim.2007.09.010

15. Hassan AN, Frank JF. Starter cultures and their use. In: Marth EH, Steele JL, editors. Applied Dairy Microbiology. 2nd ed. New York, NY: Marcel Dekker Inc; 2001. p. 151–206.

16. Gomes MP Ana, Malcata FX. Bifidobacterium spp. and Lactobacillus acidophilus: biological, biochemical, technological and therapeutical properties relevant for use as probiotics. Trends in Food Science and Technology. 1999;10:139–157.

17. Barrangou R, Lahtinen SJ, Ibrahim F, Ouwehand AC. Genus Lactobacillus. In: Lahtinen SJ, Ouwehand AC, Salminen S, von Wright A, editors. Lactic Acid Bacteria. Microbiological and Functional Aspects. 4th ed. Boca Raton: CRC Press; 2012. p. 77–92.

18. Albesharat R, Ehrmann MA, Korakli M, Yazaji S, Vogel RF. Phenotypic and genotypic analyses of lactic acid bacteria in local fermented food, breast milk and faeces of mothers and their babies. Systematic and Applied Microbiology. 2011;34:148–155. DOI: 101016/ j.syapm.2010.12.001

19. Solís G, De Los Reyes-Gavilan CG, Fernández N, Margolles A, Gueimonde M. Establishment and development of lactic acid bacteria and bifidobacteria microbiota in breast-milk and the infant gut. Anaerobe. 2010;16:307–310. DOI: 101016/ j.anaerobe.2010.02.004

20. Mitsou EK, Kirtzalidou E, Oikonomou I, Liosis G, Kyriacou A. Fecal microflora of Greek healthy neonates. Anaerobe. 2008;14:94–101. DOI: 101016/ j.anaerobe.2007.11.002

21. Liptáková D, Hornická M, Valík Ľ. Powdered infant formulas fortified with probiotics. Farmaceutický obzor. 2015; 84:241–245.

22. Fernández L, Langa S, Martín V, Maldonado A, Jiménez E, Martín R, Rodríguez JM. The human milk microbiota: origin and potential roles in health and disease. Pharmacological Research. 2013;69:1–10. DOI: 10.1016/j. phrs.2012.09.001

23. Martín R, Langa S, Reviriego C, Jiménez E, Marín MA, Olivares M, Boza J, Jiménez J, Fernández L, Xaus J, Rodríguez JM. The commensal microflora of human milk: new perspectives for food bacteriotherapy and probiotics. Trends in Food Science and Technology. 2004;15:121–127. DOI: 101016/ j.tifs.2003.09.010

24. Martín S, Maldonado-Baragán A, Moles L, Rodriguez-Ba'noz M, del Campo R, Fernández L, Rodríguez JM, Jiménez E. Sharing of bacterial strains between breast milk and infant faeces. Journal of Human Lactation. 2012;28:36–44. DOI: 10.1177/08903344I I 424729

25. Jost T, Lacroix C, Braegger CP, Rochat F, Chassard C. Vertical mother-neonate transfer of maternal gut microflora. Environmental Microbiology. 2014;16:2891–2904. DOI: 10.1111/1462-2920.12238

26. Liptáková D, Koreňová J, Hornická M, Valík Ľ. Microbiological analysis of breast milk. Lekársky obzor. 2016; 65:227–232.

27. Görner F, Valík Ľ. Applied food microbiology. 1st ed. Bratislava: Malé Centrum; 2004. 528 p.

28. Tamine AY, Robinson's RK. Yoghurt. 3rd ed. Cambridge: Woodhead Publishing in Food Science, Technology and Nutrition; 2007. 791 p.

29. Ward LJH, Davey GP, Heap HA, Kelly WJ. Lactococcus lactis. In: Roginski H, Fuquay JW, Fox PF, editors. Encyclopedia of Dairy Sciences. 1st ed. Oxford: Academic Press; 2003. p. 1511–1516.

30. Matejčeková Z, Liptáková D, Valík L. Fermentation of milk- and water-based amaranth mashes. Acta Chimica Slovaca. 2015;8:140–145. DOI: 10.1515/acs-2015-0024

31. Cho S-LS, Nam W, Yoon J-H, Lee J-S, Sukhoom A, Kim W. Lactococcus chungangensis sp. nov., a lactic acid bacterium isolated from activated sludge foam. International Journal of Systematic and Evolutionary Microbiology. 2008;58:1844–1849. DOI: 10.1099/ijs.0.65527-0

32. Cavanagh D, Fitzgerald FG, McAuliffe OA. From field to fermentation: the origins of Lactococcus lactis and its domestication to the dairy environment. Food Microbiology. 2015;47:45–61. DOI: 10.1016/j.fm.2014.11.001

33. Curioni P, Bosset J. Key odorants in various cheese types as determined by gas chromatography - olfactometry. International Dairy Journal. 2002;12:959–984. DOI: 10.1016/S0958-6946(02)00124-3

34. Sadiq S, Imran M, Hassan MN, Iqbal M, Zafar Y, Hafeez FY. Potential of bacteriocinogenic Lactococcus lactis subsp. lactis inhabiting low pH vegetables to produce nisin variants. LWT–Food Science and Technology. 2014;59:204–210. DOI: 10.1016/j.lwt.2014.05.018

35. Huys G, Leisner J, Björkroth J. The lesser LAB gods: Pediococcus, Leuconostoc, Weissella, Carnobacterium, and affiliated genera. In: Lahtinen SJ, Ouwehand AC, Salminen S, Von Wright A, editors. Lactic Acid Bacteria. Microbiological and Functional Aspects. 4th ed. Boca Raton: CRC Press; 2012. p. 93–122.

36. Liu S-Q, Holland R. Leuconostoc spp. In: Roginski H, Fuquay JW, Fox PF, editors. Encyclopedia of Dairy Sciences. 1st ed. Oxford: Academic Press; 2003. p. 1539–1544.

37. de Bruyne K, Schillinger U, Caroline L, Boehringer B, Cleenwerck I, Vancanneyt M, de Vuyst L, Franz MAP Charles, Vandamme P. Leuconostoc holzapfelii sp. nov., isolated from Ethiopian coffee fermentation and assessment of sequence analysis of housekeeping genes for delineation of Leuconostoc species. International Journal of Systematic and Evolutionary Microbiology. 2007;57:2952–2959. DOI: 10.1099/ijs.0.65292-0

38. Ogier JC, Casalta E, Farrokh C, Saïhi A. Safety assessment of dairy microorganisms: The Leuconostoc genus. International Journal of Food Microbiology. 2008;126:286–290. DOI: 10.1016/j.ijfm.2007.08.012

39. Franz MAP Charles, Vancanneyt M, Vandemeulebroecke K, de Wachter M, Cleenwerck I, Hoste B, Schillinger U, Holzapfel WH, Swings J. Pediococcus stilesii sp. nov., isolated from maize grains. International Journal of Systematic and Evolutionary Microbiology. 2006;56:329–333. DOI: 10.1099/ijs.0.63944-0

40. Garai-Ibabe G, Dueñas MT, Irastorza A, Sierra-Filardi E, Werning ML, López P, Corbí AL, Fernández de Palencia P. Naturally occurring 2-substituted (1,3)-β-d-glucan producing Lactobacillus suebicus and Pediococcus parvulus strains with potential utility in the production of functional foods. Bioresource Technology. 2010;101:9254–9263. DOI: 10.1016/j.biortech.2010.07.050

41. Papagianni M, Anastasiadou S. Encapsulation of Pediococcus acidilactici cells in corn and olive oil microcapsules emulsified by peptides and stabilized with xanthan in oil-in-water emulsions: Studies on cell viability under gastro-intestinal simulating conditions. Enzyme and Microbial Technology. 2009;45:514–522. DOI: 10.1016/j.enzmictec.2009.06.007

42. Sherman JM. The Streptococci. Bacteriological Reviews. 1937;1:3–97.

43. Schleifer KH, Killper-Bälz R. Transfer of Streptococcus faecalis and Streptococcus faecium to the genus Enterococcus nom. rev. as Enterococcus faecalis com. nov. and Enterococcus faecium comb. nov. International Journal of Systematic Bacteriology. 1984;34:31–34.

44. Bentley RW, Leigh JA, Collins MD. Intrageneric structure of Streptococcus based on comparative analysis of small subunit rRNA sequences. International Journal of Systematic Bacteriology. 1991;41:487–494.

45. Facklam R. What happened to the Streptococci: overview of taxonomic and nomenclature changes. Clinical Microbiology Reviews. 2002;15:613–630. DOI: 10.1128/CMR.15.4.613-630.2002

46. Köhler W. The present state of species within the genera Streptococcus and Enterococcus. International Journal of Medical Microbiology. 2007;297:133–150. DOI: 10.1016/j.ijmm.2006.11.008

47. Rossi F, Marzotto M, Cremonese S, Rizzotti L, Torriani S. Diversity of Streptococcus thermophilus in bacteriocin production; inhibitory spectrum and occurrence of thermophilin genes. Food Microbiology. 2013;35:27–33. DOI: 10.1016/j.fm.2013.02.006

48. Tagg JR, Wescombe PA, Burton JP. Streptococcus: a brief update on the current taxonomic status of the genus. In: Lahtinen SJ, Ouwehand AC, Salminen S, Von Wright A, editors. Lactic Acid Bacteria. Microbiological and Functional Aspects. 4th ed. Boca Raton: CRC Press; 2012. p. 23–146.

49. Kabuki T, Uenishi H, Seto Y, Yoshioka T, Nakajima H. A unique lantibiotic, thermophilin 1277, containing a disulfide bridge and two thioether bridges. Journal of Applied Microbiology. 2009;106:853–862. DOI: 10.1111/j.1365-2672.2008.04059.x

50. Erkus O, Okuklu B, Yenidunya AF, Harsa S. High genetic and phenotypic variability of Streptococcus thermophilus strains isolated from artisanal

Yuruk yoghurts. LWT – Food Science and Technology. 2014;58:348–354. DOI: 10.1016/j.lwt.2013.03.007

51. Biavati B, Vescovo M, Torriani S, Bottazzi V. Bifidobacteria: history, ecology, physiology and applications. Annals of Microbiology. 2000;59:117–131.

52. Ishizuka A, Tomizuka KI, Aoki R, Nishijima T, Saito Y, Inoue R, Ushida K, Mawatari T, Ikeda T. Effects of administration of Bifidobacterium animalis subsp. lactis GCL2505 on defecation frequency and bifidobacterial microbiota composition in humans. Journal of Bioscience and Bioengineering. 2012;113:587–591. DOI: 10.1016/j.jbiosc.2011.12.016

53. Picard C, Fioramonti J, Francois A, Robinson T, Neant F, Matuchansky C. Review article: bifidobacteria as probiotic agents – physiological effects and clinical benefits. Alimentary Pharmacology and Therapeutics. 2005;22:495–512. DOI: 0.1111/j.1365-2036.2005.02615.x

54. Matsuki T, Watanabe K, Tanaka R, Fukuda M, Oyaizu H. Distribution of bifidobacterial species in human intestinal microflora examined with 16S-rRNA-genetargeted species-specific primers. Applied and Environmental Microbiology. 1999;65:4506–4512.

55. Ventura M, van Sinderen D, Fitzgerald GF, Zink R. Insights into the taxonomy, genetics and physiology of bifidobacteria. Antonie van Leeuwenhoek. 2004;86:205–223.

56. Ventura M, Turroni F, van Sinderen D. Bifidobacteria: general overview on ecology, taxonomy, and genomics. In: Lahtinen SJ, Ouwehand AC, Salminen S, von Wright A, editors. Lactic Acid Bacteria. Microbiological and Functional Aspects. 4th ed. Boca Raton: CRC Press; 2012. p. 147–164.

57. Shah NP. Functional cultures and health benefits. International Dairy Journal. 2007;17:1262–1277. DOI: 10.1016/j.dairyj.2007.01.014

58. Sathe SJ, Nawani NN, Dhakephalkar PK, Kapadnis BP. Antifungal lactic acid bacteria with potential to prolong shelf-life of fresh vegetables. Journal of Applied Microbiology. 2007;103:2622–2628. DOI: 10.1111/j.1365-2672.2007.0 3525.x

59. Batish VK, Lal R, Grover S. Effect of nutritional factors. Australian Journal of Dairy Technology. 1990;45:74–76.

60. Dalié DKD, Deschamps AM, Richard-Forget F. Lactic acid bacteria – Potential for control of mould growth and mycotoxins: a review. Food Control. 2010;21:370–380. DOI: 10.1016/j.foodcont.2009.07.011

61. Nes IF, Kjos M, Diep DB. Antimicrobial components of lactic acid bacteria. In: Lathinen S, Ouwehand AC, Salminen S, Wright AV, editors. Lactic Acid Bacteria. 4th ed. Boca Raton: CRC Press; 2011. p. 285–330.

62. Ray RC, Joshi VK. Fermented foods: past, present and future. In: Ray RC, Montet D, editors. Microorganisms and Fermentation of Traditional Foods. London: CRC Press; 2014. p. 1–36.

63. Breidt F, Hayes JS, McFeeters RF. Independent effects of acetic acid and pH on survival of Escherichia coli in simulated acidified pickle products. Journal of Food Protection. 2004;67:12–18.

64. Liptáková D, Hudecová A, Valík L, Medvedová A. Interaction between dairy yeasts and Lactobacillus rhamnosus GG in milk. Journal of Agricultural Science and Technology. 2010;4:88–95.

65. Greifová M, Marunová E, Greif G, Zimanová M. Antifugálna aktivita kyseliny D,L-fenylmliečnej. Mlékařské Listy. 2014;142:6–10.

66. Liptáková D, Valík L, Lauková A, Strompfová V. Characterization of Lactobacillus rhamnosus VT1 and its effect on the growth of Candida maltosa YP1. Czech Journal of Food Sciences. 2007;25:272–282.

67. Plocková M, Stiles J, Chumchalová J, Halfarová R. Control of mould growth by Lactobacillus rhamnosus VT1 and Lactobacillus reuteri CCM 3625 on milk agar plates. Czech Journal of Food Sciences. 2001;19:46–50.

68. Schnürer J, Magnusson J. Antifungal lactic acid bacteria as biopreservatives. Trends in Food Science and Technology. 2005;16:70–78. DOI: 10.1016/j.tifs.2004.02.014

69. Fitzsimmons N, Berry DR. Inhibition of Candida albicans by Lactobacillus acidophilus: evidence for the involvement of a peroxidase system. Microbios. 1994;80:125–133.

70. Lozo J, Vukasinovic M, Strahinic I, Topisirovic L. Characterization and antimicrobial activity of bacteriocin 217 produced by natural isolate Lactobacillus paracasei subsp. paracasei BGBUK2-16. Journal of Food Protection. 2004;67:2727–2734.

71. Güven K, Benlikaya N. Acid pH produced by lactic acid bacteria prevent the growth of Bacillus cereus in boza, a traditional fermented Turkish beverage. Journal of Food Safety. 2005;25:98–108. DOI: 10.1111/j.1745-4565.2005.00568.x

72. Valdés-Stauber N, Scherer S. Isolation and characterization of linocin M18, a bacteriocin produced by Brevibacterium linens. Applied and Environmental Microbiology. 1994;60:3809–3814.

73. Corsetti A, Settanni L, Van Sinderen D. Characterization of bacteriocin-like inhibitory substances (BLIS) from sourdough lactic acid bacteria and evaluation of their in vitro and in situ activity. Journal of Applied Microbiology. 2004;96:521–534. DOI: 10.1111/j.1365-2672.2004.02171.x

74. Björkroth KJ, Geisen R, Schillinger U, Weiss N, de Vos P, Holzapfel WH, Korkeala HJ, Vandamme P. Characterization of Leuconostoc gasicomitatum sp. nov., associated with spoiled raw tomato-marinated broiler meat strips packaged under modified atmosphere conditions. Applied and Environmental Microbiology. 2000;66:3764–3772.

75. Nieto-Lozano JC, Reguera-Useros JI, del Peláez-Martínez M, Sacristián-Pérez-Minayo G, Gutíerrez-Fernández AJ, de la Torre AH. The effect of the pediocin PA – 1 produced by Pediococcus acidilactici against Listeria monocytogenes and Clostridium perfringens in Spanish dry – fermented sausages and frankfurters. Food Control. 2010;21:679–685. DOI: 10.1016/j.foodcont.2009.10.007

76. Gurira OZ, Buys EM. Characterization and antimicrobial activity of Pediococcus species isolated from South African farm – style cheese. Food Microbiology. 2005;22:159–168. DOI: 10.1016/j.fm.2004.08.001

77. Altuntas EG, Cosansu S, Ayhan K. Some growth parameters and antimicrobial activity of a bacteriocin – producing strain Pediococcus acidilactici 13. International Journal of Food Microbiology. 2010;141:28–31. DOI: 10.1016/j.ijfoodmicro.2010.04.024

78. Khalid K. An overview of lactic acid bacteria. International Journal of Biosciences. 2011;1:1–13.

79. Ouwehand AC, Salminen S, Isolauri E. Probiotics: an overview of beneficial effects. Journal of Microbiology. 2002;82:279–289.

80. Prado FC, Parada JL, Pandey A, Soccol CR. Trends in non – dairy probiotic beverages. Food Research International. 2008;41:111–123. DOI: 10.1016/j.foodres.2007.10.010

81. Saad N, Delattre C, Urdaci M, Schmitter JM, Bressollier P. An overview of the last advances in probiotic and prebiotic field. LWT – Food Science and Technology. 2013;50:1–16. DOI: 10.1016/j.lwt.2012.05.014

82. Da Cruz AG, Buriti CA Flávia, De Souza HB Cínthia, Faria AF José, Saad MI Susana. Probiotic cheese: health benefits, technological and stability aspects. Trends in Food Science and Technology. 2009;20:344–354. DOI: 10.1016/j.tifs.2009.05.001

83. Vasiljevic T, Shah NP. Probiotics – from Metchnikoff to bioactives. International Dairy Journal. 2008;18:714–728. DOI: 10.1016/j.idairyj.2008.03.004

84. Vijaya Kumar B, Vijayendra SV Naga, Reddy OV Sarathi. Trends in dairy and non-dairy probiotic products- a review. Journal of Food Science and Technology. 2015;52:6112–6124. DOI: 10.1007/s13197-015-1795-2

85. Tripathi MK, Giri SK. Probiotic functional foods: survival of probiotics during processing and storage. Journal of Functional Foods. 2014;9:225–241. DOI: 10.1016/j.jff.2014.04.030

86. Lee NK, Son SH, Jeon EB, Jung GH, Lee JY, Paik HD. The prophylactic effect of probiotic Bacillus polyfermenticus KU3 against cancer cells. Journal of Functional Foods. 2015;14:513–518. DOI: 10.1016/j.jff.2015.02.019

87. Liptáková D, Petruláková M, Pelikánová J, Valík L, Krištúfková K. Growth and survival of probiotic lactobacilli in pressed buckwheat product. Chemické Listy. 2016;110:149–152.

88. Mortazavian AM, Ehsani MR, Mousavi SM, Sohrabvandi S, Reinheimer J. Effect of refrigerated storage temperature on the viability of probiotic microorganisms in yoghurt. International Journal of Dairy Technology. 2007a;59:123–127. DOI: 10.1111/j.1471-0307.2007

89. Mortazavian AM, Razavi SH, Ehsani MR, Sohrabvandi S. Principles and methods of microencapsulation of probiotic microorganisms. Iranian Journal of Biotechnology. 2007;5:1–18.

90. Liptáková D, Valík L, Janovčíková L. Lactobacillus acidophilus we can be in daily contact with it. Farmaceutický Obzor. 2008b;77:217–221.

91. Valík L, Liptáková D, Medvedová A. Content of probiotic bacteria L. acidophilus in fermented milk determined at the end of the recommended "use by date". Farmaceutický Obzor. 2012;79:49–53.

92. Commane D, Hughes R, Shortt C, Rowland I. The potential mechanisms involved in the anti - carcinogenic action of probiotics. Mutation Research. 2005;591:276–289. DOI: 10.1016/j.mrfmmm.2005.02.02

93. Isolauri E, Salminen S, Ouwehand AC. Probiotics. Best Practise and Research Clinical Gastroenterology. 2004;18:299–313. DOI: 10.1053/ybega.2004.443

94. Liptáková D, Valík L, Görner F. Nutrition and health benefits of yoghurts. Farmaceutický Obzor. 2006;75:159–162.

95. Saavedra JM, Bauman NA, Oung I, Perman JA, Yolken RH. Feeding of Bifidobacterium bifidum and Streptococcus thermophilus to infants in hospital for prevention of diarrhoea and shedding of rotavirus. Lancet. 1994;344:1046–1049.

96. Guandalini S, Pensabene L, Zikri MA, Dias JA, Casali LG. Hoekstra. Lactobacillus rhamnosus GG administered in oral rehydration solution to children with acute diarrhea: a multicenter European trial. Journal of Pediatric Gastroenterology and Nutrition. 2000;30:214–216.

97. Zajewska H, Skórka A, Ruszcynski M, Gieruszczak-Biatek D. Metaanalysis: Lactobacillus GG for treating acute diarrhoea in children. Alimentary and Pharmacology Therapeutics. 2007;25:871–881. DOI: 10.1111/apt.12403

98. Gismondo MR, Drago L, Lombardi A. Review of probiotics available to modify gastrointestinal flora. International Journal of Antimicrobial Agents. 1999;12:287–292.

99. Elmer GW, McFarland LV, Surawicz CM, Danko L, Greenberg RN. Behaviour of Saccharomyces boulardii in recurrent Clostridium difficile disease patients. Alimentary Pharmacology and Therapeutics. 1999;13:1663–1668. DOI: 10.1046/j.1365-2036.1999.00666.x

100. Surawicz CM, McFarland LV, Greenberg RN, Rubin M, Fekety R, Mulligan ME, Garcia RJ, Brandmarker S, Bowen K, Borjal D, Elmer GW. The search for a better treatment for recurrent Clostridium difficile disease: use of high-dose vancomycin combined with Saccharomyces boulardii. Clinical Infectious Diseases. 2000;31:1012–1017.

101. Czerucka D, Rampal P. Experimental effects of Saccharomyces boulardii on diarrheal pathogens. Microbes and Infection. 2002;4:733–739.

102. Cenci G, Rossi J, Throtta F, Caldini G. Lactic acid bacteria isolated from dairy products inhibit genotoxic effect of 4 - nitroquinoline - 1 - oxide in SOS - chromtest. Systematic and Applied Microbiology. 2002;25:483–490.

103. Faghfoori Z, Gargari BP, Gharamaleki AS, Bagherpour H, Khosroushahi AY. Cellular and molecular mechanisms of probiotics effects on colorectal cancer. Journal of Functional Foods. 2015;18:463–472. DOI: 10.1016/j.jff.2015.08.013

104. Charalampopoulos D, Wang R, Pandiella SS, Webb C. Application of cereals and cereal components in functional foods: a review. International Journal of Food Microbiology. 2002;79:131–141. DOI: 10.1016/S0168-1605(02)00187-3

105. Huggett AC, Schliter B. Research needs for establishing the safety. Nutrition Reviews. 1996;54:143–148. DOI: 10.1111/j.1753-4887.1996.tb03835.x

106. Helland MH, Wicklund T, Narvhus JA. Growth and metabolism of selected strains of probiotic bacteria in milk- and water-based cereal puddings. International Dairy Journal. 2004;14:957–965. DOI: 10.1016/j.idairyj.2004.03.008

107. Kedia G, Wang, R, Patel H, Pandiella SS. Use of mixed cultures for the fermentation of cereal-based substrates with potential probiotic properties. Process Biochemistry. 2007;42:65–70. DOI: 10.1016/j.procbio.2006.07.011

108. Prugar J. Kvalita Rostlinných Produktu na Prahu 3. Tisíciletí. Praha: VÚPS; 2008. 327 p.

109. Pelikánová J, Liptáková D, Valík L, Stančeková K. Evaluation of the growth of selected lactobacilli in pseudocereal substrate. Potravinárstvo. 2011;5:53–57. DOI: 10.5219/169

110. FAO. World Food Situation [Internet]. 2016. Available from: http://www.fao.org/worldfoodsituation/csdb/en/ [Accessed: 2016-03-10].

111. Kowieska A, Lubowicki R, Jaskowska I. Chemical composition and nutritional characteristics of several cereal grain. Acta Scientiarum Polonorum Zootechnica. 2011;10:37–50.

112. Wrigley C. Cereals. In: Wrigley C, Corke H, Walker CH, editors. Encyclopedia of Grain Science. 1st ed. Oxford: Elsevier Academic Press; 2004. p. 187–201.

113. Mckewith B. Nutritional aspects of cereals. Nutrition Bulletin. 2004;29:111–142. DOI: 10.1111/j.1467-3010.2004.00418.x

114. Žajová A, Porubská M. Obilniny vo Výžive Zdravých i Chorých Ludí. Obilniny. Pieštany: VÚRV; 1997. 400 p.

115. Kučerová J. Technologie Cereálií. 1st ed. Brno: MZLU; 2004. 141 p.

116. Marko A, Rakická M, Mikušová L, Valík L, Šturdík E. Lactic acid fermentation of cereal substrates in nutritional perspective. International Journal of Research in Chemistry and Environment. 2014;4:80–92.

117. Lambo AM, Öste R, Nyman MEGL. Dietary fibre in fermented oat and barley β-glucan rich concentrates. Food Chemistry. 2005;89:283–293. DOI: 10.1016/j.foodchem.2004.02.035

118. Shewry PR. Improving the protein content and composition of cereal grain. Journal of Cereal Science. 2007;46:239–250. DOI: 10.1016/j.jcs.2007.06.006

119. Kadlec P, Melzoch K, Voldřich M. Přehled Tradičních Potravinářských Výrob-Technologie Potravin. 1st ed. Ostrava: Key Publishing; 2012. 569 p.

120. Gänzle MG. Enzymatic and bacterial conversion during sourdough fermentation. Food Microbiology. 2014;37:2–10. DOI: 10.1016/j.fm.2013.04.007

121. Antony U, Sripriya G, Chandra TS. The effect of fermentation on the primary nutrients in foxtail millet (Seteria italica). Food Chemistry. 1996;56:381–384. DOI: 10.1016/0308-8146(95)00186-7

122. Cui L, Li D, Liu C. Effect of fermentation on the nutritive value of maize. International Journal of Food Science and Technology. 2012;47:755–760. DOI: 10.1111/j.1365-2621.2011.02904.x

123. Koehler P, Wieser H. Chemistry of cereal grains. In: Gobbetti M, Gänzle M, editors. Handbook on Sourdough Biotechnology. New York, NY: Springer Science; 2013. p. 11–45. DOI: 10.1007/978-1-4614-5425-0

124. Volkov A, Liavonchanka A, Kamneva O, Fiedler T, Goebel C, Kreikemeyer B, Feussner I. Myosin cross-reactive antigen of Streptococcus pyogenes M49 encodes a fatty acid double bond hydratase that plays a role in oleic acid

detoxification and bacterial virulence. Journal of Biological Chemistry. 2010;285:10353–10361. DOI: 10.1074/jbc.M109.081851

125. Belitz HD, Grosch W, Schieberle P. Food Chemistry. 4th ed. Berlin: Springer; 2004. 1070 p.

126. Poutanen K, Flander L, Katina K. Sourdough and cereal fermentation in a nutritional perspective. Food Microbiology. 2009;26:693–699. DOI: 10.1016/j.fm.2009.07.011

127. Němečková I, Dragounová H, Pechačová M, Rysová J, Roubal P. Fermentation of vegetable substrates by lactic acid bacteria as a basis of functional foods. Czech Journal of Food Sciences. 2011;29:42–48.

128. Keresteš J. et al. Zdravie a Výživa Ludí. Bratislava: CAD Press; 2011. 1040 p.

129. Kohajdová Z, Karovičová J. Fermentation of cereals for specific purpose. Journal of Food and Nutrition Research. 2007;46:51–57.

130. Ross RP, Morgan S, Hill C. Preservation and fermentation: past, present and future. International Journal of Food Microbiology. 2002;79:3–16. DOI: 10.1016/S0168-1605(02)00174-5

131. Adams M, Mitchell R. Fermentation and pathogen control: a risk assessment approach. International Journal of Food Microbiology. 2002;79:75–83. DOI: 10.1016/S0168-1605(02)00181-2

132. Vogelmann SA, Hertel CH. Impact of ecological factors on the stability of microbial associations in sourdough fermentation. Food Microbiology. 2011;28:583–589. DOI: 10.1016/j.fm.2010.11.010

133. Petruláková M, Valík L. Legumes as potential plants for probiotic strain Lactobacillus rhamnosus GG. Acta Universitatis Agriculturae et Silviculturae Mendelianae Brunensis. 2015;63:1505–1511.DOI: 10.11118/actaun201563051505

134. Font de Valdez G, Rollán G, Gerez CL, Torino MI. Microbial applications in the biopreservation of cereal products. In: Lacroix CH, editor. Protective Cultures, Antimicrobial Metabolites and Bacteriophages for Food and Beverage Biopreservation. Philadelphia, PA: Woodhead Publishing; 2011. p. 348–359.

135. Kocková M, Valík L. Development of new cereal-, pseudocereal-, and cereal-leguminous-based probiotic foods. Czech Journal of Food Science. 2014;32:391–397.

136. Haard NF, Odunfa SA, Lee CH, Quintero-Ramirez R, Lorence-Quiñones A, Wacher-Radarte C. Fermented cereals. A global perspective. FAO Agricultural Services Bulletin. 1999;138:1–114.

137. Bressani R. Amaranth. In: Caballero B, Trugo L, Finglas P, editors. Encyclopedia of Food Sciences and Nutrition. Oxford: Academic Press; 2003. p. 166–173.

138. Sterr Y, Weiss A, Schmidt H. Evaluation of lactic acid bacteria for sourdough fermentation of amaranth. International Journal of Food Microbiology. 2009;136:75–82. DOI: 10.1016/j.ijfoodmicro.2009.09.006

139. Léder I. Buckwheat, amaranth, and other pseudocereal plants. In: Füleky G, editor. Encyclopedia of Life Support Systems. 1st ed. Ramsey: EOLSS Publishers Co Ltd; 2009. p. 1–17.

140. Aghamirzaei M, Heydari-Dalfard A, Karami F, Fathi M. Pseudo-cereals as a functional ingredient: effects on bread nutritional and physiological properties-Review. International Journal of Agriculture and Crop Sciences. 2013;5:1574–1580.

141. Meo De B, Freeman G, Marconi O, Booer C, Perrettti G, Fantozzi P. Behaviour of malted cereals and pseudocereals for gluten- free beer production. Journal of the Institute of Brewing. 2011;117:541–546.

142. Písaříková B, Zralý Z, Kráčmar S, Trčková M, Herzig I. Nutritional value of amaranth (genus Amaranthus L.) grain in diets for broiler chickens. Czech Journal of Animal Science. 2005;50:568–573.

143. Amicarelli V, Camaggio G. Amaranthus: A crop to rediscover. Forum Ware International. 2012;2:4–11.

144. Bhat A, Satpathy G, Gupta KR. Evaluation of nutraceutical properties of Amaranthus hypochondriacus L. grains and formulation of value added cookies. Journal of Pharmacognosy and Phytochemistry. 2015;3:51–54.

145. Mlakar GS, Turinek M, Jakop M, Bavec M, Bavec F. Grain amarant as an alternative and perspective crop in temperate climate. Journal for Geography. 2010;5-1:135–145.

146. Matejčeková Z, Liptáková D, Valík Ľ. Evaluation of the potential of amaranth flour for lactic acid fermentation. Journal of Pharmacy and Nutrition Sciences. 2016;6:1–6. DOI: 10.6000/1927-5951.2016.06.01.1

147. Kocková M, Valík Ľ. Suitability of cereal porridges as substrate for probiotic strain Lactobacillus rhamnosus GG. Potravinárstvo. 2013;7:22–27. DOI: 10.5219/242

148. Jubete-Alvarez L, Arendt EK, Gallagher E. Nutritive value of pseudocereals and their increasing use as functional gluten-free ingredients. Trends in Food Science and Technology. 2010;2:106–113. DOI: 10.1016/j.tifs.2009.10.014

149. Kreft I, Fabjan N, Germ M. Rutin in buckwheat – protection of plants and its importance for the production of functional food. Fagopyrum. 2003;20:7–11.

13 UTILIZING SOLID-STATE CULTURE FOR THE PRODUCTION OF LIGNOCELLULASES

1. INTRODUCTION: AN OVERVIEW

Laccases and xylanases are crucial lignocellulases utilized in several industrial processes. Xylanases (EC 3.2.1.8) are enzymes that break down xylan, a primary polysaccharide found in the cell walls of several cereals, into its individual monomers. These enzymes are mostly utilized in the textile, pulp and paper, bread making, feed sectors, as well as in the preparation of juice and fruit extracts [1, 2]. Laccases (EC 1.10.3.2) are enzymes with multiple copper ions that facilitate the oxidation of several substances including mono-, di-, and polyphenols, aminophenols, methoxyphenols, aromatic amines, and ascorbic acid. They have multiple industrial applications, including the breakdown of harmful fungal metabolites and phenolic compounds. They are also utilized in the creation of biosensors for detecting phenols in wastewater, the advancement of biofuel cells, and the bleaching and delignification processes in the pulp and paper industry. Additionally, they are employed in the production of innovative paper products [3]. Xylanases and laccases can work together to enhance the bleaching process of many types of pulps, resulting in a more environmentally friendly procedure with significantly reduced use of harmful chemicals [4, 5]. Fungi are the primary source of these enzymes, which can be produced through either submerged or solid-state cultures. The former option is highly favored because to its ability to effectively control operating parameters, provide high productivity, facilitate easy downstream processing, ensure homogeneity of the culture and pH, and enable better regulation of oxygen supply and agitation speed [6]. Nevertheless, solid-state fermentation may be more advantageous for the production of these enzymes, as it mimics the natural circumstances that fungi encounter while invading lignocellulosic material. In relation to this matter, our research group has made progress in enhancing the components of culture media to achieve the maximum levels of xylanases and laccases activities and yields produced by Aspergillus sp and Trametes versicolor, respectively [7]. The enzymatic activities have been assessed for pulp pretreatment, yielding favorable outcomes [8, 9]. Nevertheless, the behavior of the optimized cultures has not undergone thorough analysis, despite the potential utility of this information in the development of a scalable bioprocess. Alternatively, the simultaneous utilization of fungi that generate xylanase or laccase can be explored to create a coculture system. This system can yield a combination of enzymes that possess both xylanase and laccase activity. Such a mixed enzyme preparation can then be employed in biopulping and biobleaching procedures. This approach will yield economic benefits due to the decrease in the overall production costs [10].

Hence, the aim of this study is to analyze the properties of solid-state monocultures in terms of their development, substrate utilization, and production of xylanases or laccases. Additionally, we will evaluate the potential of a coculture to simultaneously produce both enzymes by solid-state fermentation.

2. METHODOLOGY

2.1. Microorganisms

Trametes versicolor CDBB-H-1051 and *Aspergillus* sp were used in these experiments. Stock cultures were maintained on malt-extract agar (malt extract 2%, agar 1.8%) or PDA slants at 4°C with a periodic transfer.

2.2. Culture conditions for solid-state monocultures

For developing the corresponding monoculture, each microorganism was cultivated in solid-state fermentation (SSF) using 4 g of wheat bran and sugar cane bagasse (1:1 w/w) as support and substrate. For doing this, the support was moistened with water and autoclaved in 250 mL beakers at 121°C for 20 min.

For laccases production monoculture, an appropriate quantity (around 4 mL) of Kirk medium was added to maintain the desired moisture level of the support (50%) for several experimental units. Each beaker was then inoculated with 4 mycelial plugs (50 mm diameter) taken from the periphery if a *T. versicolor* colony grown on malt extract agar at 30°C for 240 h, withdrawing two experimental units every 24 h for quantifying growth, residual substrate content and oxidases activities, and employing the analytical techniques described in the following sections.

Kirk basal medium composition (in g/L) was as follows: sodium tartrate, 0.275; $MgSO_4 \cdot 7H_2O$, 0.55; K_2HPO_4, 2.2; $CaCl_2 \cdot 2H_2O$, 0.145; $(NH_4)_2SO_4$, 0.44; Glucose, 8.2; $CuSO_4 \cdot 5H_2O$, 0.28; trace elements, 11 mL (in g/L: $MnSO_4 \cdot H_2O$, 0.5; NaCl, 1; $FeSO_4 \cdot 7H_2O$, 0.1; $CoCl_2 \cdot 6H_2O$, 0.185; $ZnSO_4 \cdot 7H_2O$, 0.11; $Na_2MO_4 \cdot 2H_2O$, 0.011; H_3BO_3, 0.011). The pH of the medium was adjusted at 5.0 using 1 M HCl before sterilization.

Inoculum from *Aspergillus* sp was obtained from several 5 day old PDA plates incubated at 37°C. The spores in the agar surface were gently scraped and blended in 10 mL sterile saline and used as spore suspension. The spores were enumerated under microscope using a Neubauer chamber. The experimental units used for developing the *Aspergillus* monocultures for obtaining hydrolytic activities were inoculated using 1×10^8 spores/$g_{substrate}$ and incubated at 37°C for 96 h. Two experimental units were taken from incubation every 24 h, for quantifying growth, residual substrate content and hydrolytic activities, employing the analytical techniques described further.

Basal medium composition employed for this monoculture (in g/L) was as follows: K_2HPO_4, 2; KH_2PO_4, 2; $(NH_4)_2SO_4$, 5.

2.3. Cocultures developing

For developing the cocultures in solid state, the same support employed for monocultures was used. This was moisturized and sterilized as indicated before. In this case, the support was moisturized with Kirk basal medium, and both inoculums (*T. versicolor* and *Aspergillus* sp.) were added, using the same quantities of the corresponding monocultures (see Section 2.2). Several experimental units were subsequently incubated at 30 or 37°C for 240 h. In this case oxidative and hydrolytic activities were quantified for each culture after 240 h.

2.4. Extraction and storage of crude enzymes

After incubation, 80 mL of 100 mM sodium acetate buffer pH 5 was added to each experimental unit, homogenizing with a Multi Braun® mixer and a posterior constant agitation in an ice bath. Afterwards, the enzymatic extracts were recovered by centrifugation at 4000 rpm and stored at 4°C until analyzed. The solids obtained after centrifugation were used for estimate biomass and residual substrate content, as explained below.

2.5. Enzymatic activities quantification

Laccase (Lac) activity was determined by measuring the increase in A_{470} ($\varepsilon_{470\ nm}$ = 26,600 M^{-1} cm^{-1}) due to the oxidation of 10 mM guaiacol in 0.1 M Na-acetate buffer (pH 5.0) after incubation with the crude extract at 25°C for 10 min. One unit of laccase activity (U) was defined as the amount of enzyme required to oxidize 1 µmol of guaiacol per minute of reaction [11].

Xylanolytic (Xyl) activity was assayed by incubating at 45°C for 20 min using the crude enzyme with 1% (w/v) xylan dissolved in 100 mM acetate buffer pH 5.0, and the release of reducing sugars as xylose was monitored at 575 nm after stopping the reaction by the addition of DNS. The optical density obtained was compared with a 1 g/L xylose standard curve. One unit of xylanolytic activity was defined as the enzyme necessary for the release of 1 µmol of xylose under the described conditions [12].

Lignin peroxidase (LiP) activity was estimated by incubating at 25 °C for 20 min the crude enzyme with 4 mM veratryl alcohol diluted in 100 mM tartrate buffer pH 3.5 and 0.4 mM H_2O_2. The formation of veratraldehyde was monitored at 310 nm ($\varepsilon_{310\ nm}$ = 9.3331 mM^{-1} cm^{-1}). One unit of lignin peroxidase activity was defined as the enzyme required for oxidize 1 μmol of veratryl alcohol per minute of reaction [13].

Manganese peroxidase (MnP) activity was assayed by Incubating at 25 °C for 5 min the crude enzyme with 0.1 mM $MnSO_4$, 1 mM H_2O_2, and 1 mM 2,6-dimethoxyphenol (DMP) as substrate diluted in 0.1 M tartrate buffer pH 4.5, measuring the increase in A_{469} nm ($\varepsilon_{469\ nm}$ = 27,500 mM^{-1} cm^{-1}). One unit of manganese peroxidase activity (U) was defined as the amount of enzyme required to oxidize one μmol of DMP per minute of reaction [14].

Glucoamylase (Glcamyl) activity was estimated by incubating at 60 °C for 15 min the crude enzyme with 1% (w/v) starch dissolved in 0.15 M sodium chloride buffer pH 5.0. The released reducing sugars as glucose were monitored at 575 nm after stopping the reaction by the addition of DNS. The optical density obtained was compared with a 1 g/L glucose standard curve. One unit of glucoamylase activity was defined as the enzyme necessary for the release of 1 μmol of glucose under the reaction conditions [15].

α-Amylase (α-amyl) activity was determined by incubating at 37 °C for 20 min the crude enzyme with 1% (w/v) starch dissolved in 0.15 M sodium chloride buffer pH 5. The photometric disappearance of starch was quantified after stopping the reaction by the addition of an iodine (I2/IK) mother solution, and the resultant optical density at 580 nm was registered. One unit of α-amylase activity was defined as the enzyme necessary for hydrolyze 0.1 mg of starch [15].

Invertase (Inv) activity was determined by incubating at 30 °C for 30 min the crude enzyme with 0.1 M sucrose dissolved in 0.15 M acetate buffer pH 5.5. The reducing sugars as fructose were quantified after stopping the reaction with DNS. The optical density obtained was compared with a 1 g/L glucose standard curve. One unit of invertase activity was described as the enzyme necessary for the release of 1 μmol of reducing sugars per minute of reaction [16].

Carboxymethyl cellulase (CMCase) activity was quantified By incubating at 50 °C for 5 min the crude enzyme with (1% w/v) carboxymethyl cellulase low viscosity in 50 mM citrate buffer (pH 5.0). The reducing sugars as glucose were quantified after stopping the reaction with DNS. The optical density obtained was compared with a 1 g/L glucose standard curve. One unit of CMCase activity was described as the enzyme necessary for the release of 1 μmol of reducing sugars per minute of reaction [17].

The total cellulase activity (filter paper activity, FPAse) was assayed by using a 0.5 × 6 cm string of filter paper as the substrate. This was incubated using 100 mM acetate buffer (pH 5.0) and the crude enzyme for 5 min at 45°C, stopping the reaction with DNS. The optical density obtained was compared with a 1 g/L glucose standard curve. One unit of FPAse activity was defined as the amount of enzyme used for the release of 1 μmol of glucose under the assayed conditions [18].

2.6. Biomass quantification

The quantification of fungal biomass was made after quantifying ergosterol content of the biomass in each sample. For doing this, the solid content of each experimental unit was resuspended in 10 mL of water by vigorous agitation. From this homogeneous solid suspension, 1 mL was withdrawn and filtered, to recover the solids. To this, 3 mL of an alcoholic basic solution (25% w/v of KOH dissolved in methanol) was added, boiling the resultant mixture for 5 h. Afterwards, 1 mL of distilled water and 5 mL of *n*-heptane were added, for mixing at the vortex for 3 min. The tubes were let to settle until the phase separation (around 1 h), for recording the corresponding optical density at 230 nm (for detecting the presence of 24(28) DHE, an intermediary of ergosterol pathway) and 280 nm (for detecting the ergosterol presence) of the organic phase. Ergosterol content of the biomass was estimated using the following equation:

$$Ergosterol(\%) = \frac{O.D.at280nm}{290} - \frac{O.D.at230nm}{518}$$

where 290 and 518 correspond to the molar extinction coefficient ($M^{-1}\ cm^{-1}$) of crystalline ergosterol and 24(28) DHE, respectively. A 10 g/L mycelium (*Aspergillus* sp or *T. versicolor*) standard curve was developed, for correlate g/L of biomass with ergosterol %. In fact, the calculated ergosterol % contained on each sample was compared with this curve in order to estimate biomass content [19].

2.7. Substrate quantification

The solids obtained from each culture were dried at 60°C for 12 h. The biomass content (estimated as explained before) was subtracted to the corresponding dry weight in order to obtain the real substrate content of each sample.

All the experiments were performed in duplicate, and the results are expressed as the mean of these repetitions and the corresponding standard deviation.

3. RESULTS AND DISCUSSION

3.1. Study of the behavior of monocultures in solid-state fermentation for the synthesis of enzymes

On previous work, we optimized the culture media components in order to obtain high xylanase and laccase productions on solid-state fermentation, using *Aspergillus* sp or *T. versicolor*, respectively [7]. However, these experiments were made at very low volumes, in which only the information of final enzyme production was obtained at 72 or 240 h, respectively. As these enzymes were used successfully for jonote pulp bleaching [8], we decided to increase the volume of the reaction unit, and analyze the behavior of each culture with respect to fungal growth, substrate consumption and xylanase or laccase production. By doing this, we could obtain the basic information to scale up the process for producing both enzymatic activities in large amounts. Also, it could be possible to design an efficient bioprocess in which both enzymatic units can be produced simultaneously.

In this regard, it is worth mentioning that biomass quantification on solid-state cultures is a complicated work. Fungal growth is not easy to quantify because fungus grows as hyphal filaments that cannot be quantified by the usual enumeration techniques, and specifically on cultures in which an insoluble material is used as the only carbon source, complete recovery of fungal biomass from the substrate is very difficult, because the fungal hyphae tend to penetrate into and binds tightly to the solid substrate particles [20]. It is important therefore to have reliable and convenient methods for measuring fungal growth. For this reason, we used the ergosterol content methodology [19] for quantifying biomass evolution along the culture, and employed a simple mass balance for knowing the corresponding residual substrate. Therefore, this is one of the main contributions of the present work.

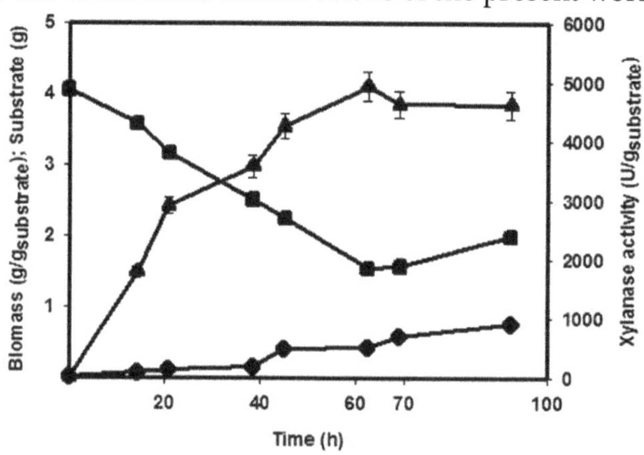

Figure 1. Fungal growth (♦), substrate consumption (▪), and xylanase production (▴) by Aspergillus sp along the solid-state monoculture developed on wheat bran:sugar cane bagasse 1:1 (w/w) as substrate for solid-state fermentation, incubated at 37°C for 96 h.

With respect to monocultures developed with *Aspergillus* sp., growth seemed to start increasing when the culture was among 20 and 40 h, reaching finally 0.7 g of mycelium/g of substrate at the end of culture. In this case, the fungus showed an approximate growth rate of 0.03 h^{-1}. Substrate consumption showed a constant rate during the first 60 h, stopping after that. Almost 2.5 g of substrate seems to have been consumed during the culture, at a consumption rate of -0.04 $g_{substrate}$/h. With respect to xylanases production, this started with the same trend as that of fungal growth, and increased constantly, although when the growth stopped xylanolytic activity decreased slightly (Figure 1).

With respect to monocultures developed with *T. versicolor* for laccase production (Figure 2), it can be seen that in this case there was an adaptation phase of around 50 h, afterwards the fungus started to grow. In this case, it was obtained about 0.6 $g_{mycelium}/g_{substrate}$, with a constant specific growth rate of 0.11 h^{-1}. However, substrate consumption was really slow (r_s = -0.0006 $g_{substrate}$/h), and the fungi only consumed approximately 1.5 g along the culture, which represents 40% of the initial substrate. Laccase production and growth started together, increasing at constant rate, but the enzyme production remained even when biomass stopped growing.

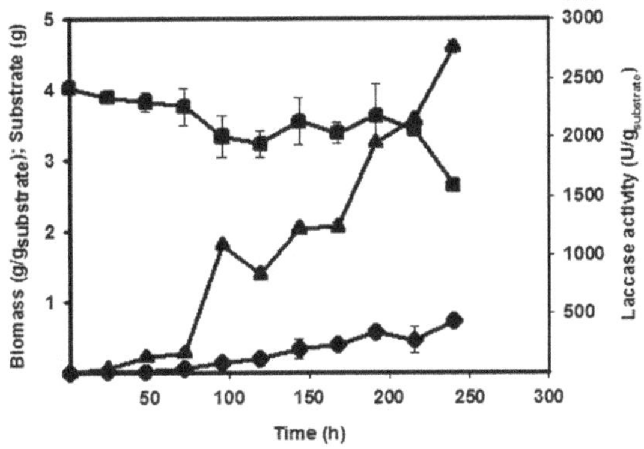

Figure 2. Fungal growth (♦), substrate consumption (▪), and laccase production (▴) by T. versicolor in monoculture developed on wheat bran:sugar cane bagasse 1:1 (w/w) as support/substrate for solid-state fermentation, incubated at 37°C for 96 h.

Several studies developed on liquid or submerged fermentation have reported that laccases production has a secondary metabolite behavior; it means the activity is produced mainly during the secondary metabolism [21]. On solid-state culture, this relationship is not well known due to the difficulty of quantifying accurately the total biomass grown in the substrate along the culture. However, other studies in which fungal biomass has been quantified by indirect methods, like that used in the present work, have shown that laccases production is growth related, as happened for laccases produced by *Streptomyces psammoticus* on rice straw [22]. And those results are in accordance with the *T. versicolor* profile obtained in the present work.

The titers obtained on each monoculture for xylanases and laccase activities were high. These results show that solid-state culture is a good alternative for producing high oxidative or hydrolytic activities, in order to employ them for several industrial bioprocesses. At this respect, the results obtained in the present work would represent one basis for developing this process on a full- scale. Thereby, for further characterization of the monocultures, in the next section the production profiles of other oxidative or hydrolytic activities obtained were analyzed.

3.2. Enzymatic profiles of crude enzymes

Total enzyme production obtained along *Aspergillus* sp. and *T. versicolor* solid-state monocultures was determined by quantifying different activities: xylanases, oxido-reductases and cellulases, since it is known that these enzymatic families are involved in invasion of lignocellulosic material [23]. Furthermore, due to wheat bran and sugar cane bagasse composition [24], amylase and invertase activities were also measured.

Along *Aspergillus* monoculture, xylanolytic activity was predominant; a peak of this activity was reached after 60 h, followed by a constant level until the end of culture. However, another hydrolases were quantified: cellulases and invertases had maximum titers during the first 20 h, after that their levels remained below 1000 U/g for FPase and 500 U/g for CMCase, while amylases reached their maximum levels after 48 h of cultivation, and these activities were kept until the end of the culture. In this monoculture, no oxidase activities were obtained.

With respect to oxidases, during the first 72 h of monocultures developed with *T. versicolor*, LiP and MnP activities were predominant; from this moment, laccase activity increased continuously to reach a peak around 2700 U/g at the end of process. This activity was almost five times higher than lignin and manganese peroxidases. After 100 h of cultivation, LiP and MnP activities stayed at a level around 250 and 800 U/g, respectively. In this *T. versicolor* monoculture none hydrolytic activity was detected. Table 1 concentrates final titers of all hydrolase and oxidase activities obtained on each of both monocultures.

Table 1. Oxidases and hydrolase activities produced by monocultures developed by *T. versicolor* or *Aspergillus* sp on solid-state monocultures.

Enzymatic activities (U/gs)	*Trametes versicolor*	*Aspergillus* sp
Laccases	2759 ± 30	ND
LiP	410 ± 10	ND
MnP	996 ± 2	ND
Xylanases	ND	4617 ± 38
FPAses	ND	746 ± 7
Carboximetil-celulases	ND	829 ± 34
Glucoamylases	ND	1762 ± 4
α-amylases	ND	1898 ± 6
Invertases	ND	1023 ± 73

ND indicates the enzyme whose activity was not detected in the corresponding monoculture.

Taking into account that enzymatic extracts obtained on solid-state cultures by *Aspergillus* sp. and *T. versicolor* had several hydrolytic or oxidative activities, there could be some other uses for them. In this respect, extracts obtained from *Aspergillus* sp. monoculture could be used for saccharification processes of a number of agro-industrial residues [25], while the extract obtained from *T. versicolor* monoculture can be proved in color removal of some industrial effluents [26].

3.3. Cocultures for joint production of lignocellulases

Solid-state cultures showed high productions of either hydrolase or oxidase activities in the corresponding monocultures. Previous work of this group has shown that bleach boosting of kraft [9] and jonote [8] pulp can be improved with the employment of the combined action of xylanases and laccases, produced by solid-state cultures as those described in the present work. So, a greater enzyme production would be desirable in order to have a more efficient bioprocess. At this respect, it has been suggested that the cocultivation of microbes in fermentation can increase the quantity of the desirable components on a cellulose complex [27]. On the other side, some reports have shown that laccases or xylanases production can increase in a coculture mode, as happens with *P. ostreatus*, which increased fivefold its laccase production in a coculture with *Trichoderma viride* in submerged fermentation [28]. As discussed earlier, it became interesting to probe if a coculture of *Aspergillus* sp and *T. versicolor* was feasible for obtaining higher xylanase and laccase activities at the same time considering that both fungi used wheat bran and sugar cane bagasse as substrate. In addition, this coculture strategy would provide economic advantages because of reduction in overall cost production. This is why both fungi were inoculated at the same time in the support and two incubation temperatures were probed: 30°C (the best for *T. versicolor*) and 37°C (the best for *A. niger*), developing independent experiments. Figure 3 shows the enzyme proportions obtained in cocultures developed at 30 or 37°C with respect to the control conditions, obtained in the respective monocultures of each fungi.

With respect to hydrolytic activities proportion, it can be seen that while on monocultures xylanase activity was the predominant hydrolytic activity, glucoamylases were in greater proportion for cocultures developed at 30°C, and α-amylases highlighted on cocultures developed at 37°C (Figure 3 A).

Comparing both cocultures, one can see that the highest hydrolytic activities were obtained on that developed at 37°C, which means that temperature could be affecting seriously the *Aspergillus* sp. behavior along this culture. On the other hand, invertase and cellulase activity proportions were low on monocultures, but decreased on both cocultures. The diminution of invertase activity production on coculture modes can be due to the presence of glucose in the media, as Kirk medium was used to moisturize the support in these experiments, and this mineral medium contains glucose. A similar diminution on invertase activity production due to glucose presence in solid-state fermentation was reported previously for *A. ochraceus* in similar culture conditions, which employed wheat bran or sugar cane bagasse as substrates [29].

Laccase activities decreased on both cocultures with respect to those values obtained on monocultures. This behavior could be considered unusual, because in general the coculture strategy is used in order to increase these activities. In fact, the addition of soil microorganisms to white rot fungi cultures has increased laccase and other oxidases production [30]. In this case, the diminution on laccase activity can be explained from different points of view. First of all, in the present study both fungi were inoculated at the same time to the support. The decrease in oxidases and hydrolases production could be provoked by problems in fungal growth, considering the differences in specific growth rate of each fungal species. This is because mineral medium used in these cocultures contained glucose, and it has been reported as laccase inductor for *T. versicolor* [31]. So, even when *Aspergillus* grows slower in its respective monoculture, in this case it could be consuming the available glucose of the medium faster than *T. versicolor* could, reducing the ability of *T. versicolor* for producing laccases and the another oxidases. However, to probe this hypothesis, some additional experiments would be necessary.

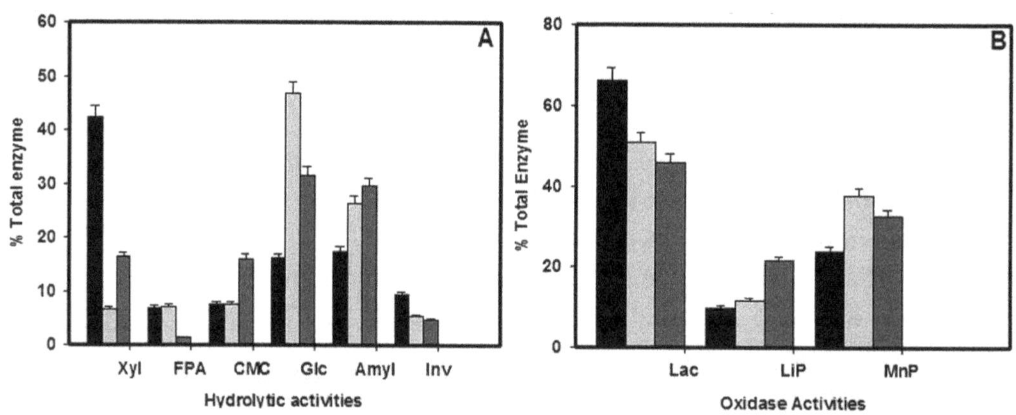

Figure 3. Enzyme proportion of hydrolytic (A) or oxidase (B) activities obtained on monocultures (▪), and cocultures developed at 30°C () or 37°C ().

On the other hand, we must remember that enzymes are produced at different rates along the cultures, as described in Section 3.2. In this case, activities analyzed for this experimental phase are those obtained after 10 days, but the enzyme evolution along both cocultures was not registered. This is a great limitation for obtaining valid conclusions. It could be possible that one of these enzymes has been increased at any time during the fermentations, but as we did not quantify the enzyme evolution, we could not know if the reported activity was the highest obtained for the corresponding enzyme in this experiment. So, we can only analyze the enzyme activities at the end of the coculture, and these could not be the highest activities obtained in this case.

Finally, even when some hydrolytic or oxidative enzymes decreased on coculture mode, these extracts can be employed on several processes. Agricultural by-products typically vary in their chemical composition and nutritional value, and sometimes are also higher in low-quality fiber, so a specific enzyme complex is required to break it down in order to be used in ruminant feed. Besides, their nutritional value could be increased by biodegradation methods of fiber in the rumen, through efficient delignification [32]. Therefore, the filtrates obtained on both cocultures could be a good alternative for being employed for using agricultural by-products for ruminant feed.

4. CONCISE SUMMARY

The indirect method employed to measure fungal biomass content proved to be valuable, making a significant contribution to the analysis of solid-state cultures. It is evident from this observation that both fungi exhibit distinct behaviors during the cultivation process. T. versicolor demonstrates a faster growth rate compared to Aspergillus sp., resulting in a higher rate of substrate consumption. Nevertheless, the monocultures yielded significant quantities of xylanases and laccases. In this instance, the production of xylanases appeared to be correlated with growth, whereas laccases began creating during the development phase and maintained production even after the fungus ceased to grow. Furthermore, the enzymatic extracts exhibited the capability to be utilized in many industrial bioprocesses, as evidenced by the existence of additional hydrolases and oxidases activities. The implementation of coculture mode resulted in a reduction in the production of xylanase and laccase. The incubation temperature appears to have an impact on xylanase synthesis. Additionally, the slower growth of Aspergillus sp. may be attributed to its consumption of glucose in Kirk mineral medium, which in turn delays the generation of laccases by T. versicolor. Modifications to the inoculation process are necessary to enhance the production of these enzymes through coculture. From this information, it can be inferred that solid-state fermentation, either on its own or in combination with other cultures, has the potential to be a very effective and environmentally friendly method of manufacturing various lignocellulosic enzymes from agricultural waste. These enzymes can then be utilized in a range of industrial applications, all at a reduced cost.

REFERENCES

1. Jin N, Ma S, Liu Y, Yi X, He R, Xu H, Qiao DR, Cao Y. Thermophilic xylanase production by Aspergillus niger in solid state fermentation using wheat straw and corn cob. African Journal of Microbiology Research 2012;6(10):2387–2390. doi: 10.5897/AJMR11.1433
2. Mohamed SA, Al-Malki AL, Khan JA, Kabli SA, Al-Garni SM. Solid state production of polygalacturonase and xylanase by Trichoderma species using cantaloupe and watermelon rinds. Journal of Microbiology. 2013;51(5):605–611. doi: 10.1007/s12275-013-3016-x
3. Fernández-Fernández M, Sanromán Má, Moldes D. Recent developments and applications of immobilized laccase. Biotechnology Advances. 2013;31(8):1808–1825. doi: 10.1016/j.biotechadv.2012.02.013
4. Record E, Asther M, Sigoillot C, Pages S, Punt PJ, Delattre M, Haon M, Van den Hondel CA, Sigoillot JC, Lesage-Meessen L, Asther M. Overproduction of the Aspergillus niger feruloyl esterase for pulp bleaching application. Applied Microbiology and Biotechnology. 2003;62(4):349–355. doi: 10.1007/s00253-003-1325-4
5. Kapoor M, Kapoor RK, Kuhad RC. Differential and synergistic effects of xylanase and laccase mediator system (LMS) in bleaching of soda and waste pulps. Journal of Applied Microbiology. 2007;103(2):305–317. doi: 10.1111/j.1365-2672.2006.03251.x
6. Xu P, Ding ZY, Qian Z, Zhao CX, Zhang KC. Improved production of mycelial biomass and ganoderic acid by submerged culture of Ganoderma lucidum SB97 using complex media. Enzyme and Microbial Technology. 2008;42(4):325–331. doi: 10.1016/j.enzmictec.2007.10.016
7. Domínguez-Morales D. Optimización de la producción de xilanasas y lacasas fúngicas mediante el empleo de la Metodología de Superficie de Respuesta [Thesis]. Tecnológico de Estudios Superiores de Ecatepec; 2013.
8. Licona-Soto S.R. Bioproceso integrado para el tratamiento biológico de la pulpa de jonote [thesis]. Tecnológico de Estudios Superiores de Ecatepec; 2014.
9. García-Rivero M, Membrillo-Venegas I, Vigueras-Carmona SE, Zafra-Jiménez G, Zárate-Segura PB, Martínez-Trujillo MA. Enzymatic pretreatment to enhance chemical bleaching of a kraft pulp. Revista Mexicana de Ingeniería Química. 2015;14(2):335–345.
10. Dwivedi P, Vivekanand V, Pareek N, Sharma A, Singh RP. Co-cultivation of mutant Penicillium oxalicum SAU E-3.510 and Pleurotus ostreatus

for simultaneous biosynthesis of xylanase and laccase under solid-state fermentation. New biotechnology. 2011;28(6):616–626. doi: 10.1016/j.nbt.2011.05.006

11. Bertrand B, Martínez-Morales F, Tinoco-Valencia R, Rojas S, Acosta-Urdapilleta L, Trejo-Hernández MR. Biochemical and molecular characterization of laccase isoforms produced by the white-rot fungus Trametes versicolor under submerged culture conditions. Journal of Molecular Catalysis B: Enzymatic. 2015;122:339–347.

12. Membrillo Venegas I, Fuentes-Hernández J, García-Rivero M, Martínez-Trujillo A. Characteristics of Aspergillus niger xylanases produced on rice husk and wheat bran in submerged culture and solid?state fermentation for an applicability proposal. International Journal of Food Science and Technology. 2013;48(9):1798–1807. doi: 10.1111/ijfs.12153

13. Jung H, Xu F, Li K. Purification and characterization of laccase from wood-degrading fungus Trichophyton rubrum LKY-7. Enzyme and Microbial Technology. 2002;30(2):161–168. doi: 10.1016/S0141–0229(01)00485-9

14. Palma C, Martınez AT, Lema JM, Martınez MJ. Different fungal manganese-oxidizing peroxidases: a comparison between Bjerkandera sp. and Phanerochaete chrysosporium. Journal of Biotechnology. 2000;77(2):235–245. doi: 10.1016/S0168-1656(99)00218-7

15. Matute L, Bertsch A, Díaz I. Evaluación de la actividad amilolítica de Aspergillus niger ANM-1 en fermentaciones en estado sólido y sumergido para la obtención y caracterización de aditivos enzimáticos. Revista de la Facultad de Agronomía (UCV). 2011;38(1): 9–17.

16. Romero-Gómez SJ, Augur C, Viniegra-González G. Invertase production by Aspergillus niger in submerged and solid-state fermentation. Biotechnology Letters. 2000;22(15):1255–1258. doi: 10.1023/A:1005659217932

17. Miller GL, Blum R, Glennon WE, Burton AL. Measurement of carboxymethylcellulase activity. Analytical Biochemistry. 1960;1(2):127–132. doi: 10.1016/0003-2697(60)90004-X

18. Ponce-Noyola T, De la Torre M. Isolation of a high-specific-growth-rate mutant of Cellulomonas flavigena on sugar cane bagasse. Applied Microbiology and Biotechnology. 1995;42(5):709–712. doi: 10.1007/BF00171949

19. Bhosle SR, Sandhya G, Sonawane HB, Vaidya JG. Ergosterol content of several wood decaying fungi using a modified method. International Journal of Pharmacy and Life Sciences. International Journal of Pharmacy and Life Science (IJPLS). 2011;2(7):916–918.

20. Abd-Aziz S, Hung GS, Hassan MA, Karim MI, Samat N. Indirect method for quantification of cell biomass during solid-state fermentation of palm kernel cake based on protein content. Asian Journal of Scientific Research 2008;1(4):385–393.

21. Brijwani K, Rigdon A, Vadlani PV. Fungal laccases: production, function, and applications in food processing. Enzyme Research. 2010 Sep 21;2010. http://dx.doi.org/10.4061/2010/149748

22. Niladevi KN, Sukumaran RK, Prema P. Utilization of rice straw for laccase production by Streptomyces psammoticus in solid-state fermentation. Journal of Industrial Microbiology and Biotechnology. 2007;34(10):665–674. doi: 10.1007/s10295-007-0239-z

23. Hasunuma T, Okazaki F, Okai N, Hara KY, Ishii J, Kondo A. A review of enzymes and microbes for lignocellulosic biorefinery and the possibility of their application to consolidated bioprocessing technology. Bioresource Technology. 2013;135:513–522. doi: 10.1016/j.biortech.2012.10.047

24. Veana F, Martínez-Hernández JL, Aguilar CN, Rodríguez-Herrera R, Michelena G. Utilization of molasses and sugar cane bagasse for production of fungal invertase in solid state fermentation using Aspergillus niger GH1. Brazilian Journal of Microbiology. 2014;45(2):373–377. http://dx.doi.org/10.1590/S1517-83822014000200002

25. Mostafa FA, Ahmed SA, Helmy WA. Enzymatic saccharification of pretreated lemon peels for fermentable sugar production. Journal of Applied Sciences Research. 2013;9(3):2301–2310.

26. Zeng X, Cai Y, Liao X, Zeng X, Li W, Zhang D. Decolorization of synthetic dyes by crude laccase from a newly isolated Trametes trogii strain cultivated on solid agro-industrial residue. Journal of Hazardous Materials. 2011;187(1):517–525. doi: 10.1016/j.jhazmat.2011.01.068

27. Kaushal R, Sharma N, Tandon D. Cellulase and xylanase production by co-culture of Aspergillus niger and Fusarium oxysporum utilizing forest waste. Turkish Journal of Biochemistry/Turk Biyokimya Dergisi. 2012;37(1).

28. Flores C, Casasanero R, Trejo-Hernández MR, Galindo E, Serrano-Carreón L. Production of laccases by Pleurotus ostreatus in submerged fermentation in co?culture with Trichoderma viride. Journal of Applied Microbiology. 2010;108(3):810–817. doi: 10.1111/j.1365-2672.2009.04493.x

29. Alegre AC, Polizeli MD, Terenzi HF, Jorge JA, Guimarães LH. Production of thermostable invertases by Aspergillus caespitosus under submerged or solid state fermentation using agroindustrial residues as carbon source. Brazilian

Journal of Microbiology. 2009;40(3):612–622. http://dx.doi.org/10.1590/S1517-83822009000300025

30. Chan Cupul W, Heredia Abarca G, Martínez Carrera D, Rodríguez Vázquez R. Enhancement of ligninolytic enzyme activities in a Trametes maxima-Paecilomyces carneus co-culture: key factors revealed after screening using a Plackett-Burman experimental design. Electronic Journal of Biotechnology. 2014;17(3):114–121. http://dx.doi.org/10.1016/j.ejbt.2014.04.007

31. Piscitelli A, Giardina P, Lettera V, Pezzella C, Sannia G, Faraco V. Induction and transcriptional regulation of laccases in fungi. Current Genomics. 2011;12(2):104–112. doi: http://dx.doi.org/10.2174/138920211795564331

32. Graminha EB, Goncalves AZ, Pirota RD, Balsalobre MA, Da Silva R, Gomes E. Enzyme production by solid-state fermentation: application to animal nutrition. Animal Feed Science and Technology. 2008;144(1):1–22. doi: 10.1016/j.anifeedsci.2007.09.029

14

ANAEROBIC DIGESTION: I. AN UBIQUITOUS PROCESS FACILITATING THE TRANSFER OF ENERGY A ND THE CYCLING OF MATTER IN ECOSYSTEMS II. A DEVICE DESIGNED FOR THE GENERATION OF GASEOUS BIOFUELS

1. INTRODUCTION: AN OVERVIEW

The process of anaerobic digestion of biomass at mesophilic temperatures (microbial breakdown of organic matter in the absence of oxygen), results in the production of methane and carbon dioxide. This process plays a crucial role in the energy flow and cycling of matter in ecosystems. The global carbon cycle is facilitated by the action of various kinds of microbes, making it a crucial process. Anaerobic digestion is a process that frequently happens in natural environments that lack oxygen, such as freshwater sediments, wetlands, marshlands, paddy fields, and the deeper areas of marine sediments. The process of methane synthesis also occurs in the digestive systems of animals, including ruminants and termites. Biological methanogenesis is believed to account for over 70% of the total methane emissions worldwide [1, 2]. Anaerobic breakdown of biomass into carbon dioxide and methane happens only in settings that lack oxygen and have a low potential for oxidation-reduction reactions. This means that the concentrations of other electron acceptors, such as nitrate, oxidized forms of metals like Mn(IV) and Fe(III), or sulphate, are low in these situations. The redox potential determines the extent to which nitrate, oxidized metal ions, and sulphate hinder anaerobic digestion. Figure 1 demonstrates that a decrease in redox potential results in a shift in the prevailing form of anaerobic respiration towards less energy-producing activities. The presence of specific final electron acceptors in an environment plays a crucial role in determining the ecological habitats suitable for specific bacteria.

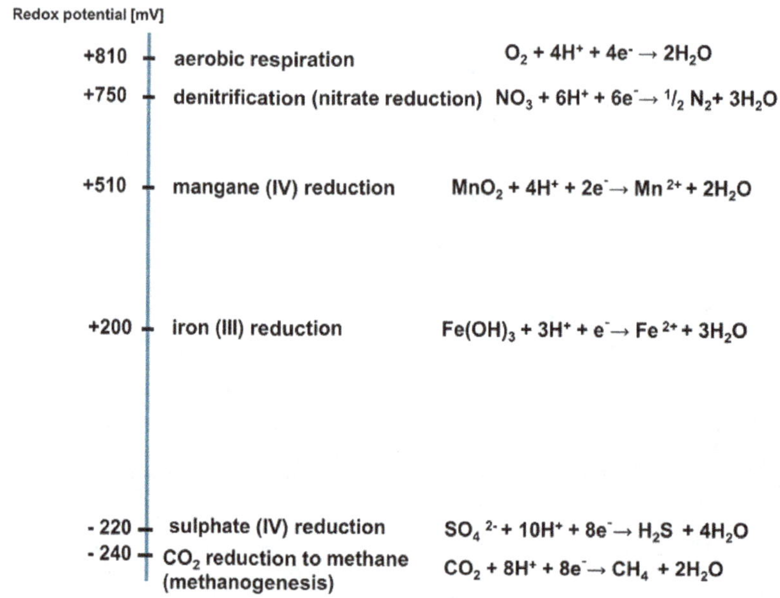

Figure 1. Redox potential for different types of final electron acceptors in anaerobic respiration and energy gain for microbial cells.

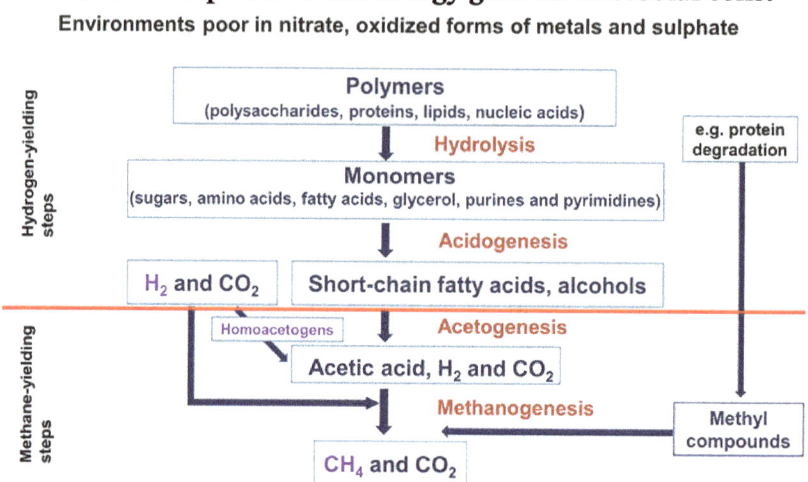

Figure 2. Scheme of anaerobic digestion of polymeric organic matter to methane and carbon dioxide.

The overall process of anaerobic digestion is widely understood (Figure 2). The process is intricate and facilitated by the interaction of multiple groups of microorganisms, consisting of four main phases. One of the processes is the hydrolysis of intricate organic polymers into individual monomers. During the acidogenesis phase, hydrogen, carbon dioxide, and non-gaseous fermentation products such as low-molecular weight organic acids (short-chain fatty acids) and alcohols are formed. During the third stage, referred to as acetogenesis, the non-gaseous substances are further oxidized through syntrophic breakdown mechanisms to produce hydrogen, carbon dioxide, and acetate. Methanogenesis is the fourth step. The latter two stages, acetogenesis and methanogenesis, are tightly interconnected and entail symbiotic relationships between acetogenic bacteria that produce hydrogen and methanogens that consume hydrogen. These linkages maintain a sufficiently low hydrogen partial pressure, enabling acetogenesis to become thermodynamically favorable. The process, known as interspecies electron transfer, involves the transfer of hydrogen and formate. Acetate serves as a primary source for the production of methane and can also undergo a process called syntrophic oxidation to produce hydrogen and carbon dioxide [3–8].

Anaerobic digestion is prevalent in landfill sites and anaerobic wastewater treatment plants. Biogas production is a regularly employed method for generating alternative energy by decomposing biomass, such as energy crops or organic agro-waste, through anaerobic processes. There is currently significant enthusiasm for the advancement of novel technologies aimed at modernizing landfills and wastewater treatment plants. These technologies aim to effectively manage the emission of biogas and capture methane for use as a fuel source. Furthermore, in order to develop advanced technologies that rely on microbial processes, it is advantageous to construct contemporary biogas facilities where the stages of anaerobic digestion that produce hydrogen (hydrolysis and acidogenesis) and methane (acetogenesis and methanogenesis) are kept separate. This separation allows for the optimized production of hydrogen and methane in controlled environments. To achieve efficient production of methane or hydrogen and methane from organic matter, it is crucial to have a comprehensive understanding of anaerobic digestion at the molecular level. This includes knowledge about the composition and variety of microbial communities, as well as the metabolic pathways involved in converting the organic substrate into the desired gaseous products.

2. META-OMICS APPROACHES FOR EXPLORING MICROBIAL COMMUNITIES

Current knowledge of microbial ecology and physiology, derived from culture-dependent techniques, is limited and incomplete because the majority of microorganisms have not been cultivated. It has been predicted that only 1% or less of all microorganisms present in natural ecosystems may be cultivated as a pure culture using standard methods [9]. Moreover, syntrophy is believed to be common in microbial communities, and syntrophic bacteria cannot be grown as a monoculture. However, culture-dependent techniques have permitted the isolation and characterization of some species involved in specific metabolic processes during anaerobic digestion, and numerous genomes have been sequenced. Data from genome sequence analyses supported by the results of physiological and biochemical studies on cultivated bacteria have provided hints as to which physiological groups of microorganisms are responsible for the key steps of anaerobic digestion. Information on methane-yielding microbial communities is now being obtained using culture-independent analytical techniques (Figure 3).

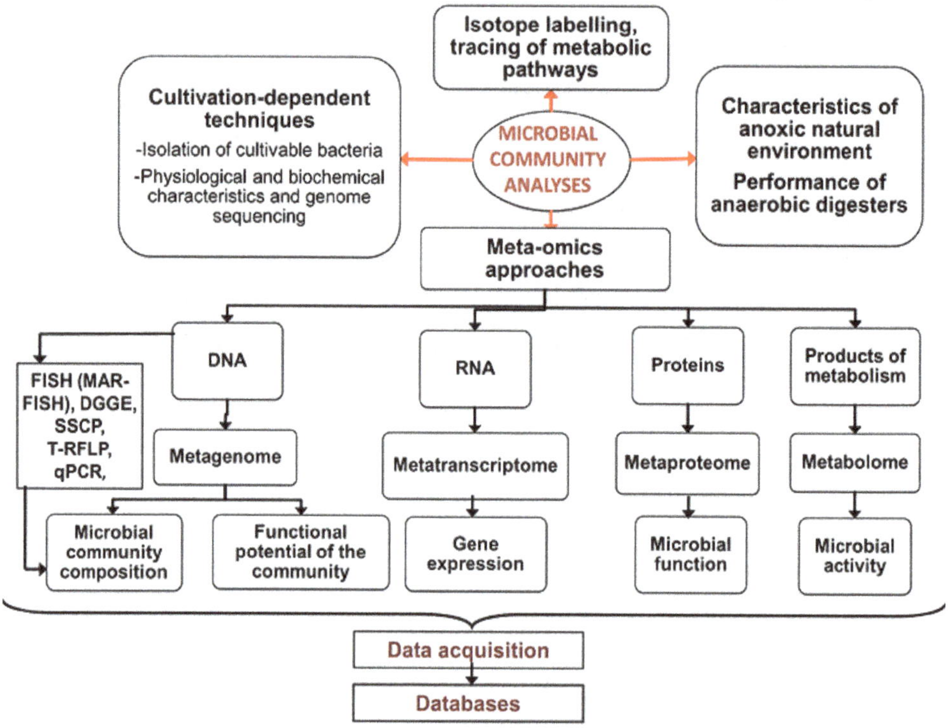

Figure 3. Culture-independent approaches to analyse methane-yielding microbial communities. FISH, fluorescence in-situ hybridization; MAR-FISH, microautoradiography combined FISH; DGGE, denaturing gradient gel electrophoresis; SSCP, single-strand conformation polymorphism; T-RFLP, terminal restriction fragment length polymorphism; qPCR, real-time quantitative PCR.

The recent increase in the number of culture-independent molecular biology techniques and bioinformatic tools for exploring microbial communities has helped to develop the field of meta-omics. Meta-omics encompasses metagenomics, metatranscriptomics, metaproteomics and metabolomics, based on analyses of, respectively, total DNA, mRNA, total proteins and metabolites isolated from the microbial communities [10–14]. Metagenomics shows microbial potential by describing the genes present in a microbial community or ecosystem. Metatranscriptomics analyses gene expression and thus represents potential microbial function. Messenger RNA (mRNA) can be sequenced directly or used to generate cDNA (by reverse transcription) that is subsequently sequenced using metagenomics platforms. Metaproteomics is focused on microbial function—it investigates proteins expressed within a microbiome. Metabolomics analyses the intermediates and end-products of metabolism and then shows microbial activity.

The data generated by these novel methodologies have provided significant insights into the structure and function of microbial communities in both natural environments and man-made systems. However, meta-omics-based approaches do suffer from certain limitations: the variable extraction efficiency of DNA/RNA/protein/metabolites may affect the results, and reference databases used for comparative analyses often contain false or missing assignments of DNA and protein sequences or chromatography/mass spectrometry data. For example, metagenomic analyses always generate large numbers of sequences that are of low complexity, unclassified, not assigned or show no hits. Such unidentified reads usually constitute a significant proportion of the total reads, as discussed by Chojnacka et al. [15]. It is noteworthy that the limited number of microorganisms that can be propagated as pure cultures determines the number of sequenced reference genomes available for genomic studies. So far, only five genomes of syntrophic bacteria involved in acetogenesis have been sequenced: *Syntrophus aciditrophicus*, *Syntrophus wolfei*, *Syntrophobacter fumaroxidans*, *Pelotomaculum thermopropionicum* and *Syntrophothermus lipocalidus*, as discussed by Li et al. [16].

Using metagenomic sequence data and genomic assembly procedures, it is possible to reconstruct genomes of bacteria that have not been cultivated. One example is a reconstruction of the genome of *Candidatus* Cloacimonas acidaminovorans [17]. This is a representative of the *Cloacimonetes*, a sub-dominant group of bacteria found in anaerobic mesophilic digesters and gut microflora. They are regarded as syntrophs capable of amino acid fermentation, propionate and butyrate oxidation as well as cellulose degradation and have never been grown in pure culture.

In the case of anaerobic digestion, the combined use of meta-omic approaches and isotope labelling techniques in both natural anoxic environments and bioreactors plus the analysis of reactor performance data will allow us to develop a fundamental understanding of the processes leading to methane production. Meta-omic data can also be used to validate commonly accepted theses.

Other cultivation-independent techniques include isolation of total DNA from microbial communities, amplification, cloning and sequencing of marker genes, most frequently 16S rRNA or others such as *gyrB* or *mcrA* for methanogenic *Archaea*; fluorescence in-situ hybridization (FISH) and its derivatives such as microautoradiography combined FISH (MAR-FISH); community fingerprinting by denaturing gradient gel electrophoresis (DGGE); single-strand conformation polymorphism (SSCP); terminal restriction fragment length polymorphism (T-RFLP) and real-time quantitative PCR (qPCR), as discussed by Dziewit et al. [18].

3. HYDROLYSIS AND ACIDOGENESIS ARE THE STAGES OF ANAEROBIC DIGESTION THAT RESULT IN THE PRODUCTION OF SHORT-CHAIN FATTY ACIDS AND HYDROGEN.

3.1. Hydrolysis

Hydrolysis is the first step in the anaerobic decomposition of organic matter. It involves the conversion of polymeric organic matter (e.g., polysaccharides, lipids, proteins) to monomers (e.g., sugars, fatty acids, amino acids) by hydrolases secreted to the environment by microorganisms. Three key groups of hydrolases are involved in the process of anaerobic digestion: esterases, glycosidases and peptidases, which catalyse the cleavage of ester bonds, glycoside bonds and peptide bonds, respectively [19]. The bacteria most commonly associated with hydrolysis include representatives of the *Firmicutes* (*Clostridia, Bacilli*), *Bacteroidetes* and *Gammaproteobacteria* [20–22]. Usually, the same bacteria are also able to conduct acidogenesis, the second step in the decomposition of organic matter.

Metaproteomic analysis of microbial communities mediating the decomposition of dead plant material in forest leaf litter revealed fungi to be the main producers of extracellular hydrolytic enzymes, the most prominent of which are cellulolytic enzymes: exo- and endo-glucanases as well as β-glucosidases. Other hydrolases involved include phosphatases, pectinases, xylanases, lipases, amylases, chitinases and oxidoreductases. Moreover, the species of fungi – the main cellulase producers – changed depending on the season. In a sample collected in February, *Leotiomycetes* dominated, whereas in samples collected in May, *Eurotiomycetes, Dothideomycetes, Leotiomycetes* and *Sordariomycetes* were the most abundant fungal phyla. Interestingly, no bacterial hydrolases were detected [23].

3.2. Acidogenesis

3.2.1. Fermentation of sugars

During acidogenesis, the hydrolysis products undergo conversion into non-gaseous short-chain fatty acids, alcohols, aldehydes, as well as gases like carbon dioxide and hydrogen [3]. The type of fermentation is determined by the major end-products of the fermentation process (Figure 4A). The primary hydrogen-producing fermentations that occur under moderate temperature circumstances are butyric acid fermentation (also known as Clostridium-type fermentation) and mixed-acid fermentation (also known as Enterobacteriaceae-type fermentation). The initial and most frequent phase is glycolysis, specifically the Embden-Meyerhof-Parnas pathway, which involves the conversion of glucose into pyruvate and the formation of NADH (as shown in Figure 4B). Hydrogenases are accountable for the liberation of hydrogen in both forms of fermentation. Hydrogenases are metalloenzymes that can be classified into two primary categories based on the type of metal present in the active site's prosthetic group: [FeFe] and [FeNi] hydrogenases [24, 25].

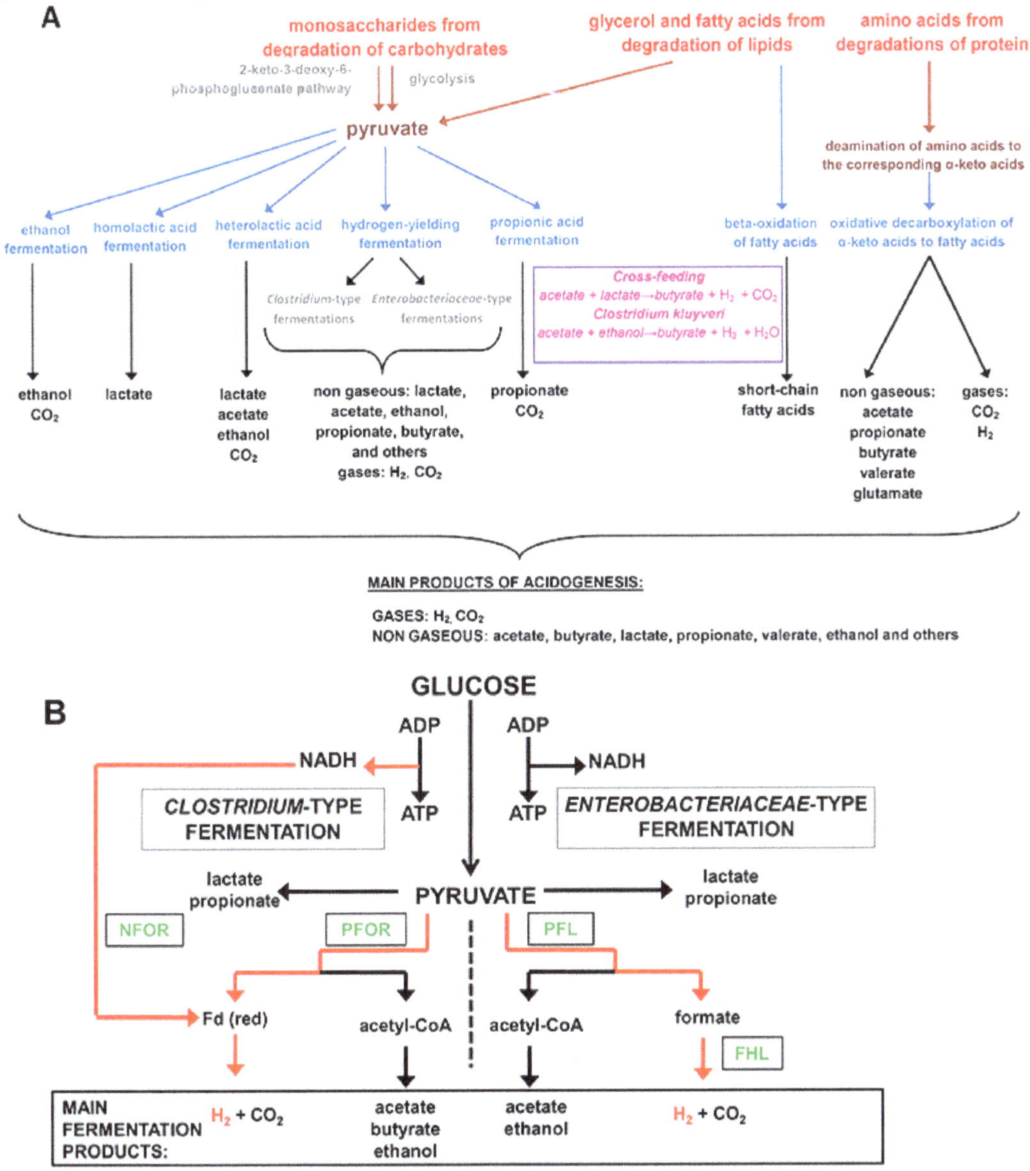

Figure 4. Metabolic pathways of acidogenesis: (A) general overview and (B) glycolytic hydrogen-yielding fermentations. PFOR, pyruvate:ferredoxin oxidoreductase; NFOR, NADH:ferredoxin oxidoreductase; PFL, pyruvate formate-lyase; FHL, formate hydrogen-lyase complex.

During Clostridium-type fermentation, the process of pyruvate oxidation to acetyl-CoA is carried out by an enzyme called pyruvate:ferredoxin oxidoreductase (PFOR), with the assistance of a molecule called ferredoxin (Fd). NADH:ferredoxin oxidoreductase (NFOR) catalyzes the formation of reduced ferredoxin through a reaction with NADH. Hydrogen is mostly emitted through the action of [FeFe] hydrogenases, which facilitate the reduction of protons by utilizing electrons derived from ferredoxin. A variety of enzymes catalyze the conversion of Acetyl-CoA into non-gaseous end-products, which consist of short-chain fatty acids (acetate, butyrate, lactate, propionate), alcohols (mostly ethanol, but also butanol and propanol), and ketones (such as acetone). The synthesis of hydrogen during acidogenesis is reduced by an elevated quantity of non-gaseous fermentation byproducts. The concentration of hydrogen controls the relative activity of PFOR and NFOR. An hydrogen partial pressure exceeding 60 Pa suppresses NFOR activity and promotes the production of non-gaseous end-products. However, PFOR remains functional even when exposed to hydrogen concentrations as high as 30,000 Pa [3, 26–29].

In the mixed-acid fermentation (also known as formic acid fermentation), pyruvate is converted to acetyl-CoA and formic acid by pyruvate formate-lyase (PFL). The formic acid can then be degraded into hydrogen and carbon dioxide by formate hydrogen-lyase (FHL) complex. One of the FHL subunits is the [FeNi] hydrogenase Hyd-3. There are two types of mixed-acid fermentation: the *Escherichia coli* type and the *Enterobacter* type. In the *Enterobacter*-type fermentation, hydrogen can also be generated through oxidation of NADH by NFOR in reactions similar to those described for the *Clostridium*-type fermentation. Non-gaseous products of the *Enterobacteriaceae*-type fermentation can include ethanol, short-chain fatty acids (formate, acetate, lactate, succinate) as well as acetoin and 2,3-butanediol [30–32].

Besides glycolysis, other pathways of pyruvate formation exist, e.g., the 2-keto-3-deoxy-6-phosphogluconate (Entner-Doudoroff) pathway. Two intermediates of glycolysis, glyceraldehyde-3-phosphate and fructose-6-phosphate, are also formed in the pentose phosphate pathway. Monosaccharides other than glucose can enter glycolysis or other pathways leading to pyruvate formation. Pyruvate can also be formed from glycerol [33].

In addition to the hydrogen-yielding fermentations, other fermentations occur during acidogenesis, including lactic, propionic and ethanol fermentations. Two types of lactic acid fermentation are distinguished: homolactic and heterolactic, whose products are, respectively, lactate only or lactate, ethanol, acetate and carbon dioxide.

3.2.2. Fermentation of amino acids

Members of the *Clostridiales* (families *Clostridiaceae, Eubacteriaceae, Peptococcaceae, Peptostreptococcaceae*), *Fusobacterales, Synergistetes* (*Aminobacterium colombiense*) and *Cloacimonetes* (*Candidatus* Cloacimonas acidaminovorans) are capable of amino acid fermentation [4, 7]. Amino acids are generally degraded to acetate, propionate, hydrogen and carbon dioxide, with the formation of butyrate and ammonia. This process involves NAD(P)- or FAD-dependent deamination of amino acids to the corresponding α-keto acids with subsequent oxidative decarboxylation of α-keto acids to fatty acids with CoA and phosphate derivatives (Figure 4A). The pathways of fermentation differ depending on the amino acid type [4]. Amino acid mixtures are often degraded by coupled fermentation of pairs of amino acids through the Stickland reaction (e.g., alanine and glycine in *Clostridium sporogenes*). Oxidation of one amino acid is coupled to the reduction in another in a single cell:

$$\text{alanine} + 2H_2O \rightarrow \text{acetate}^- + CO_2 + NH_4^+ + 2H_2$$

$$\text{glycine} + 2H_2O + H^+ \rightarrow 2CO_2 + NH_4^+ + 3H_2$$

$$\text{serine} + H_2O \rightarrow \text{acetate}^- + CO_2 + NH_4^+ + H_2$$

$$\text{threonine} + H_2O \rightarrow \text{propionate}^- + CO_2 + NH_4^+ + H_2$$

$$\text{histidine} + 4H_2O + H^+ \rightarrow \text{glutamate}^- + CO_2 + 2NH_4^+ + H_2$$

$$\text{proline} + 2H_2O \rightarrow \text{glutamate}^- + H^+ + 2H_2$$

$$\text{glutamate}^- + 2H_2O + H^+ \rightarrow \text{propionate}^- + 2CO_2 + NH_4^+ + 2H_2$$

$$\text{glutamate}^- + 2H_2O \rightarrow 2\ \text{acetate}^- + CO_2 + NH_4^+ + H_2$$

$$\text{aspartate}^- + 2H_2O + H^+ \rightarrow \text{acetate}^- + 2CO_2 + NH_4^+ + 2H_2$$

Notably, glutamate may be fermented through five different pathways by various bacterial species: Pathway 1—through 3-methylaspartate; Pathway 2—through 3-methylaspartate to acetate, propionate, carbon dioxide and ammonium; Pathway 3—through 2-hydroxyglutarate to acetate, butyrate, hydrogen, carbon dioxide and ammonium; Pathway 4—through 4-aminobutyrate to acetate, butyrate and ammonium and Pathway 5—through 5-aminovalerate to acetate, propionate, valeriate and ammonium [34].

3.2.3. Transformation of lipids during acidogenesis

The products of lipid hydrolysis are glycerol and long-chain fatty acids (Figure 4A). Glycerol can enter (i) a reductive pathway and be converted to 1,3-propanediol or (ii) an oxidative pathway and be transformed to phosphoenolopyruvate in a four-step process. Phosphoenolopyruvate can then be converted to succinate and propionate and/or to pyruvate. In the latter case, further transformations of pyruvate occur through glycolytic fermentations as described for sugars [33, 35]. Significant hydrogen production was observed when *Enterobacter aerogenes* [36] and *Klebsiella pneumoniae* [37] were grown on glycerol-rich media.

Long-chain fatty acids are transformed to acetate and hydrogen through the beta-oxidation pathway, requiring syntrophic cooperation between acetogens and methanogens (described in Section 2.3). However, long-chain fatty acids have an inhibitory effect on anaerobic digestion due to their adherence to microbial cell walls, which can block the passage of nutrients through the cell membrane and/or cause flotation of the cells.

3.2.4. Cross-feeding

Symbiotic interactions between lactic acid bacteria and butyrate-producing bacteria involving clostridia, called "cross-feeding", have been detected in the gastrointestinal tract (Figure 4A). Numerous observations in different animal models have described lactate and acetate conversion to butyrate by butyrate-producing intestinal bacteria, stimulated by lactic acid bacteria (for review, see Ref. [38]). The incubation of human microflora in media containing ^{13}C-labelled lactate revealed that butyrate was the major net product of lactate conversion [39]. Other studies performed using ^{2}H-labelled acetate and ^{13}C-labelled lactate showed that acetic and lactic acids are important precursors of butyrate production in human faecal samples [40]. The metabolic pathway of lactate and acetate utilization to produce butyrate proposed for *Eubacterium hallii* and *Anaerostipes caccae* involves the conversion of lactate to pyruvate by lactate dehydrogenase [41, 42]. The next steps are typical of hydrogen-yielding *Clostridium*-type fermentation. Pyruvate is oxidized to acetyl coenzyme A (acetyl-CoA), which is subsequently routed to acetate and butyrate. Additional acetate is converted to acetyl-CoA. Hydrogen can be produced by both PFOR and NFOR complexes and hydrogenases. The conversion of lactate to butyrate is an important factor in maintaining homeostasis in gastrointestinal tracts.

It is commonly accepted that anaerobic digestion requires symbiotic interactions between specific groups of microorganisms. Some studies have indicated that lactic acid bacteria (LAB), often detected within mesophilic hydrogen-producing microbial communities, may support hydrogen production during acidogenesis. Based on our own research and the findings of other groups, we have considered the true role of LAB in bioreactors and their influence on hydrogen producers [38]. Our metagenomic survey of microbial communities in anaerobic bioreactors, performed using 454-pyrosequencing, revealed that *Clostridiaceae, Enterobacteriaceae* and heterolactic fermentation bacteria, mainly *Leuconostocaeae,* were the most dominant bacteria in hydrogen-producing consortia under optimal conditions for gas production. Furthermore, the complete consumption of lactic acid and predominance of butyric acid in the acidic effluent were observed [43].

An analysis of the hydrogen-yielding granular sludge using the FISH technique [44] revealed that *Streptococcus* spp. cells are located inside the granules, surrounded by *Clostridium* cells. This finding indicates the importance of *Streptococcus* spp. in sludge granule formation and the positive role they play within these microbial communities by stimulating hydrogen production.

Others researchers have examined the effects of lactic acid on hydrogen production by communities of fermentative bacteria. In one study, the complete consumption of lactic acid increasing hydrogen production and butyric acid formation was observed [45]. Subsequently, another group demonstrated that lactic acid increased the efficiency of hydrogen production [46]. FISH analysis revealed that *Clostridium* spp. were the dominant hydrogen producers in the examined system.

Many studies have examined the conversion of lactate and acetate to butyrate and hydrogen by clostridial species, and all point to pH as a critical factor for this process. It is noteworthy that the results of studies on gastrointestinal microflora indicate that acidity is a key regulatory factor in lactate metabolism. The pH may influence both bacterial growth and the development of specific groups of bacteria, as well as fermentation processes affecting the relative proportions of short-chain fatty acids (for review, see Ref. [38]).

A phenomenon analogous to cross-feeding observed in the gastrointestinal tract may occur in hydrogen-producing bioreactors and natural environments [38, 43] (Figure 4A).

Clostridium kluyveri ferments ethanol and acetate to butyrate and hydrogen (Figure 4A; for review, see Ref. [47]).

4. ACETOGENESIS

4.1. The essence of acetogenesis

The two final steps of anaerobic digestion, acetogenesis (Stage III) and methane formation (Stage IV), are tightly connected. Acetogenesis supplies substrates for methanogens. Three groups of substrates for methane production and three types of methanogenic pathways have been recognized: (i) splitting of acetate (aceticlastic/acetotrophic methanogenesis); (ii) reduction in CO_2 with H_2 or formate and rarely ethanol or secondary alcohols as electron donors (hydrogenotrophic methanogenesis) and (iii) reduction in methyl groups of methylated compounds such as methanol, methylated amines or methylated sulphides (hydrogen-dependent and hydrogen-independent methylotrophic methanogenesis) [2, 48–51].

Due to the limited number of substrates for methanogenesis, methanogens are strictly dependent on partner microbes with which they form syntrophic systems. Syntrophy is a special type of symbiotic cooperation between two metabolically different types of microorganisms, which depend on one another for the degradation of a certain substrate, typically through the transfer of one or more metabolic intermediate. In this case, the partner microbes oxidize non-gaseous products of acidogenesis to acetate, carbon dioxide, hydrogen and formate that are directly utilized by the methanogens, making the entire syntrophic metabolism efficient and thermodynamically favourable. This is the essence of acetogenesis. The process of hydrogen or formate transfer (interspecies hydrogen/formate transfer) between acetogenic bacteria and methanogenic *Archaea* is an excellent example of syntrophy [4, 6, 7].

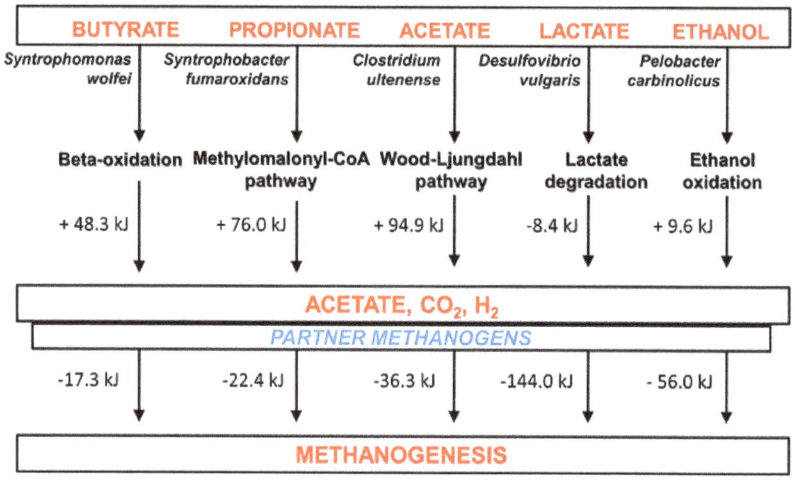

Figure 5. Syntrophic metabolism during acetogenesis—oxidation of non-gaseous products of acidogenesis based on reverse electron transfer in syntrophy with hydrogen-consuming methanogens. The ΔG°' values for acetate, butyrate, propionate and ethanol come from the study of Kamagata [53] and those for lactate oxidation, coupled or uncoupled with the methanogen partner, from the study of McInerney and Bryant [54].

Under standard conditions, the oxidation of butyrate, propionate, acetate, ethanol and other non-gaseous products of acidogenesis, coupled to hydrogen or formate production, is endergonic, as demonstrated by the positive change in Gibbs free energy. However, when the oxidation processes are coupled to methane production, the conversion is energetically feasible (exergonic) due to the very low hydrogen partial pressure ensured by hydrogen-consuming methanogens (Figure 5). Oxidation of non-gaseous products of acidogenesis during acetogenesis is based on reverse electron transfer: the energetically unfavourable movement of electrons that requires the input of energy to drive the oxidation/reduction reaction (Figure 5). This involves multiple systems, most of which are membrane-located, comprising formate dehydrogenases (FDHs), ferredoxin:NAD+ oxidoreductase, hydrogenases, *c*-type cytochromes, quinone reactive complexes, flavoprotein:quinone oxidoreductases and confurcating hydrogenases. Electron confurcation is a key process in reverse electron transfer. It involves a combined biochemical reaction using two dissimilar electron donors to generate a single product. Confurcating hydrogenases couple hydrogen production from reduced ferredoxin with hydrogen production from NADH [7]. The process responsible for energy conservation in syntrophically growing acetogens is called flavin-based electron bifurcation. Electron bifurcation is the reverse process whereby two products are formed, e.g., NADH and reduced ferredoxin from butyryl-CoA (see Section 4.2) [47, 52].

The second known mechanism of interspecies electron transfer in methanogen-yielding communities is direct transfer. This was described between *Geobacter* and *Shewanella* species as the electron donor and methanogen (the electron acceptor), respectively, in environments lacking Mn(IV) and Fe(III) compounds. In this case, pili and outer membrane *c*-type cytochromes are involved in the cell-to-cell electron transfer. Interspecies electron transfer in syntrophic methanogenic microbial communities has been recently reviewed [55].

Our current understanding of the microbial ecology and physiology associated with anaerobic digestion is restricted to culture-dependent techniques and thus is incomplete. The majority of microorganisms involved in the process of anaerobic digestion have yet to be cultivated. It is noteworthy that acetogenic bacteria are unable to grow without their syntrophic partners and cannot be cultivated as a monoculture. Thus, the mechanisms of acetogenesis are poorly characterized at the molecular level. Data derived using recently developed meta-omics approaches are likely to give a deeper insight into syntrophic metabolic pathways of anaerobic digestion.

4.2. Biochemistry of syntrophic oxidation of non-gaseous products of acidogenesis

The metabolic pathways utilized for syntrophic oxidation of common non-gaseous products of acidogenesis include beta-oxidation for butyrate, the methylmalonyl-CoA pathway for propionate, the Wood-Ljungdahl pathway for acetate, the pathway of lactate oxidation recognized in *Desulfovibrio* in the absence of sulphate and the pathway of ethanol oxidation recognized in the genera *Pelobacter* and *Desulfovibrio* in the absence of other electron acceptors.

In the first reaction of butyrate oxidation, butyrate is activated with acetyl-CoA to butyryl-CoA by butyrate-CoA transferase. This is followed by the conversion of butyryl-CoA to crotonyl-CoA catalysed by butyryl-CoA dehydrogenase, to release electrons as hydrogen or formate, which requires ATP. This process is only possible by a reverse electron transport through electron transfer flavoprotein EtfAB and a membrane-anchored DUF224 protein to the menaquinone pool in the membrane, cytochromes and other electron transfer complexes, terminating at the formate dehydrogenase and hydrogenase/formate dehydrogenase complexes. Crotonyl-CoA is transformed to 3-hydroxy-butyryl-CoA by crotonase and then to aceto-acetyl-CoA by 3-hydroxybutyryl-CoA dehydrogenase. The latter reaction also yields electrons as hydrogen or formate due to reverse electron transfer and the activity of the NADH:hydrogenase/formate dehydrogenase complex. Aceto-acetyl-CoA is split into two moieties of acetyl-CoA by acetyl-CoA acetyltransferase: one is used for butyrate activation and the second is transformed to acetate by phosphotransacetylase and acetate kinase activity, accompanied by the release of ATP [52, 56, 57].

In the first reaction of propionate oxidation, propionate is activated with acetyl-CoA to propionyl-CoA by propionate-CoA transferase. This is then transformed to (S) methylmalonyl-CoA, (M) methylmalonyl-CoA, succinyl-CoA and succinate by, respectively, methylmalonyl-CoA decarboxylase, methylmalonyl-CoA epimerase, methylmalonyl-CoA mutase and succinyl-CoA synthetase. The final step generates ATP. The next reaction is the conversion of succinate to fumarate by fumarate reductase, which releases electrons. This is the first key reaction that requires reverse electron transport. Fumarate is transformed to malate by fumarate hydratase. Malate is then converted to oxaloacetate by malate dehydrogenase in the second key reaction coupled to reverse electron transport. Pyruvate formed from oxaloacetate by pyruvate carboxylase is then transformed to acetyl-CoA by pyruvate:ferredoxin oxidoreductase. Finally, acetyl-CoA is converted to acetate in the third step generating electrons during propionate oxidation. The oxidation of oxaloacetate to fumarate involves coupling menaquinone reduction, proteins encoded by cytochrome c gene homologues, cytochrome b:quinone oxidoreductases, formate dehydrogenases, and hydrogenases including confurcating [FeFe]-hydrogenases [7, 52, 55, 57].

Acetogens that synthesize acetate from hydrogen and carbon dioxide use the reductive carbon monoxide dehydrogenase/acetyl-CoA synthase pathway (reductive CODH/ACS) known as the Wood-Ljungdahl pathway. Acetate-oxidizing syntrophs use the same pathway in reverse (oxidative CODH/ACS). Electrons as hydrogen or formate are released in the reactions catalysed by the carbon monoxide dehydrogenase/acetyl-CoA synthase, methylene-tetrahydrofolate (methylene-THF) reductase and methylene-THF dehydrogenase formate dehydrogenase. Reverse electron transfer during acetate oxidation has yet to be confirmed. It is likely that the same electron transfer mechanism is used in both pathways (reductive and oxidative) [52, 58].

It is believed that ethanol is oxidized to acetaldehyde coupled to NADH formation. Subsequently, acetaldehyde is oxidized to acetate and reduced ferredoxin is formed. Ethanol-oxidizing *Pelobacter carbinolicus* possesses genes encoding membrane-bound ion-translocating ferredoxin:NAD+ oxidoreductase and a confurcating hydrogenase that could directly catalyse the oxidation of NADH and reduced ferredoxin to form hydrogen [7, 59].

The key reaction of syntrophic lactate oxidation in *Desulfovibrio* spp. is the conversion of lactate to acetyl-CoA (via pyruvate) by lactate dehydrogenase followed by pyruvate:ferredoxin oxidoreductase, in a reaction that requires reverse electron transfer. The membrane-bound Qmo (quinone-interacting membrane-bound oxidoreductase) complex, cytochromes (involving Hmc, high-molecular-weight cytochrome c complex), menaquinone, hydrogenases (Hyn, Hyd, Hys) and formate dehydrogenases are responsible for reverse electron transport and final hydrogen and formate release. Acetyl-CoA is further processed to acetate by phosphate acetyltransferase and acetate kinase or to ethanol by alcohol dehydrogenase [60].

Worm and co-workers analysed the genomes of the butyrate- or propionate-oxidizing syntrophs *Syntrophus aciditrophicus, Syntrophus wolfei, Syntrophobacter fumaroxidans, Pelotomaculum thermopropionicum* and *Syntrophothermus lipocalidus,* and identified six syntrophy-specific functional domains [52]. These include the extra-cytoplasmic formate dehydrogenase (FDH) alpha subunit, as well as an FdhE-like protein and FDH accessory protein. The functions of the latter two proteins are tightly connected with FDH. This finding points to the important role of formate in interspecies electron transfer. The fourth domain was detected in CapA, a protein involved in capsule or biofilm formation that may facilitate syntrophic growth. The fifth domain is characteristic of FtsW, RodA and SpoVE proteins involved in membrane integration, cell division, sporulation and shape determination. The final domain was detected in a conserved site of ribonuclease P involved in tRNA maturation.

In the same study, functional domains involved in electron transfer were also identified [52]. These were found in the following proteins: cytoplasmic FDH, extra-cytoplasmic FDH, formate transporter, Fe-Fe hydrogenase, NiFe hydrogenase, Rnf complex, Ech complex, Etf alpha, Etf beta, Bcd, cytochromes c, cIII, b561 and b5 and the DUF224 protein complex.

Notably, the genomes of sulphate-reducing non-syntrophs were found to lack the syntrophy-specific domains. However, these domains are present in other sulphate reducers that have never been tested for syntrophy: *Desulfobacterium autotrophicum*, *Desulfomonile tiedjei* and *Desulfosporosinus meridiei* [52].

4.3. A model of methane-yielding granules

Based on the methane-yielding granule model proposed over 25 years ago, acetotrophic methanogens form the center core of the granule. They are surrounded by acetogenic bacteria and hydrogenotrophic methanogens. The outside layer consists of microorganisms responsible for acidogenesis. The minimum physical distances required for energetically favorable hydrogen transfer between acetogenic bacteria and hydrogenotrophic methanogens have been determined based on research conducted on the Pelotomaculum thermopropionicum, a syntroph that oxidizes propionate, propanol, and ethanol, and the Methanothermobacter thermoautotrophicus, a hydrogenotrophic methanogen. The optimal distance required for effective transfer of hydrogen between different species varied depending on the substrate. It was determined to be approximately 2, 16, and 32 μm for the oxidation of propionate, ethanol, and propanol, respectively. It is important to note that for the less energetically favorable syntrophic process, a closer relationship between syntrophic partners is necessary. This means that the partners need to be physically closer to each other. The specific energy values for propionate, ethanol, and propanol oxidation are +76.0 kJ, +9.6 kJ, and +3.0 kJ, respectively. For more information, refer to references [53, 55].

4.4. Syntrophic relationships between acetogenic bacteria and methanogens during anaerobic digestion

The most well-studied examples of syntrophic metabolism in methanogenic communities are described below. The *Syntrophomonadaceae*, a family from the order *Clostridiales*, are highly specialized syntrophic microbes found in methanogenic consortia that can oxidize butyric, propionic and long-chain fatty acids to acetic and formic acids with the production of hydrogen and carbon dioxide—the basic substrates for their partner methanogens [61–63].

The most frequently recognized butyrate oxidizers are representatives of the *Syntrophomonadaceae*—*Syntrophomonas wolfei, S. bryantii, S. curvata, S. sapovorans, S. palmitatica, S. cellicola, S. saponavida, S. erecta, S. zehnderi; Syntrophothermus lipocalidus, Thermosyntropha lipolytica* and representatives of the *Syntrophobacterales (Deltaproteobacteria)* and *Syntrophus acidotrophicus.* Proteins expressed specifically during syntrophic growth of *S. wolfei* with butyrate have been investigated by proteomic analysis [4, 6, 7, 56].

The propionate-oxidizing bacteria are members of the *Syntrophomonadaceae*—genus *Syntrophobacter* (*S. fumaroxidans, S. wolinii*) and *Smithella propionica* and of the *Peptococcaceae*—*Desulfotomaculum* (*D. thermocisternum, D. thermobenzoicum* subs. *thermosyntrophicum*) and *Pelotomaculum* (*P. thermopropionicum*) genera [4, 6, 7].

Li and co-workers [16] developed specific PCR assays for propionate-CoA transferase genes (*pct*) to identify and analyse propionate oxidizers in the methane-yielding microbial communities in anaerobic digesters treating various food industry wastes. In addition to *Syntrophobacter fumaroxidans*, six other distinct clusters of putative *pct* genes were detected. The diversity and abundance of the *pct* genes were determined by the nature of the feedstocks of the anaerobic digesters. There was little difference between the *pct* gene profiles of the granular sludge and the liquid phase in the same digester. These authors postulated that the feedstock is a critical factor influencing propionate metabolism in anaerobic digesters.

It is noteworthy that such PCR assays may also be used to examine anaerobic decomposition of organic matter in natural environments.

Acetate is the major intermediate product during anaerobic digestion of organic matter to methane and carbon dioxide. It can be directly transformed to methane and carbon dioxide by acetoclastic methanogens (Section 2.4) or syntrophically oxidized to hydrogen and carbon dioxide. The latter reaction requires the participation of two microbial partners: an acetate-oxidizing bacterium and a hydrogenotrophic methanogen. Recognized acetate-oxidizing bacteria include members of the *Clostridia—Thermoacetogenium phaeum, Clostridium ultunense, Clostridium sporomusa, Syntrophaceticus schinkii, Tepidanaerobacter syntrophicus, Tepidanaerobacter acetatoxydans, Candidatus* Syntrophonatronum acetioxidans and *Moorella* sp., as well as *Deltaproteobacteria—Geobacter* spp. and *Thermotogae—Thermotogae lettingae* [4, 52]. Analyses using culture-independent techniques have revealed many other uncultured acetate oxidizers.

Ito and co-workers [64] used MAR-FISH combined with phylogenetic analysis of ^{13}C-labelled bacterial 16S rRNA and tracing of [2-^{14}C]-labelled acetate degradation to study metabolic pathways of acetate transformation in methanogenic sludge from an anaerobic digester fed with mineral medium containing powdered whole milk. These analyses identified *Synergistes* Group 4, belonging to the phylum *Synergistetes*, as the major acetate-utilizing group of bacteria. Moreover, acetate oxidizers were shown to win the competition with acetoclastic methanogens from the genus *Methanosaeta* for the utilization of acetate. At high acetate concentrations, the *Synergistetes* showed a lower affinity for acetate and higher utilization rate in comparison with *Methanosaeta.*

Lee and co-workers [8] presented evidence that in anaerobic digesters fed with a medium containing acetate as the sole carbon source, *Spirochaetes* syntrophically oxidize this substrate with hydrogenotrophic *Methanomicrobiales.* Quantitative PCR (qPCR) and reverse transcription quantitative PCR (RT-qPCR) targeting the 16S rRNA genes of cluster II *Spirochaetes* and methanogens (*Methanosaetaceae, Methanosarcinaceae, Methanomicrobiales* and *Methanobacteriales*) revealed that an increase in the former was correlated with higher numbers of *Methanomicrobiales.* High concentrations of hydrogen inhibited the activity of the *Spirochaetes.*

Synergistetes and *Spirochaetes* are frequently found in anaerobic digesters and natural environments, but little is known about their role in anaerobic digestion besides the fact that the latter are thought to be capable of glucose fermentation.

Interestingly, current knowledge concerning the oxidation of lactate in methanogenic consortia is limited to members of the *Desulfovibrio* genus. These species are capable of syntrophic growth on lactate and ethanol with hydrogenotrophic methane-producing partners in the absence of sulphate. As methanogenesis is thermodynamically unfavourable, such syntrophic metabolism is possible only when other electron acceptors such as sulphate are absent. Otherwise, sulphate reduction occurs. Lactate can also act as a substrate for the non-methanogen *Archaeoglobus*, a known sulphate reducer capable of oxidizing lactate to carbon dioxide [7, 60].

Recent studies on anaerobic digestion of molasses wastewater in an upflow anaerobic sludge blanket (UASB) reactor revealed the significant contribution of *Lactococcus* and *Methanosaeta* and their close interaction in methane production [65]. These authors analysed cDNA obtained by reverse transcription of RNA isolated from methane-yielding sludge samples. They proposed lactate as the major fermentation product, which is subsequently oxidized to acetate, a substrate for *Methanosaeta.*

Chojnacka and co-workers [15] hypothesized that a symbiotic interaction between lactic acid bacteria and clostridia, known as lactate cross-feeding (described in Section 2.2.3.), may also occur in methanogenic communities. Butyrate and hydrogen are the products of lactate transformation. The hydrogen and the products of further syntrophic butyrate oxidation constitute substrates for methanogenesis.

Ethanol is also effectively utilized by the methane-yielding microbial communities [15, 66]. Apart from *Desulfovibrio* species, other well-recognized syntrophic ethanol oxidizers are representatives of the *Deltaproteobacteria, Geobacter* and *Pelobacter*—well-known Fe(III) reducers. To be an energetically effective reaction, the oxidation of ethanol to carbon dioxide and hydrogen also requires strict cooperation with hydrogenotrophic methanogens [4].

Members of the orders *Clostridiales* (*Clostridiaceae, Eubacteriaceae, Peptococcaceae, Peptostreptococcaceae* families), *Fusobacteriales* and phylum *Synergistetes* (*Aminobacterium colombiense*) are capable of amino acid fermentation [4, 7]. However, the transformation of amino acids to produce methane is only energetically possible in syntrophic association with hydrogenotrophic methanogens that scavenge hydrogen.

The *Cloacimonetes*, including Waste Water of Evry 1 (WWE1), are a sub-dominant group of bacteria found in mesophilic anaerobic digesters and gut microflora. So far, all attempts to cultivate representatives of the *Cloacimonetes* have failed, probably due to their need for obligatory symbiotic relationships with other microorganisms. However, the genome of a representative bacterium *Candidatus* Cloacimonas acidaminovorans has been reconstructed using metagenomic sequence data and genomic assembly procedures [93]. The candidate division WWE1 bacteria are regarded as syntrophs capable of amino acid fermentation, propionate and butyrate oxidation as well as cellulose degradation [7, 66].

Actinobacteria, Chloroflexi and *Plantomycetes* are often among the bacterial phyla detected in methane-producing anaerobic digesters and wastewater treatment plants. Their functional activities in methanogenic communities have not been well characterized. *Actinobacteria* and *Chloroflexi* are thought to hydrolyse and ferment carbohydrates. The contribution of *Chloroflexi* and *Plantomycetes* to butyrate oxidation was identified in experiments performed with [^{13}C]-labelled butyrate [62].

We have examined the microbial community processing an acidic effluent from molasses fermentation to methane in a UASB bioreactor [15]. Total DNA isolated from the methanogenic community formed in the reactor was sequenced by 454-pyrosequencing. The results revealed that the biodiversity of methanogenic sludge is significantly higher than that of the hydrogen-producing community. The ratio of *Bacteria* to *Archaea* in the methanogenic community was 4:1. The domain *Bacteria* was dominated by *Firmicutes* (~24%), *Bacteroidetes* (~21%), *Proteobacteria* (~9%), *Cloacimonetes* (~7.5%) and *Spirochaetes* (~7%). The *Firmicutes* were dominated by *Clostridia*, which constituted approximately 14% of all bacterial reads. The *Proteobacteria* were mostly represented by the delta and gamma subdivisions (~9 and ~1.5%, respectively), whereas the alpha and beta subdivisions were poorly represented (~0.5%). Other minor groups were *Actinobacteria* (~2%), *Chlamydiae* (~1%), *Synergistetes* (~1%) and *Chloroflexi* (~0.5%). A small number of reads were sequences from *Armatimonadetes, Negativicutes* and *Plantomycetes*. The low level of unfermented sugars and the abundance of *Clostridia* and *Bacteroidetes* suggested that these bacteria play a previously unrecognized role in acetogenesis, involving syntrophic oxidation of non-gaseous products of hydrogen-yielding fermentation. Moreover, an analysis of short-chain fatty acids revealed that butyric and lactic acids were the main substrates utilized in the methanogenic step.

Some of the aforementioned bacterial phyla are capable of oxidizing other compounds, including 1-propanol, benzoate, hydroxybenzoate, phenol and phthalates.

5. METHANE FORMATION

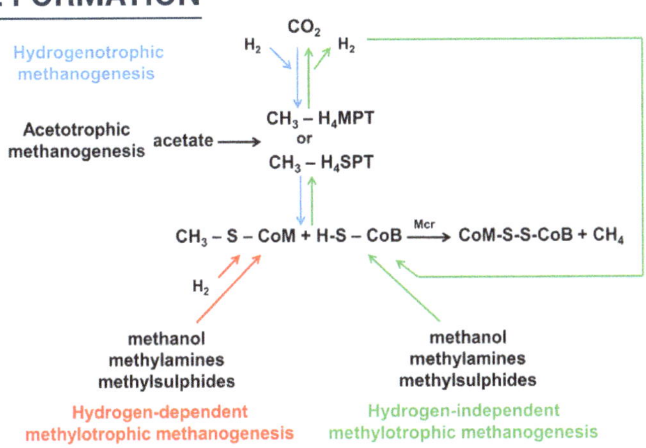

Figure 6. Pathways of methanogenesis. H-S-CoM, coenzyme M; H-S-CoB, coenzyme B; H_4MPT, tetrahydromethanopterin; H_4SPT, tetrahydrosarcinapterin.

Methane formation, Stage IV of anaerobic digestion, is a complex process requiring specific enzymes and cofactors not found in other microorganisms. The course of the reaction depends on the substrates utilized by the methanogens. Three groups of substrates are recognized: (i) acetate, (ii) CO_2 and H_2 or formate, and rarely ethanol or secondary alcohols and (iii) methylated compounds including methanol, methylated amines and methylated sulphides. These substrates are, respectively, processed through three recognized pathways of methanogenesis: aceticlastic/acetotrophic, hydrogenotrophic and methylotrophic (hydrogen-dependent and hydrogen-independent) (Figure 6) [67]. Irrespective of the substrate, the final step in each methanogenic pathway is the reaction of methyl-coenzyme M (CH_3-S-CoM) and coenzyme B to produce heterodisulphide CoM-S-S-CoB and methane:

$$CH_3-S-CoM+Co\,B\rightarrow CoM-S-S-CoB+CH_4$$

This reaction is catalysed by methylcoenzyme M reductase (Mcr), the key enzymatic complex of the methanogenic process. It possesses a unique prosthetic group, coenzyme F_{430}, containing nickel. CoM-S-S-CoB acts as the final electron acceptor during anaerobic respiration and is the key compound for energy gain by methanogens. Methane is a by-product of methanogen metabolism. The pathways of methanogenesis are in fact pathways of CoM-S-S-CoB synthesis.

Splitting of acetate (acetotrophic methanogenic pathway) involves the formation of acetyl-CoA, the transfer of methyl groups to tetrahydrosarcinopterin (H_4SPT) and the formation of methyl tetrahydrosarcinopterin CH_3-H_4STP. CH_3-S-CoM is formed in the reaction of CoM with CH_3-H_4STP. The electrons required to reduce CH_3-S-CoM to methane come from oxidation of the carboxyl group of acetate.

The formation of CH_3-S-CoM by the reduction in CO_2 with H_2, formate or alcohols constitutes the hydrogenotrophic methanogenic pathway. This pathway is comprised of the following steps: (i) the formation of formylmethanofuran (formyl-MFR) from methanofuran (MFR) and CO_2, (ii) the reaction of formyl-MFR and tetrahydromethanopterin (H_4MPT) to produce formyl tetrahydromethanopterin (formyl H_4MPT), (iii) the formation of methylene H_4MPT that in reaction with F_{420}, a derivative of 5′ dezaflavin, produces methyl H_4MPT and (iv) the reaction of methyl H_4MPT with CoM to generate CH_3-S-CoM. The electrons required to reduce CH_3-S-CoM to methane come from hydrogen, formate or alcohols.

In the methylotrophic pathway of methanogenesis, CH_3-S-CoM is formed by the direct transfer of methyl groups from methylated compounds to CoM. One methyl group bound to CoM is oxidized to CO_2 and hydrogen (in the form of $F_{420}H_2$ and reduced ferredoxin) to reverse the hydrogenotrophic pathway. The reducing equivalents are used to reduce CH_3-S-CoM to methane.

In the recently discovered process of hydrogen-dependent methylotrophic methanogenesis, CH_3-S-CoM is also formed through the direct transfer of methyl groups from methylated compounds to CoM. However, the electrons required to reduce CH_3-S-CoM to methane come from externally supplied hydrogen. Genomic analysis revealed that organisms generating methane by this process lack genes encoding the enzymes of hydrogenotrophic methanogenesis [50, 51].

The known cultured methanogens are strict anaerobes and comprise seven orders in the class *Euryarchaeota* of the domain *Archaea*: *Methanobacteriales*, *Methanococcales*, *Methanomicrobiales*, *Methanosarcinales*, *Methanopyrales* [2, 48], *Methanocellales* [49] and *Methanomassiliicoccales* [50, 51]. Surprisingly, only two known genera, *Methanosarcina* and *Methanosaeta*, members of the order *Methanosarcinales*, are capable of methane production from acetate. Moreover, only *Methanosaeta* is strictly acetoclastic, whereas *Methanosarcina* is able to produce methane from acetate, CO_2 and H_2 and from methylated compounds. The recognized methylotrophic methanogens belong to the order *Methanosarcinales*. All other known methanogens produce methane by the reduction in CO_2 [1, 2, 48, 49, 67]. The known members of the *Methanomassiliicoccales* order are H_2-dependent methylotrophs. They use methylated compounds (mono-, di-, tri-methylamine and dimethylsulphide) as substrates for methanogenesis, and the methyl group is reduced by hydrogen [50, 51].

It has been estimated that 70% of methane is produced from acetate. When biomass is transformed into methane under mesophilic conditions in anaerobic digesters or natural environments, it is first fermented to acetate, carbon dioxide and hydrogen and formate, as well as short-chain fatty acids during acidogenesis. The theoretical maximum hydrogen yield during dark fermentation occurs with the conversion of one-third of the substrate to hydrogen and carbon dioxide and two-thirds of the substrate to acetate. Therefore, it follows that two-thirds of methane originates from acetate and one-third is from hydrogen, formate and carbon dioxide [2].

Culture-independent analyses of methanogenic communities (mainly from anaerobic digesters) based on cloning and sequencing of 16S rRNA and *mcrA* gene fragments or high-throughput DNA sequencing technologies have revealed that the contribution of methanogens performing the aceticlastic or hydrogenotrophic pathways depends on the substrate and the process conditions. *Methanomicrobiales*, represented by *M. marisnigri*, often predominate in methanogenic communities in biogas plants, indicating that methane is produced through the hydrogenotrophic pathway. This finding does not support the thesis that methane is produced primarily from acetate through the acetoclastic pathway [68]. We found that the hydrogenotrophic pathway of methane production was predominant in the bioreactor processing an acidic effluent from molasses fermentation to methane, and the order *Methanomicrobiales* dominated the archaeal community, constituting about 59%. The most abundant genus within this order was *Methanoculleus* represented by *M. marsigni* and *M. bourgensis*, while the second and the third most abundant genera were *Methanocorpusculum* and *Methanofollis*. Other representatives of this order were members of the genus *Methanoplanus* including the species *Methanoregula formicica, Methanosphaerula palustris* and *Methanospirillum hungatei*. Among the identified hydrogenotrophic methanogens were representatives of the *Methanobacteriales* including the genera *Methanobacterium, Methanococcales* and *Methanocellales*. *Archaea* conducting the aceticlastic pathway of methane production included the *Methanosarcinales* (~3.5%), represented by the genera *Methanosaeta* and *Methanosarcina*. Metagenomic analysis revealed a relatively large contribution of sequences assigned to the genus *Methanomassiliicoccus*, including *Methanomassiliicoccus luminyensis, Candidatus* Methanomassiliicoccus intestinalis and *Candidatus* Methanomethylophilus alvus.

It should be noted that the acetoclastic pathway provides only a small amount of energy available for growth:

$$CH_3COO^- + H^+ \rightarrow CO_2 + CH_4 \left(\Delta G^{0\prime} = -36\ kJ/mol\right)$$

In comparison, the hydrogenotrophic pathway produces fourfold more energy:

$$4H_2 + CO_2 \rightarrow CH_4 + H_2O\left(\Delta G^{0\prime} = -131\ kJ/mol\right)$$

$$4HCOO^- + 4H^+ - CH_4 + 3CO_2 + H_2O\ (\Delta G^{0\prime} = -144.5\ kJ/mol)\ [67]$$

Thus, the hydrogenotrophic pathway is much more energetically effective, and this may be one of the reasons for the dominance of the *Methanomicrobiales* order in the analysed communities. Moreover, as it was mentioned previously, acetate oxidizers such as *Synergistetes* successfully compete with acetoclastic methanogens belonging to the *Methanosaeta* for acetate [64].

An analysis of the substrate preferences of the recognized methanogenic *Archaea* revealed that hydrogen and carbon dioxide, methyl compounds and acetate are utilized by 74.5, 33 and 8.5% of the methanogens, respectively [69].

In all methanogenic microbial communities examined by high-throughput DNA sequencing, the contribution of unidentified sequences is usually high. As phylogenetic analyses are dependent on comparison with DNA sequences present in databases and the majority of the recognized genera of methanogens produce methane through the hydrogenotrophic pathway, it is possible that acetoclastic methanogens are hidden among the unidentified sequences. Therefore, the apparent dominance of hydrogenotrophic methanogens such as *Methanomicrobiales* may only be due to our limited knowledge of methanogenic *Archaea*.

Recently, Dziewit and co-workers [18] described four novel molecular markers—other than 16S rRNA and *mcrA*—for the metagenomic analysis of methanogenic communities, with a particular focus on methylotrophic methanogens. These are the *mcrB, mcrG, mtaB* and *mtbA* genes encoding beta and gamma subunits of the methyl-CoM reductase, methanol-5-hydroxybenzimidazolylcobamide Co-methyltransferase and methylated [methylamine-specific corrinoid protein]:coenzyme M methyltransferase, respectively.

It is commonly recognized that methanogenic granular sludge is rich in minerals, mainly ferric sulphide and Ca-, Mg-, Na-, K- or Al-containing compounds. They constitute between 10 and 90% of the dry mass, depending on the composition of the wastes and nature of the methanogenic process [70]. The inorganic components of the extracellular matrix of methanogenic granules may inhibit some metabolic pathways and thus determine the processes leading to methane production by the microbial community. Both Al and K are undesirable elements in the methanogenic sludge due to their competition with other essential metals, inhibiting microbial growth and consequently their adverse effect on the methanogenic process. In contrast, Ca and Mg have a positive effect due to their promotion of the granulation process. Sodium plays a role in the formation of ATP and oxidation of NADH and then is essential for the growth of methanogens. However, high concentrations of Ca^2+, Mg^2+ and Na+ ions cause inhibitory effects on methanogen activity. The optimum concentration of Ca^2+ and Na+ ions for methane synthesis from acetate was found to be 200 and 230 mg/L, respectively, whereas a concentration of 8000 mg/L of either ion inhibited the process [71]. Interestingly, the combination of various elements can mitigate the toxicity of others, e.g., magnesium, sodium and ammonium counteract potassium toxicity. It is noteworthy that the acetoclastic pathway of methanogenesis and the oxidation of propionate are particularly sensitive to raised levels of certain minerals [71]. Moreover, it has been observed that inhibition of the acetotrophic pathway of methane formation is usually accompanied by inhibition of propionate oxidation [61].

6. THE PROCESS OF GENERATING HYDROGEN AND METHANE THROUGH A TWO-STAGE ANAEROBIC DIGESTION.

There is currently great interest in the development of new technologies for the production of energy from renewable sources, of which fermentation processes generating methane and hydrogen show great promise. Hydrogen-yielding fermentation is considered to be one of the most attractive alternative biological methods of hydrogen production. However, there are two major drawbacks: low productivity of the process and the formation of large amounts of environmentally unfriendly non-gaseous fermentation products [29, 72]. The theoretical maximum hydrogen yield during *Clostridium*-type fermentation is four moles of hydrogen per mole of glucose, when all of the substrates are converted to acetic acid according to the following equation:

$$C_6H_{12}O_6 + 2\,H_2O \rightarrow 4\,H_2 + 2\,CO_2 + 2\,CH_3COOH$$

This gives the highest possible yield of hydrogen during dark fermentation. The complete oxidation of glucose provides 12 moles of hydrogen per mole of glucose:

$$C_6H_{12}O_6 + 6\,H_2O \rightarrow 12\,H_2 + 6\,CO_2$$

Theoretically, only one-third of the biomass can be converted to hydrogen by the process of hydrogen-yielding fermentation. In practice, this value is lower due to the formation of non-gaseous products such as organic acids and alcohols. For example, when the glucose is converted to butyrate, the hydrogen yield drops to two moles. It is estimated that the efficiency of hydrogen production must reach 60–80% to be economically attractive [73, 74]. This level of efficiency may be attained by using two-stage systems to achieve the transformation of substrates into hydrogen and methane. In such systems, the hydrogenic (hydrolysis and acidogenesis) and methanogenic (acetogenesis and methanogenesis) steps are performed separately under controlled conditions to favour biohydrogen and biomethane production, respectively. In the first stage, hydrogen-rich fermentation gas is produced, while in the second, the non-gaseous products of hydrogen fermentation act as substrates for methanogenic consortia. These two processes are carried out in separate bioreactors that differ in design and have different pH conditions and hydraulic retention times.

There is an increasing amount of literature documenting the utilization of two-stage systems for the synthesis of hydrogen and methane. These systems have demonstrated potential in laboratory and pilot experiments employing different materials such as organic wastes, plant biomass, food industry by-products, and pure hydrocarbons [66, 75–90]. These systems have been able to obtain energy recovery increases of up to 20-30% compared to one-stage biogas-producing bioreactors [76, 78, 85, 90]. The economic viability of hydrogen production by fermentation could be enhanced by efficiently producing biomethane from non-gaseous fermentation products. According to estimates, biohydrogen production may reach an industrial size by 2040 [91]. The concept of utilizing two-phase anaerobic digestion as an efficient means of breaking down biomass into methane and carbon dioxide is not a recent development [92]. The new feature is the simultaneous synthesis of hydrogen and methane through co-production. Previous research on the anaerobic digestion of biomass to produce hydrogen and methane has mostly examined the overall process's performance and efficiency, neglecting a thorough investigation of the microbial communities present in the bioreactors responsible for the two phases. Understanding the composition and variety of microbial communities that work together in converting organic matter into hydrogen and methane gases is important for improving the efficiency of two-stage systems for co-producing these gases. Our laboratory has been conducting research on two-stage anaerobic digestion for numerous years. We have created and explained a small-scale anaerobic digestion system in the laboratory. This system has two stages: in the first stage, it creates hydrogen, and in the second stage, it produces methane. The system uses sucrose-rich by-products from the sugar beet refining sector as the main source of energy. The process takes place under moderate temperature settings. Hydrogen is first produced in a three-litre packed bed reactor (PBR) by acidogenesis processes. This is done by a microbial population that ferments molasses and yields hydrogen. Afterwards, the non-gaseous organic substances resulting from the initial phase are directed to a 3.5-litre UASB reactor, where a colony of microorganisms that create methane (biogas) is present. The researchers conducted a thorough analysis of the molecular composition of a two-stage anaerobic digestion system that produces hydrogen and methane from sugar beet molasses. They used improved methods for extracting DNA and high-throughput pyrosequencing (454 Roche) to obtain precise results [15, 43]. The two-stage system for hydrogen and methane synthesis mentioned earlier has been successfully increased in size by a factor of 10 and is currently undergoing testing in a sugar factory in Poland.

REFERENCES

1. Thauer RK, Kaster A-K, Seedorf H, Buckel W, Hedderich R. Methanogenic Archaea: ecologically relevant differences in energy conservation. Nat Rev Microbiol. 2008;6:579–591. DOI:10.1038/nrmicro1931
2. Liu Y, Whitman WB. Metabolic, phylogenetic, and ecological diversity of the methanogenic Archaea. Ann N Y Acad Sci. 2008;1125:171–189. DOI:10.1196/annals.1419.019
3. Angenent LT, Karim K, Al-Dahhan MH, Wrenn BA, Domiguez-Espinosa R. Production of bioenergy and biochemicals from industrial and agricultural wastewater. Trends Biotechnol. 2004;22:477–485. DOI:10.1016/j.tibtech.2004.07.001
4. Schink B, Stams AJM. Syntrophism among prokaryotes. In: Dworkin M, editor. The Prokaryotes. 3rd ed. New York: Springer-Verlag; 2006. p. 309–335. DOI:10.1007/978-3-642-30123-0_59
5. Demirel B, Scherer P. The roles of acetotrophic and hydrogenotrophic methanogens during anaerobic conversion of biomass to methane: a review. Rev Environ Sci Biotechnol. 2008;7:173–190. DOI:10.1007/s11157-008-9131-1
6. Stams AJM, Plugge CM. Electron transfer in syntrophic communities of anaerobic bacteria and Archaea. Nat Rev Microbiol. 2009;7:568–577. DOI:10.1038/nrmicro2166

7. Sieber JR, McInerney MJ, Gunsalus RP. Genomic insight into syntrophy: the paradigm for anaerobic metabolic cooperation. Annu Rev Microbiol. 2012;66:429–452. DOI: 10.1146/annurev-micro-090110-102844

8. Lee S-H, Park J-H, Kim S-H, Yu BJ, Yoon J-J, Park H-D. Evidence of syntrophic acetate oxidation by Spirochaetes during anaerobic methane production. Bioresour Technol. 2015;190:543–549 DOI:10.1016/j.biortech.2015.02.066

9. Amann RI, Ludwig W, Schleifer KH. Phylogenetic identification and in situ detection of individual microbial cells without cultivation. Microbiol Rev. 1995;59:143–169. PMCID: PMC239358

10. Vanwonterghem I, Jensen PD, Ho DP, Batstone DJ, Tyson GW. Linking microbial community structure, interactions and function in anaerobic digesters using new molecular techniques. Curr Opin Biotechnol. 2014;27:55–64. DOI:10.1016/j.copbio.2013.11.004

11. Koch C, Müller S, Harms H, Harnisch F. Microbiomes in bioenergy production: from analysis to management. Curr Opin Biotechnol. 2014;27:65–72. DOI:10.1016/j.copbio.2013.11.006

12. Abram F. Systems-based approaches to unravel multi-species mirobial community functioning. Comp Struct Biotechnol J. 2015;13:24–32. DOI:10.1016/j.csbj.2014.11.009

13. Ravin NV, Mardanov AV, Skryabin KG. Metagenomics as a tool for the investigation of uncultured microorganisms. Russ J Genet. 2015;51:431–439. DOI:10.1134/S1022795415050063

14. Heyer R, Kohrs F, Reichl U, Benndorf D. Metaproteomics of complex microbial communities in biogas plants. Microbial Biotechnology. 2015;8:749–763. DOI:10.1111/1751-7915.12276

15. Chojnacka A, Szczęsny P, Błaszczyk MK, Zielenkiewicz U, Detman A, Salamon A, Sikora A. Noteworthy facts about a methane-producing microbial community processing acidic effluent from sugar beet molasses fermentation. PLoS One. 2015;10(5):e0128008. DOI:10.1371/journal.pone.0128008

16. Li Y-F, Wei S, Yu Z. Feedstocks affect the diversity and distribution of propionate CoA-transferase genes (pct) in anaerobic digesters. Microb Ecol. 2013;66:351–362. DOI:10.1007/s00248-013-0234-z

17. Pellietier E, Kreimeyer A, Bocs S, Rouy Z, Gyapay G, et al. "Candidatus Cloacamonas Acidaminovorans": genome sequence reconstruction provides a first glimpse of a new bacterial division. J Bacteriol. 2008;190:2572–2579. DOI:10.1128/JB.01248-07

18. Dziewit L, Czarnecki J, Prochwicz E, Wibberg D, Schlüter A, Pühler A, Bartosik D. Genome-guided insight into the methylotrophy of Paracoccus aminophilus JCM 7686. Front Microbiol. 2015;6:852. DOI:10.3389/fmicb.2015.00852

19. The Enzyme List Class 3—Hydrolases. Nomenclature Committee of the International Union of Biochemistry and Molecular Biology (NC-IUBMB) Generated from the ExplorEnz database. September 2010. http://www.enzyme-database.org.

20. Wirth R, Kovacs E, Maroti G, Bagi Z, Rakhely G, Kovacs KL. Characterization of a biogas-producing microbial community by short-read next generation DNA sequencing. Biotechnol Biofuels. 2012;5:41. DOI:10.1186/1754-6834-5-41

21. Li A, Chu Y, Wang X, Ren L, Yu J, et al. A pyrosequencing-based metagenomic study of methane-producing microbial community in solid-state biogas reactor. Biotechnol Biofuels. 2013;6:3. DOI:10.1186/1754-6834-6-3

22. Hanreich A, Schimpf U, Zakrzewski M, Schlüter A, Benndorf D, et al. Metagenome and metaproteome analyses of microbial communities in mesophilic biogas-producing anaerobic batch fermentations indicate concerted plant carbohydrate degradation. Syst Appl Microbiol. 2013;36:330–338. DOI:10.1016/j.syapm.2013.03.006

23. Schneider T, Keiblinger KM, Schmid E, Sterflinger-Gleixner K, Ellersdorfer G, et al. Who is who in litter decomposition? Metaproteomics reveals major microbial players and their biogeochemical functions. ISME J. 2012;6:1749–1762. DOI: 10.1038/ismej.2012.11

24. Meyer J. [FeFe] hydrogenases and their evolution: a genomic perspective. Cell Mol Life Sci. 2007;64:1063–1084. DOI: 10.1007/s00018-007-6477-4

25. Forzi L, Sawers RG. Maturation of [NiFe]-hydrogenases in Escherichia coli. Biometals. 2007;20:565–578. DOI:10.1007/s10534-006-9048-5

26. Girbal L, Croux C, Vasconcelos I, Soucaille P. Regulation of metabolic shifts in Clostridium acetobutylicum ATCC 824. FEMS Microbial Rev. 1995;17:287–297. DOI:10.1016/0168-6445(95)00017-7

27. Hallenbeck PC. Fundamentals of the fermentative production of hydrogen. Water Sci Technol. 2005;52:21–29. PMID: 16180405

28. Kraemer JT, Bagley DM. Improving the yield from fermentative hydrogen production. Biotechnol Lett. 2007;29:685–695. DOI:10.1007/s10529-006-9299-9

29. Lee HS, Vermaas WFJ, Rittmann BE. Biological hydrogen production: prospects and challenges. Trends Biotechnol. 2010;28:262–271. DOI:10.1016/j.tibtech.2010.01.007

30. Nakashimada Y, Rachman MA, Kakizono T, Nishio N. Hydrogen production of Enterobacter aerogenes altered by extracellular and intracellular redox state. Int J Hydrogen Energy. 2002;27:1399–1405. DOI:10.1016/S0360-3199(02)00128-3

31. Sawers RG. Formate and its role in hydrogen production in Escherichia coli. Biochem Soc Trans. 2005;33:42–46. DOI:10.1042/BST0330042

32. Maeda T, Sanchez-Torres V, Wood TK. Enhanced hydrogen production from glucose by metabolically engineered Escherichia coli. Appl Microbiol Biotechnol. 2007;77:879–90. DOI:10.1007/s00253-007-1217-0

33. Viana MB, Freitas AV, Leitão RC, Pinto GAS, Santaella ST. Anaerobic digestion of crude glycerol: a review. Environ Technol Rev. 2012;1:81–92. DOI:10.1080/09593330.2012.692723

34. Buckel W. Unusual enzymes involved in five pathways of glutamate fermentation. Appl Microbiol Biotechnol. 2001;57:263–273. DOI:10.1007/s002530100773

35. Weng C, Jeris J. Biochemical mechanisms in the methane fermentation of glutamic and oleic acids. Water Res. 1976;10:9–18. DOI:10.1016/0043-1354(76)90151-2

36. Ito T, Nakashimada Y, Senba K, Matusi T, Nishio N. Hydrogen and ethanol production from glycerol-containing wastes discharged after biodiesel manufacturing process. J Biosci Bioeng. 2005;100:260–265. DOI:10.1263/jbb.100.260

37. Liu F, Fang B. Optimization of bio-hydrogen production from biodiesel wastes by Klebsiella pneumoniae. Biotechnol J. 2007;2:374–380. DOI:10.1002/biot.200600102

38. Sikora A, Błaszczyk M, Jurkowski M, Zielenkiewicz U. Lactic acid bacteria in hydrogen producing consortia: on purpose or by coincidence. In: Kongo M, editor. Lactic Acid Bacteria – R & D for Food, Health and Livestock Purposes. Rijeka: InTech; 2013. p. 487–514. DOI:5772/50364

39. Bourriaud C, Robins RJ, Martin L, Kozlowski, F, Tenailleau E, et al. Lactate is mainly fermented to butyrate by human intestinal microfloras but inter-individual variation is evident. J Appl Microbiol. 2005;99:201–212. DOI:10.1111/j.1365-2672.2005.02605.x

40. Morrison DJ, Mackay WG, Edwards CA, Preston T, Dodson B, Weaver LT. Butyrate production from oligofructose fermentation by the human faecal flora: what is the contribution of extracellular acetate and lactate? British J Nutr. 2006;96:570–577. DOI:10.1079/BJN20061853

41. Duncan SH, Louis P, Flint HJ. Lactate-utilizing bacteria, isolated from human feces, that produce butyrate as a major fermentation product. Appl Environ Microbiol. 2004;70:5185–5190. DOI:10.1128/AEM.70.10.5810-5817.2004

42. Munoz-Tamayo R, Laroche B, Walter E, Dore J, Duncan SH, Flint HJ, Leclerc M. Kinetic modelling of lactate utilization and butyrate production by key human colonic bacterial species. FEMS Microbiol Ecol. 2011;76:615–624. DOI:10.1111/j.1574-6941.2011.01085.x

43. Chojnacka A, Błaszczyk MK, Szczęsny P, Nowak K, Sumińska M, Tomczyk-Żak K, Zielenkiewicz U, Sikora A. Comparative analysis of hydrogen-producing bacterial biofilms and granular sludge formed in continuous cultures of fermentative bacteria. Bioresour Technol. 2011;102:10057–10064. DOI:10.1016/j.biortech.2011.08.063

44. Hung CH, Lee KS, Cheng LH, Huang YH, Lin PJ, Chang JS. Quantitative analysis of a high-rate hydrogen-producing microbial community in anaerobic agitated granular sludge bed bioreactors using glucose as substrate. Appl Microbiol Biotechnol. 2007;75:693–701. DOI:10.1007/s00253-007-0854-7

45. Baghchehsaraee B, Nakhla G, Karamanev D, Margaritis A. Effect of extrinsic lactic acid on fermentative hydrogen production. Int J Hydrogen Energy. 2009;34:2573–2579. DOI:10.1016/j.ijhydene.2009.01.01

46. Kim TH, Lee Y, Chang KH, Hwang SJ. Effects of initial lactic acid concentration, HRTs, and OLRs on bio-hydrogen production from lactate-type fermentation. Bioresour Technol. 2012;103:136–141. DOI:10.1016/j.biortech.2011.09.093

47. Buckel W, Thauer RK. Energy conservation via electron bifurcating ferredoxin reduction and proton/Na+ translocating ferredoxin oxidation. Biochim Biophys Acta. 2013;1827:94–113. DOI:10.1016/j.bbabio.2012.07.002

48. Hedderich R, Whitman WB. Physiology and biochemistry of the methane-producing Archaea. In Dworkin M, editor. The Prokaryotes. 3rd ed. New York: Springer-Verlag; 2006. p. 1050–1079. DOI: 10.1007/0-387-30742-7_34

49. Sakai S, Takaki Y, Shimamura S, Sekine M, Tajima T, et al. Genome sequence of a mesophilic hydrogenotrophic methanogen Methanocella paludicola, the first cultivated representative of the order Methanocellales. PLoS One. 2011;6(7):e22898. DOI:10.1371/journal.pone.0022898

50. Borrel G, Toole PW, Harris H, Peyeret P, Brugere JF, Gribaldo S. Phylogenomic data support a seventh order of methylotrophic methanogens and provide insights into the evolution of methanogenesis. Genome Biol Evol. 2013;5:1769–1780. DOI:10.1093/gbe/evt128

51. Borrel G, Parisot N, Harris HMB, Peyretaillade E, Gaci N, et al. Comparative genomics highlights the unique biology of Methanomassiliicoccales, a

Thermoplasmatales-related seventh order of methanogenic Archaea that encodes pyrrolysine. BMC Genomics. 2014;15:679. DOI:10.1186/1471-2164-15-679

52. Worm P, Koehorst JJ, Visser M, Sedano-Núñez VT, Schaap PJ, et al. A genomic view on syntrophic versus non-syntrophic lifestyle in anaerobic fatty acid degrading communities. Biochim Biophys Acta. 2014;1837:2004–2016. DOI:10.1016/j.bbabio.2014.06.005

53. Kamagata Y. Syntrophy in anaerobic digestion. In: Fang HP, Zhang T, editors. Anaerobic Biotechnology: Environmental Protection and Resource Recovery. London: Imperial College Press, World Scientific; 2015. p. 13–32. DOI:10.1142/p1034/suppl_file/p1034_chap02.

54. McInerney MJ, Bryant MP. Anaerobic degradation of lactate by syntrophic associations of Methanosarcina barkeri and Desulfovibrio species and effect of H_2 on acetate degradation. Appl Environ Microbiol. 1981;41:346–354. DOI:0099-2240/81/020346-09$02.00/0, the paper was published in 1981, PMID: 16345708, PMCID: PMC243697

55. Shen L, Zhao Q, Wu X, Li X, Li Q, Wang Y. Interspecies electron transfer in syntrophic methanogenic consortia: from cultures to bioreactors. Renew Sustain Energy Rev. 2016;54:1358–1367. DOI:10.1016/j.rser.2015.10.102

56. Schmidt A, Müller N, Schink B, Schleheck D. A proteomic view at the biochemistry of syntrophic butyrate oxidation in Syntrophomonas wolfei. PLoS One. 2013;8(2):e56905. DOI:10.1371/journal.pone.0056905

57. Müller N, Worm P, Schink B, Stams AJ, Plugge CM. Syntrophic butyrate and propionate oxidation processes: from genomes to reaction mechanisms. Environ Microbiol Rep. 2010;2:489–99. DOI:10.1111/j.1758-2229.2010.00147.x

58. Hattori S. Syntrophic acetate-oxidizing microbes in methanogenic environments. Microbes Environ. 2008;23:118–27. DOI:10.1264/jsme2.23.118

59. Schmidt A, Frensch M, Schleheck D, Schink B, Müller N. Degradation of acetaldehyde and its precursors by Pelobacter carbinolicus and P. acetylenicus. PLoS One. 2014;9(12):e115902. DOI:10.1371/journal.pone.0115902

60. Meyer B, Kuehl J, Deutschbauer AM, Price M, Arkin A, Sthal D. Variation among Desulfovibrio species in electron transfer systems used for syntrophic growth. J Bacteriol. 2013;195:990–1004. DOI:10.1128/JB.01959-12

61. Hatamoto M, Imachi H, Yashiro Y, Ohashi A, Harada H. Detection of active butyrate degrading microorganisms in methanogenic sludges by RNA-based stable isotope probing. Appl Environ Microbiol. 2008;74:3610–3614. DOI:10.1128/AEM.00045-08

62. Liu P, Qiu Q, Lu Y. Syntrophomonadaceae-affiliated species as active butyrate-utilizing syntrophs in paddy field soil. Appl Environ Microbiol. 2011;77:3884–3887. DOI:10.1128/AEM.00190-11

63. Struchtemeyer CG, Ducan KE, McInerney MJ. Evidence for syntrophic butyrate metabolism under sulfate reducing conditions in a hydrocarbon-contaminate aquifer. FEMS Microbiol Ecol. 2011;76:289–300. DOI:10.1111/j.1574-6941.2011.01046.x

64. Ito T, Yoshiguchi K, Ariesyady HD, Okabe S. Identification of a novel acetate-utilizing bacterium belonging to Synergistes group 4 in anaerobic digester sludge. ISME J. 2011;5:1844–1856. DOI:10.1038/ismej.2011.59

65. Kim TG, Yun J, Cho KS. The close relation between Lactococcus and Methanosaeta is a keystone for stable methane production from molasses wastewater in a UASB reactor. Appl Microbiol Biotechnol. 2015;99:8271–8283. DOI:10.1007/s00253-015-6725-8

66. Park MJ, Jo JH, Park D, Lee DS, Park JM. Comprehensive study on a two-stage anaerobic digestion process for the sequential production of hydrogen and methane from cost-effective molasses. Int J Hydrogen Energy. 2010;35:6194–6202. DOI:10.1016/j.ijhydene.2010.03.135

67. Thauer RK. Biochemistry of methanogenesis: a tribute to Marjory Stephenson. Microbiology. 1998;144:2377–2406. DOI:10.1099/00221287-144-9-2377

68. Campanaro S, Treu L, Kougias PG, Francisci D, Valle G, et al. Metagenomic analysis and functional characterization of the biogas microbiome using high throughput shotgun sequencing and a novel binning strategy. Biotechnol Biofuels. 2016;9:26. DOI:10.1186/s13068-016-0441-1

69. Jablonski S, Rodowicz P, Lukaszewicz M. Methanogenic Archaea database containing physiological and biochemical characteristics. Int J System Evol Microbiol. 2015;65:1360–1368. DOI:10.1099/ijs.0.000065

70. Chen Y, Cheng JJ, Creamer KS. Inhibition of anaerobic digestion process: a review. Bioresource Technol. 2008;99:4044–4064. DOI:10.1016/j.biortech.2007.01.057

71. Fang C, Boe K, Angelidaki I. Anaerobic co-digestion of desugared molasses with cow manure; focusing on sodium and potassium inhibition. Bioresource Technol. 2001;102:1005–1011. DOI:10.1016/j.biortech.2010.09.077

72. Hallenbeck PC, Ghosh D. Advances in fermentative biohydrogen production: the way forward? Trends Biotechnol. 2009;27:287–297. DOI:10.1016/j.tibtech.2009.02.004

73. Benemann J. Hydrogen biotechnology: progress and prospects. Nature Biotechnol. 1996;14:1101–1103. DOI:10.1038/nbt0996-1101

74. Logan BE. Extracting hydrogen and electricity from renewable resources. Environ Sci Technol. 2004;38:160A–167A. DOI:10.1021/es040468s

75. Nishio N, Nakashimada Y. High rate production of hydrogen/methane from various substrates and wastes. Adv Bochem Engin/Biotechnol. 2004;90:63–87. DOI:10.1007/b94192

76. Liu D, Liu D, Zeng RJ, Angelidaki I. Hydrogen and methane production from household solid waste in the two-stage fermentation process. Water Research. 2006;40:2230–2236. DOI:10.1016/j.watres.2006.03.029

77. Ueno Y, Fukui H, Goto M. Operation of a two-stage fermentation process producing hydrogen and methane from organic waste. Environ Sci Technol. 2007;41:1413–1419. DOI:10.1021/es062127f

78. Ueno Y, Tatara M, Fukui H, Makiuchi T, Goto M, Sode K. Production of hydrogen and methane from organic solid wastes by phase separation of anaerobic process. Bioresour Technol. 2007;98:1861–1865. DOI:10.1016/j.biortech.2006.06.017

79. Wang X, Zhao Y. A bench scale study of fermentative hydrogen and methane production from food waste in integrated two-stage process. Int J Hydrogen Energy. 2009;34:245–254. DOI:10.1016/j.ijhydene.2008.09.100

80. Chu CF, Li YY, Xu KQ, Ebie Y, Inamori Y, Kong HN. A pH- and temperature phased two-stage process for hydrogen and methane production from food waste. Int J Hydrogen Energy. 2008;33:4739–4746. DOI:10.1016/j.ijhydene.2008.06.060

81. Cheng XY, Li Q, Liu CZ. Coproduction of hydrogen and methane via anaerobic fermentation of cornstalk waste in continuous stirred tank reactor integrated with up-flow anaerobic sludge bed. Bioresour Technol. 2012;114:327–333. DOI:10.1016/j.biortech.2012.03.038

82. Giordano A, Cantu C, Spagni A. Monitoring the biochemical hydrogen and methane potential of the two-stage dark-fermentative process. Bioresour Technol. 2011;102:4474–4479. DOI:10.1016/j.biortech.2010.12.106

83. Gomez X, Fernandez C, Fierro J, Sanchez ME, Escapa A, Moran A. Hydrogen production: two stage processes for waste degradation. Bioresour Technol. 2011;102:8621–8627. DOI:10.1016/j.biortech.2011.03.055

84. Koutrouli EC, Kalfas H, Gavala HN, Skiadas IV, Stamatelatou K, Lyberatos G. Hydrogen and methane production through two-stage mesophilic anaerobic digestion of olive pulp. Bioresour Technol. 2009;100:3718–3723. DOI:10.1016/j.biortech.2009.01.037

85. Kvesitadze G, Sadunishvili T, Dudauri T, Zakariashvili N, Partskhaladze G, et al. Two-stage anaerobic process for biohydrogen and biomethane combined production from biodegradable solid wastes. Energy. 2012;37:94–102. DOI:10.1016/j.energy.2011.08.039

86. Ljunggren M, Zacchi G. Techno-economic analysis of a two-step biological process producing hydrogen and methane. Bioresour Technol. 2010;101:7780–7788. DOI:10.1016/j.biortech.2010.05.009

87. Lu Y, Lai Q, Zhang C, Zhao H, Ma K, et al. Characteristics of hydrogen and methane production from cornstalks by an augmented two- or three-stage anaerobic fermentation process. Bioresour Technol. 2009;100:2889–2895. DOI:10.1016/j.biortech.2009.01.023

88. Ruggeri B, Tommasi T, Sassi G. Energy balance of dark anaerobic fermentation as a tool for sustainability analysis. Int J Hydrogen Energy. 2010;35:10202–10211. DOI:10.1016/j.ijhydene.2010.08.014

89. Venetsaneas N, Antonopoulou G, Stamatelatou K, Kornaros M, Lyberatos G. Using cheese whey for hydrogen and methane generation in a two-stage continuous process with alternative pH controlling approaches. Bioresour Technol. 2009;100:3713–3717. DOI:10.1016/j.biortech.2009.01.025

90. Nasr N, Elbeshbishy E, Hafez H, Nakhla G, Naggar MHE. Comparative assessment of single-stage and two-stage anaerobic digestion for the treatment of thin stillage. Bioresource Technol. 2012;111:122–126. DOI:10.1016/j.biortech.2012.02.019

91. Lee DH, Lee DJ. Hydrogen economy in Taiwan and biohydrogen. Int J Hydrogen Energy. 2008;33:1607–1618. DOI:10.1016/j.ijhydene.2007.09.028

92. Schlüter A, Bekel T, Diaz NN, Dondrup M, Eichenlaub R, et al. The metagenome of a biogas-producing microbial community of a production-scale biogas plant fermenter analysed by the 454-pyrosequencing technology. J Biotechnol. 2008;136: 77–90. DOI:10.1016/j.jbiotec.2008.05.008

93. Limam RD, Chouari R, Mazéas L, Wu TD, Li T, et al. Members of the uncultured bacterial candidate division WWE1 are implicated in anaerobic digestion of cellulose. Microbiology Open. 2014;3:157–167. DOI:10.1002/mbo3.144

INDEX

www.ingramcontent.com/pod-product-compliance
Lightning Source LLC
LaVergne TN
LVHW062222200926
848484LV00006B/1451